Dielectric Material

Dielectric Material

Editor

Gustaf Kron

scitus
academics

Dielectric Material
Edited by **Gustaf Kron**

ISBN: 978-1-68117-291-0
Library of Congress Control Number: 2016937467

SCITUS Academics LLC,
www.scitusacademics.com
Box No. 4766, 616 Corporate Way,
Suite 2, Valley Cottage,
NY 10989

Printed in United States of America on Acid Free Paper

Preface

A dielectric material is an electric insulator that can be polarized with an electric field. Dielectrics are simple and plain insulators. Both words refer to the same class of materials, but they originate differently and are preferably used in different contexts. Plastic electrical cord coating is an insulator. Insulators are the glass or ceramic plates used for supplying power lines and preventing them from being cut off. In an electrical device called an insulator, a non-metallic solid is used almost at any time. Perhaps the only time the word dielectric is used for the non-conductive layer of the capacitor. When a dielectric is placed in an electric field, electric charge does not flow through the material as it does in a conductor, but only slightly from its average balance positions leading to dielectric polarization. Positive loads are moved to the field because of dielectric polarization and negative loads are moved in the opposite direction. This creates an electrical internal field that reduces the whole dielectric field itself. If a dielectric consists of weakly connected molecules, the molecules are not only polarized but also reoriented to match the field with their symmetry axes. The study of dielectric properties concerns the storage and dissipation of electrical and magnetic energy in materials. Dielectrics are important in explaining various phenomena in electronics, optics and solid state physics.

The book covers state of the art research and results on dielectric materials theory and practice for various types of industrial applications.

Table of Contents

1

CHAPTER

DIELECTRIC MATERIALS FOR COMPACT DIELECTRIC RESONATOR ANTENNA APPLICATIONS

L. Huitema and T. Monediere

[1] *University of Limoges, Xlim Laboratory, France*

1. INTRODUCTION

Dielectric resonators using high-permittivity materials were originally developed for microwave circuits, such as filters or oscillators as tuning element [1]. Indeed, in the late nineteen sixties, the development of low-loss ceramic materials opened the way for their use as high-Q elements [2-4]. Then, making use of dielectric materials to create the dielectric resonator antenna (DRA) illustrates the ingenuity of Professor S. A. Long [5], who was the first to propose such a procedure in the early nineteen eighties. Indeed, it introduced the use of a dielectric resonator as an antenna by exciting different modes using multiple feeding mechanisms. During the nineties, emphasis was placed on applying analytical or numerical techniques for determining input impedance, fields inside the resonator and Q-factor [6]. Kishk, Junker, Glisson, Luk, Leung, Mongia, Bhartia, Petosa and so on, have described a significant amount of DRAs' analyses and characterizations [7-18]. Petosa and al. proposed both in literatures and book [6,12] many of the recent advances on DRAs.

Current DRA literatures focus on compact designs to address portable wireless applications. Among them, new DRA shapes or hybrid antennas are developed to enhance the antenna impedance bandwidth [13-19] or for multiband antenna applications [20-22].

The first part will address a brief overview of the most common used DRA shapes and structures including both rectangular and cylindrical DRAs. The emphasis will be placed on better understanding what DRAs exactly are and how to develop such an antenna. This part will detail fundamental modes of

DRAs, their resonant frequencies, fields inside the resonator and radiation patterns corresponding to these modes.

A second part will focus on the relevant dielectric material properties having a significant contribution to achieve better antenna performances. It will detail the kind of materials DRAs can use, which is closely linked to the targeted application.

Multiple techniques to miniaturize such an antenna will be presented in the third part, supported by concrete examples. At the same time, everyone will be able to appreciate that dielectric material properties have a major role to play in designing a DRA. It should be noted that the material choice is even more critical when the targeted challenge is the antenna size reduction.

Therefore, depending on the intended applications, this part will enable to find the best trade-off between the material choice and its shape.

Although some wideband or multiband DRA structures have been introduced in the third part, the fourth and last part will be dedicated to a new method to design a DRA. It will address engineering design data on hybrid modes creation to enhance the bandwidth or develop multiband antennas. This part will include many references to clearly explain this research method while highlighting their contribution to expand the use of DRA in new kind of mobile handheld devices (e.g. new tablets).

2. OVERVIEW ON DRA STUDIES

The design of a DRA in any geometry must satisfy various specifications including: the resonant frequency, the impedance bandwidth, the field distribution inside the resonator and also the radiated field. The intent of this part is to provide an understanding of fundamental operation of DRAs, emphasizing both design and implementation. Thus, to provide comprehensive research method, this part will start by presenting main findings of investigations on simple-shaped DRAs. Then, it will deal with the different DRA feeding methods. Finally, this part will focus on the study of two DRA shapes: cylindrical and rectangular.

2.1. Main DRAs characteristics

A non-exhaustive list of main simple-shaped DRAs characteristics is described below:

- The main dimension of a DRA is proportional to $\lambda_0 / \sqrt{\varepsilon_r \cdot \mu_r}$ where λ_0 is the free-space wavelength at the resonant frequency, ε_r and μ_r are respectively the dielectric and the magnetic constant of the material. In a dielectric material case, $\mu_r = 1$ and the main dimension of a DRA is proportional to $\lambda_0 / \sqrt{\varepsilon_r}$.
- The radiation efficiency of the DRA is highly depending on the material losses. In case of a low-loss dielectric material, DRAs allow to achieve better efficiency than other kind of antennas because of minimal conductor losses associated with a DRA.
- For a given dielectric constant, both resonant frequency and radiated Q-factor are defined according to the resonator dimensions. That allows having a great flexibility and some degrees of freedom to design such an antenna.
- Another degree of freedom is the large spectrum of available dielectric materials. That allows doing the best trade-off between dimensions and impedance bandwidth according to the intended application.
- A number of modes can be excited within the DRA, many of them provide dipolar-like radiation characteristics.
- The most common targeted frequencies presented by the research literatures are ranging from 1GHz to 40 GHz.
- For a given DRA geometry, the radiation patterns can be made to change by exciting different resonant modes.
- A large number of DRA excitations are currently used, e.g. microstrip line, coaxial probe excitation, coplanar waveguide... The next subsection will deal with the most commonly used excitations.

2.2. Common DRAs feedings

Multiple feeding mechanisms are employed to excite different resonator modes. This subsection will summarise most widely-used excitations while giving many references in order designers to choose the most appropriate excitation.

2.2.1. Coaxial probe excitation

It can be located within the DRA or adjacent to it. Within the DRA, a good coupling can be achieved by aligning the probe along the electric field of the DRA mode as shown Figure 1.

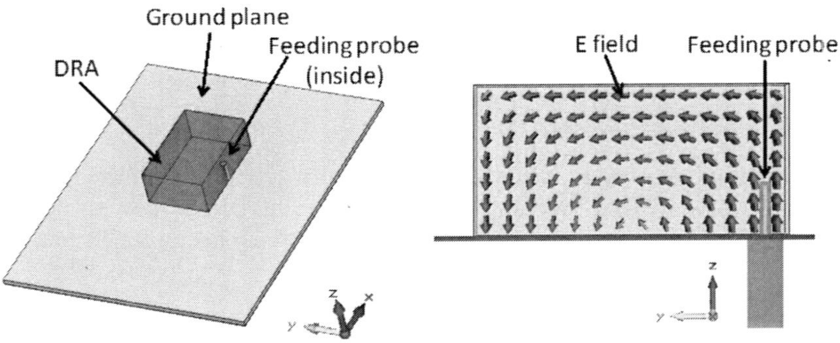

Figure 1. Coaxial probe coupling the E field

The adjacent position is currently used to couple the magnetic field of the DRA mode (Figure 2). In these both cases, the probe is exciting the TE_{111} fundamental mode of the rectangular DRA.

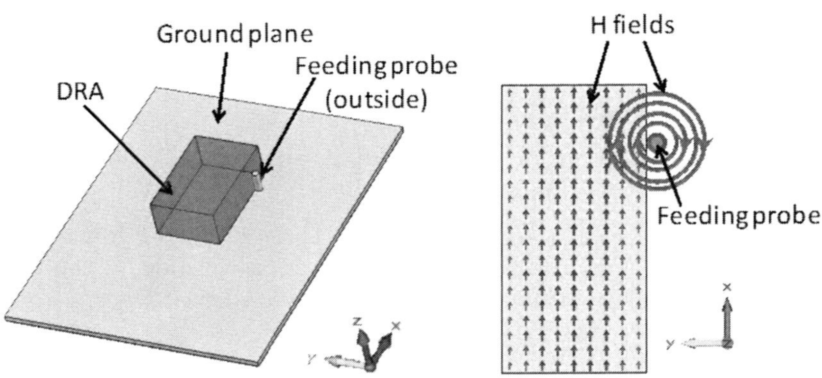

Figure 2. Coaxial probe coupling the H field

When the excitation probe is inside the resonator, particular attention has to be paid to the air gap between the probe excitation and the dielectric material. Indeed, an air gap results in a lower effective dielectric constant, which entails both a decrease in the Q-factor and a shift of the resonance frequencies [23-24]. The probe location allows choosing the intended excited mode and the coupling of the mode can be optimized by adjusting both length and height of the probe.

2.2.2. Microstrip feeding line and coplanar waveguide

The principle is similar to the probe excitation case. A microstrip line placed close to the DRA can couple the magnetic field of the DRA mode. However, this latter can affect the antenna polarisation and can thus increase the parasitic radiation. This could be reduced by placing the line under the resonator as shown Figure 3.

Another way is to replace the microstrip line by a coplanar waveguide, the Figure 3 is presenting a rectangular DRA excited by a coplanar waveguide.

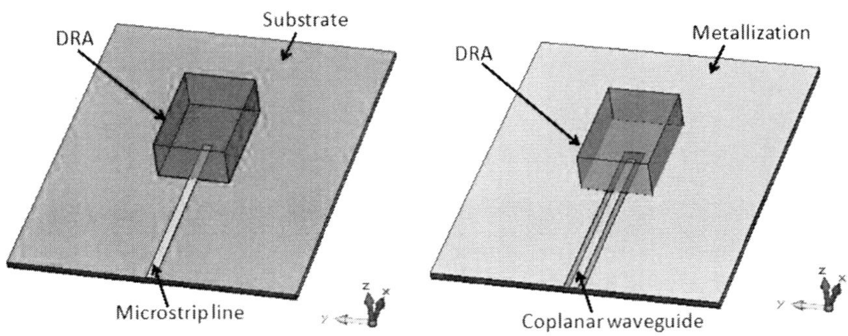

Figure 3. Microstrip feeding line and coplanar waveguide

In these both cases, the mode coupling can be optimized by changing the resonator position and/or its dielectric permittivity. For low dielectric permittivity materials (which allows obtaining a wide bandwidth), it is somewhat difficult to excite the mode. There are different solutions to obtain both miniaturization and good coupling, they will be explained in this chapter.

An important point is that these excitation methods are disturbing DRA modes by introducing electrical boundary conditions. This issue is all the more sensitive since the antenna is miniature. The last part of the chapter will show how to take advantage of this issue.

2.2.3. Aperture coupling

A common method of exciting a DRA is acting through an aperture in the ground plane. The Figure 4 shows an example of the excitation of the TE_{111} mode of a rectangular DRA with a rectangular slot. To achieve relevant coupling, the aperture has to be placed in a DRA strong magnetic area. Feeding the aperture with a microstrip line is a current approach [25-26].

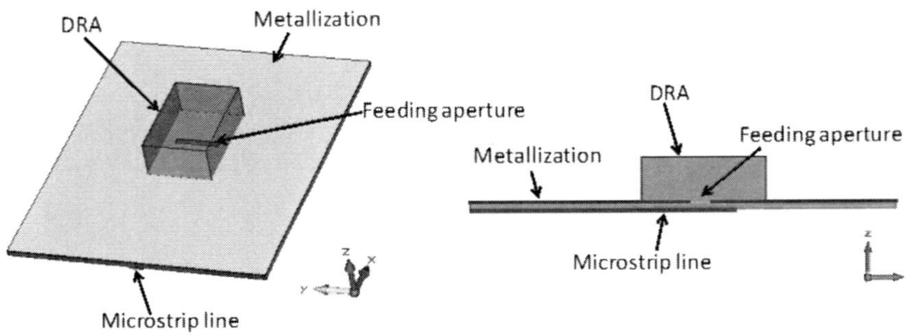

Figure 4. Aperture coupling the TE_{111} mode of the rectangular DRA

The main dimension of the aperture needs to be around $\lambda_g/2$, which is highly problematic at low frequencies.

On top of these multiple feeding methods, the choice of different DRAs shapes represents another degree of flexibility and versatility. The next subsection will deal with the cylindrical shape.

2.3. The cylindrical DRA

It offers great design flexibility, where both resonant frequency and Q-factor are depending on the ratio of radius/height. Various modes can be excited within the DRA and a significant amount of literature is devoted to their field configurations, resonant frequencies and radiation properties [27-28]. This part will present a complete and concrete study of a cylindrical DRA.

Like most realistic cases, the cylindrical DRA presented Figure 5, is mounted on a finite ground plane. Because dielectric material properties will be studied in very great depth in the third part, the dielectric permittivity ε_r of the DRA is fixed and chosen equal to 30. It is also characterized by its height d and its radius a. To keep the chapter concise while remaining comprehensive, only relevant results and equations will be given.

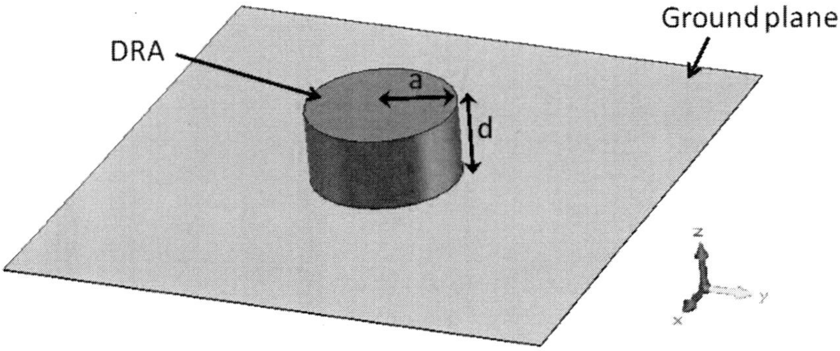

Figure 5. Cylindrical DRA

First of all, the study can begin with a modal analysis. This allows determining the most appropriate excitation method. Modes of cylindrical DRA can be divided into three types: TE, TM and hybrid modes, i.e. EH or HE [9-10,30]. The latter have a dependence on azimuth φ, while TE and TM modes have no dependence on azimuth. To identify fields' variations according φ (azimuth), r (radial) and z (axial) directions, subscripts respectively noticed n, p and m are following the mode notation. Because TE and TM modes have no azimuthal variation, n=0 for these modes. Finally, all cylindrical DRA modes can be defined such as: $TE_{0pm+\delta}$, $TM_{0pm+\delta}$, $HE_{npm+\delta}$ and $EH_{npm+\delta}$. The δ value is ranging between 0 and 1, it approaches 1 for high ε_r. It should be noted that n, m and p are natural numbers. The modal analysis of a DRA can be deducted either with analytical calculations or thanks to electromagnetic simulators like CST Microwave Studio (CST MS).

2.3.1. Modal analysis

- Analytical equations derive from the analytical calculations of a cylindrical dielectric resonator by assuming perfect magnetic and/or electric walls on resonator faces. The perfect magnetic wall boundary condition was demonstrated to be valid for high ε_r values [30], it remains accurate for lower values as well.

Fields equations inside the DRA, resonant frequencies and Q factor are detailed in [7,9] and [12]. Resonant frequencies of all modes are provided hereafter:

$$\begin{pmatrix} f_{TMnpm} \\ f_{TEnpm} \end{pmatrix} = \frac{c}{2\pi a\sqrt{\varepsilon_r \mu_r}} \sqrt{\left(\frac{X'_{np}}{X_{np}}\right)^2 + \left(\frac{(2m+1)\pi.a}{2d}\right)^2} \tag{1}$$

Where X_{np} and X'_{np} are Bessel's solutions, $(n,m,p) \in \mathrm{N}^3$, *a* and *d* are the radius and the height of the dielectric resonator.

In the case presented Figure 5, the fundamental excited mode is the $HE_{11\delta}$ and its resonant frequency equals:

$$f_{110} = \frac{3.10^8}{2\pi\sqrt{30}} \sqrt{\left(\frac{X'_{11}}{0.04}\right)^2 + \left(\frac{\pi}{2\times 0.045}\right)^2} = 503.6MHz \tag{2}$$

This method requires the resolution of both Maxwell and propagation equations. It was therefore reserved for simple-shaped DRA.

- The "Eigenmode solver" of CST MS allows viewing 3D fields of each mode and having their resonant frequencies. When considering the studied example and defining perfect magnetic boundary conditions on all DRA walls, except for the DRA bottom where perfect electric boundary condition is considered (due to the ground plane), the software gives 504 MHz for the resonant frequency value of the $HE_{11\delta}$ mode. It is also possible to see both H and E fields of this mode, they are presented Figure 6. This valuable information allows designer to choose the most appropriate excitation of the considered mode.

In light of this above, the best way to excite the $HE_{11\delta}$ mode is to integrate a coaxial probe along the E field inside the DRA.

Now that the modal analysis was explained and the excitation determined, we can go straight to the electromagnetic study of the DRA.

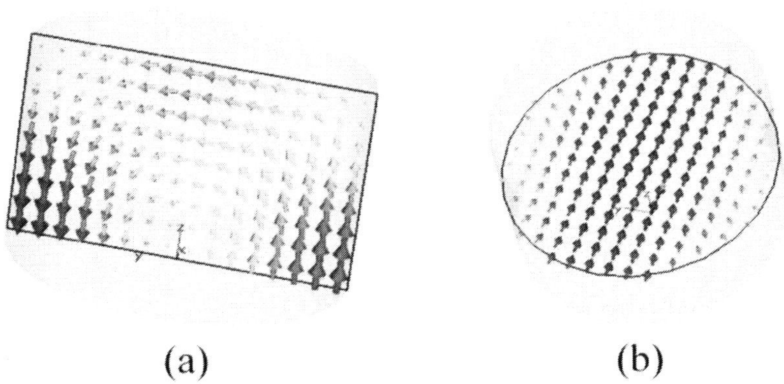

Figure 6. E field (a) and H field (b) of the $HE_{11\delta}$ mode

2.3.2. Electromagnetic study

Since input impedance and also S_{11} parameter cannot be calculated with the magnetic wall model, their study is solely possible with an electromagnetic simulator or of course can be experimentally done. Moreover, electromagnetic simulators facilitate accurate and efficient antenna analysis by providing the complete electric and magnetic fields inside and outside the antenna taking into account the finite ground plane. The impedance, the S_{11} parameter, the power radiated and the far field radiation pattern are determined everywhere and at any frequency in a single analysis thanks to the Finite Integration Temporal method.

The Figure 7 presents the cylindrical DRA excited by a coaxial probe.

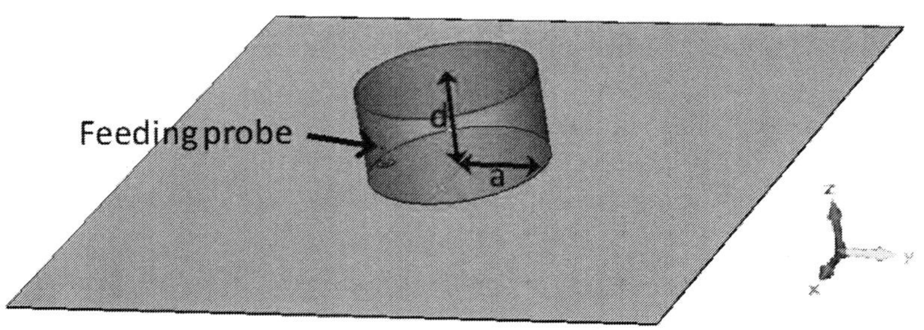

Figure 7. Excitation of the $HE_{11\delta}$ mode with a coaxial probe

Using the CST MS software, the input impedance is presented Figure 8. It shows that the fundamental mode is excited at the resonant frequency fairly corresponding to the predicted value by the modal analysis of 504 MHz. The minor shift between resonant frequencies deducted with modal analysis and electromagnetic study is due to magnetic wall model used during the modal analysis, which is not absolutely accurate.

An important data for antenna designers is the S_{11} parameter (Figure 8). It is directly deducted from the input impedance parameter.

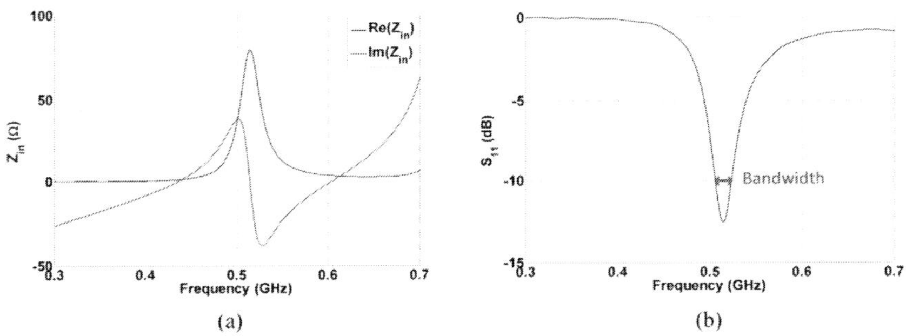

Figure 8. Input impedance (a) and S_{11} parameter (b) of the DRA

It allows knowing the matching frequency as well as the impedance bandwidth, basically corresponding to the working frequencies of the antenna. It this case, the matching frequency equals 510 MHz with 3.6% of bandwidth at -10 dB.

Another important issue is the radiation pattern. It can be expressed in spherical coordinates by using equivalent magnetic surface currents [7]. This can only be done until the DRA is mounted on an infinite ground plane. Using the electromagnetic software is another way of accessing to the 3D radiation pattern. This method is more accurate because it is taking into account the realistic structure (i.e. the finite ground plane). The Figure 9 shows the 3D radiation pattern of the cylindrical DRA using the simulator.

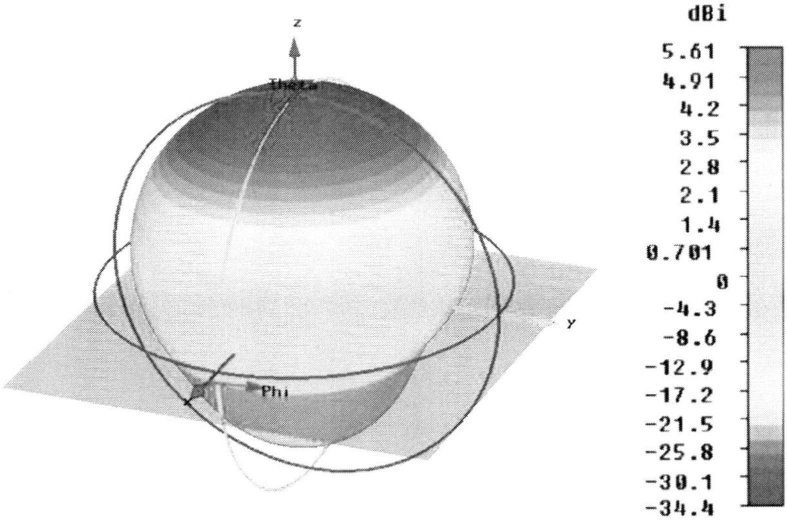

Figure 9. radiation pattern

The simulated 3D radiation pattern provides antenna information such as the radiation efficiency, which is defined as the ratio between the radiated to accepted (input) power of the antenna.

The $HE_{11\delta}$ mode of the cylindrical DRA radiates like a short horizontal magnetic dipole. Concerning other modes, the $TM_{01\delta}$ radiates like a short electric monopole, while the $TE_{01\delta}$ mode radiates like a short magnetic monopole.

2.4. The rectangular DRA

The rectangular DRA has one degree of freedom more than the cylindrical DRA. Indeed, it is characterized by three independent lengths, i.e. its length *a*, its width *b* and its height *d*. Thus, there is great design flexibility, since the large choice of both dielectric materials and different lengths ratios.

Usually the dielectric waveguide model is used to analyze the rectangular DRA [8-10]. In this approach, the top surface and two sidewalls of the DRA are assumed to be perfect magnetic walls, whereas the two other sidewalls are imperfect magnetic walls. Since the considered case is a realistic one

(Figure 10), the DRA is mounted on a ground plane, thus, an electric wall is assumed for the bottom surface.

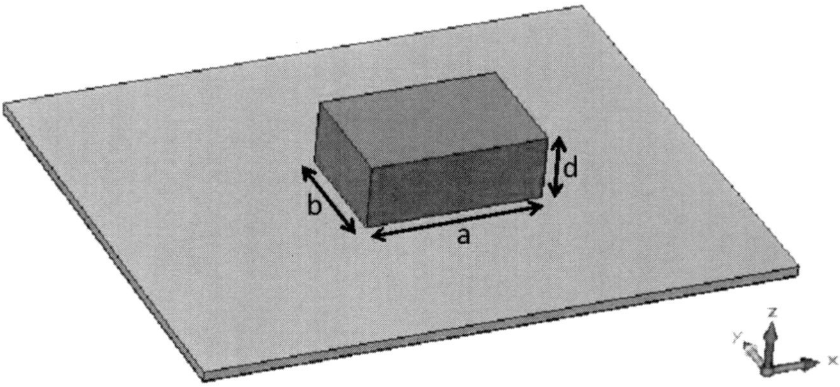

Figure 10. Rectangular DRA

The modes in an isolated rectangular dielectric resonator can be divided in two categories: TE and TM modes, but in case of the DRA mounted on a ground plane, only TE modes are typically excited. The fundamental mode is the TE_{111}. As the three dimensions of the DRA are independent, the TE modes can be along the three directions: x, y and z. By referring to the Cartesian coordinate system presented Figure 10, if the dimensions of DR are such as a>b>d, the modes in the order of increasing resonant frequency are TE^{z}_{111}, TE^{y}_{111} and TE^{x}_{111}. The analysis of all the modes is similar. The example of the TE^{z}_{111} mode is discussed in [8], the field components inside the resonator and resonant frequencies are analytically presented.

As for the cylindrical case, CST MS can be used to see both E and H fields, as presented Figure 11.

The resonant frequencies definition is reminded hereafter:

$$f_0 = \frac{c}{2\pi\sqrt{\varepsilon_r \mu_r}}\sqrt{k_x^2 + k_y^2 + k_z^2} \tag{3}$$

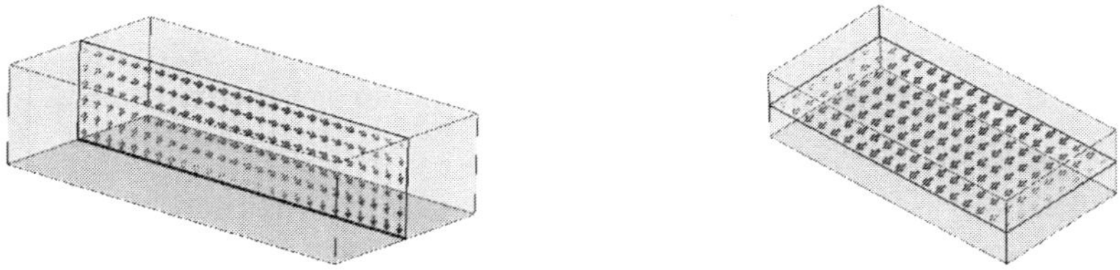

(a) (b)

Figure 11. E field (a) and H field (b) of the TE_{111} mode

It is found by solving the following transcendental equation:

$$k_z \tan(\frac{k_z a}{2}) = \sqrt{(\varepsilon_r - 1)k_0^2 - k_z^2} \quad (4)$$

where $k_x = \frac{\pi}{a}, \quad k_y = \frac{\pi}{b}, \quad k_0 = \frac{\omega_0}{v} = \frac{2\pi f_0 \sqrt{\varepsilon_r \mu_r}}{c}$ and $k_x^2 + k_y^2 + k_z^2 = k_0^2$

Values of resonant frequencies predicted by using this model are close to the measured ones for moderate to high value of ε_r. A frequency shift appears for low ε_r but it remains a good approximation method. If more accuracy is required, the electromagnetic study with CST MS (for example) presented in the cylindrical DRA case will have to be undertaken. Moreover, it allows taking into account feeding mechanism and ground plane dimensions.

Now DRA research method has been initiated, presenting resonant frequencies, fields configuration and feeding mechanisms, the next part will focus on the relevant dielectric material properties having significant influences on antenna performances.

3. ANALYSIS ON THE DIELECTRIC MATERIAL CHOICE

To supply satisfactory answers about the effects of dielectric material properties, this section will present a careful and extensive investigation into relevant cases. Indeed, properties of the dielectric material have an influence on antenna characteristics, i.e. impedance bandwidth, Q factor, resonant frequency and radiation efficiency. Thus, this part will allow the reader to correctly select a dielectric material for a targeted application.

The cylindrical DRA example presented Figure 5 with a radius a=40mm and a height h=45mm will be pursue here. However, all results included in this part are generally applicable to most shapes of DRAs. A first sub-section will detail the influence of the dielectric permittivity and a second one will be interested in the impact of the dielectric loss tangent on the DRA performances.

Dielectric material properties having an impact on antenna characteristics are dielectric permittivity values and loss tangents.

3.1. Influence of the dielectric permittivity

To show the real impact of dielectric permittivity values, this sub-section will deal with a loss less dielectric material. Analytical studies show that the Q factor of the $HE_{11\delta}$ mode is defined as

$$Q = 0.01007.\varepsilon_r.\frac{a}{h}.\left(1+100.\exp(-2.05(\frac{a}{2h}-\frac{1}{80}\left(\frac{a}{h}\right)^2\right) \quad (5)$$

It is plotted as a function of a/h for different values of ε_r in the Figure 12.

Q factor is increasing with ε_r and reaching a maximum for a/h=1.05. This Q factor can be used to estimate the fractional bandwidth of an antenna using:

$$BW = \frac{\Delta f}{f_0} = \frac{s-1}{Q\sqrt{s}} \quad (6)$$

Where Δf is the absolute bandwidth, f_0 is the resonant frequency and s the maximum acceptable voltage standing wave ratio (VSWR).

The Q factor equation is deriving from the cylindrical dielectric resonator model approach by assuming perfect magnetic and/or electric walls on resonator faces. These equations are not absolutely accurate but they offer a good starting point for the design of cylindrical DRAs.

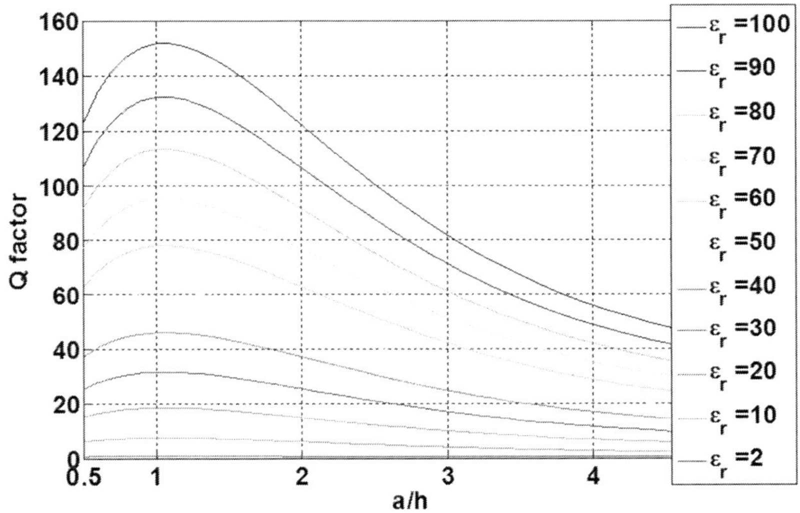

Figure 12. Q factor according the a/h values

Let's consider the electromagnetic study presented in the first section with the cylindrical DRA example (Figure 7). Previously, the dielectric permittivity was fixed and equaled 30, it is now a variable. The Figure 13 plots both resonant frequencies and impedance bandwidths according to the dielectric permittivity ε_r. Because the coupling of the mode is depending on both length and height of the probe, this latter has been optimized for each ε_r value. The Figure 13 is thus the result of a large number of simulations.

As expected (see equation 3), the resonant frequency decreases when the dielectric permittivity increases. Moreover, this Figure shows that the bandwidth is the widest for ε_r=10. Fields are less confined for a low dielectric permittivity DRA, it is thus more difficult to couple the mode inside the resonator. Indeed, for higher dielectric values (ε_r>10), strong coupling is achieved, however, the maximum amount of coupling is significantly reduced if the dielectric permittivity of the DRA is lowered. That is why the bandwidth is low for ε_r under 10. For a dielectric permittivity over 10, the Q factor is increasing and therefore the impedance bandwidth is decreasing (see equation 5).

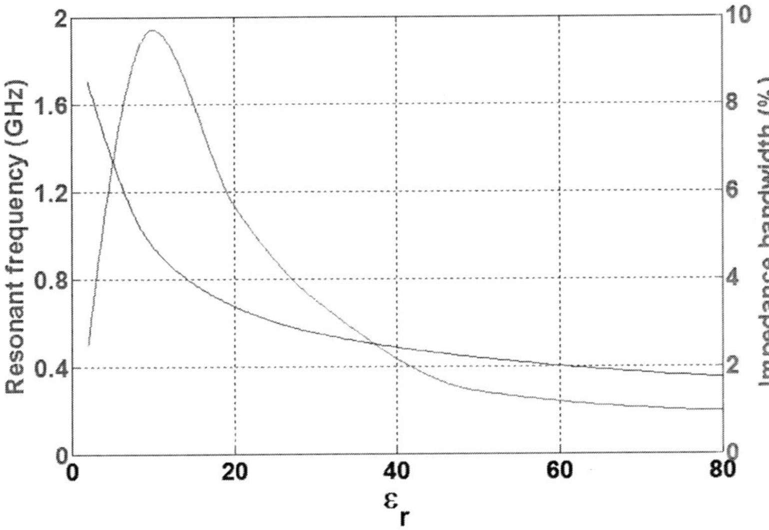

Figure 13. Resonant frequency according the dielectric permittivity values ε^r

Now that the influence of the dielectric permittivity has been shown, we can consider in the next sub-section more realistic cases by studying the impact of losses on the antenna characteristics.

3.2. Influence of dielectric material losses

Dielectric material losses directly impact the impedance bandwidth and antenna radiation efficiency. Their influence depends on the dielectric permittivity of the material.

Because analytical considerations do not take into account losses, only electromagnetic approaches are available to complete this study. To be more readable and relevant, charts will be provided to present and take into account all effects of losses.

The chosen example is still the same: the cylindrical DRA mounted on a ground plane with a radius and a height respectively equal to 40 mm and 45 mm and excited on its fundamental mode. The study of losses is done for different dielectric permittivity values.

The impedance bandwidth is studied as a first step. The Figure 14 presents the impedance bandwidth according to both dielectric permittivity and tangent loss of the dielectric material.

Several information have to be noted:

The most important loss tangents are, the widest the impedance bandwidth is.

The impedance bandwidth is the widest when for ε_r=10, whatever the losses.

The radiation efficiency is now investigated. Same simulations are done and the Figure 15 presents the antenna radiation efficiency according to both dielectric permittivity and tangent loss of the dielectric material.

Other information can be deduced from this new graph:

- Antenna radiation efficiency is all the more affected by the losses as the dielectric permittivity increase.
- For low dielectric permittivity, even in case of a high losses material, the radiation efficiency remains higher than 50%.

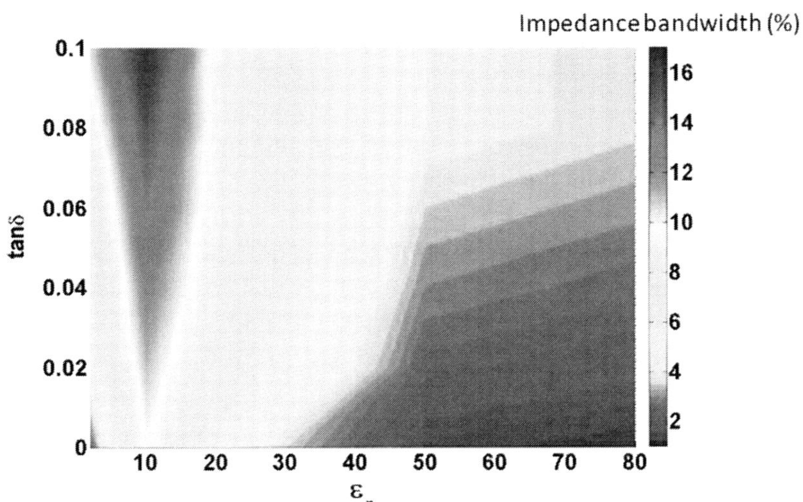

Figure 14. Impedance bandwidth according to the dielectric permittivity and loss tangent

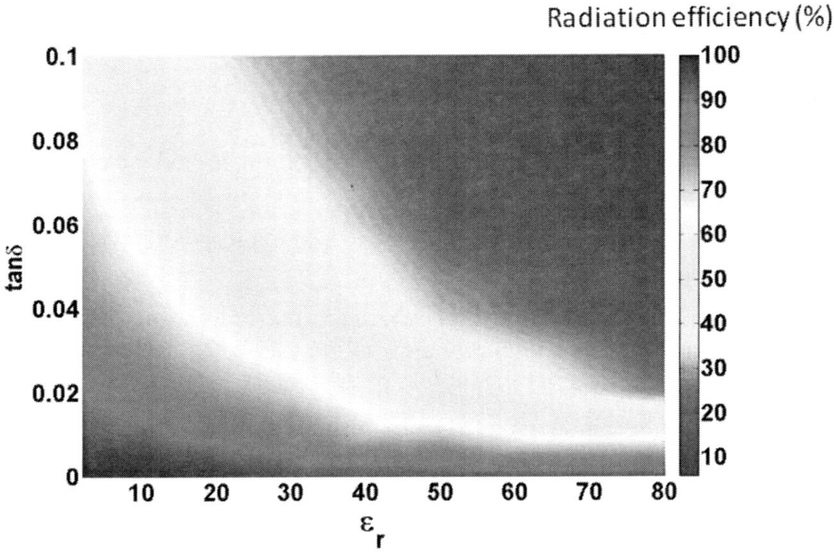

Figure 15. Radiation efficiency according to the dielectric permittivity and loss tangent

To conclude this part, a DRA designer has to choose the dielectric material according to the application for which he is aiming. If he targets a wide bandwidth application, he could choose an alumina ceramic (ε_r ~10). Depending on the radiation efficiency he aims, the chosen ceramic would have more or less losses.

Now, if he targets an ultra-miniature DRA, it will be in his interest to choose a dielectric material with a higher dielectric permittivity. In this case, the impedance bandwidth will be affected, even more if the losses are high.

The dielectric material choice is one of the most important degree of freedom in the DRA design. It is necessary to highlight the best tradeoff, keeping in mind the targeted application.

Thus, using high dielectric permittivity values is one method for achieving a compact design, but it is not the only one. The next section will deal with different techniques to miniaturize a DRA.

4. OVERVIEW OF TECHNIQUES TO MINIATURIZE DRAS

This part examines techniques to design compact DRAs. Targeted applications are mobile handsets or wireless tablet. There are several techniques to make DRAs more compact. By adding metal plates, inserting a high permittivity layer (multisegment DRA) or removing portions of the DRA, a significant size reduction can be achieved.

4.1. Addition of a metallic plate on a DRA face

The rectangular DRA shape has been studied in the first part. The perfect metallic wall implies that electric fields are normal to this conductor, while magnetic fields are tangential. E and H fields presented Figure 11 assume that a metallic plate can be inserted in the middle of the DRA according to the y-component. The principle is detailed and explained by the Figure 16. It also shows the E and H fields of the TE_{111} mode.

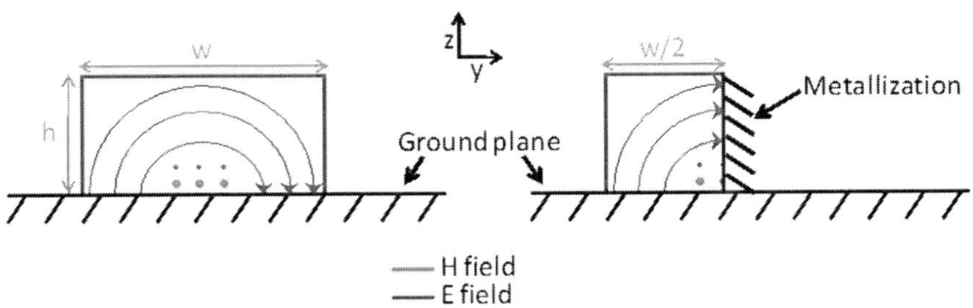

Figure 16. Integration of a metallic plate

By applying the image theory, it is possible to insert a metal plate in the y=w/2 plane. The Table 1 extracted from [12] shows the influence of the metallic plate insertion on resonant frequency and impedance bandwidth.

Table 1. Influence of the metallic plate insertion on both resonant frequency and impedance bandwidth

ε_r	w (cm)	d (cm)	h (cm)	Metallization	f_0(GHz)	Bandwidth
12	2.75	2.75	2.95	No	1.98	10%
12	2.75	2.75	2.95	Yes	1.24	5.6%

Thus, the metal plate insertion allows dividing by two the DRA size, while reducing the resonant frequency. However, as pointed by the Table 1, the metallic plate insertion involves also the decrease of the impedance bandwidth.

4.2. Multisegment DRA

Another way to decrease the DRA size is to insert different substrate layers as illustrated Figure 17.

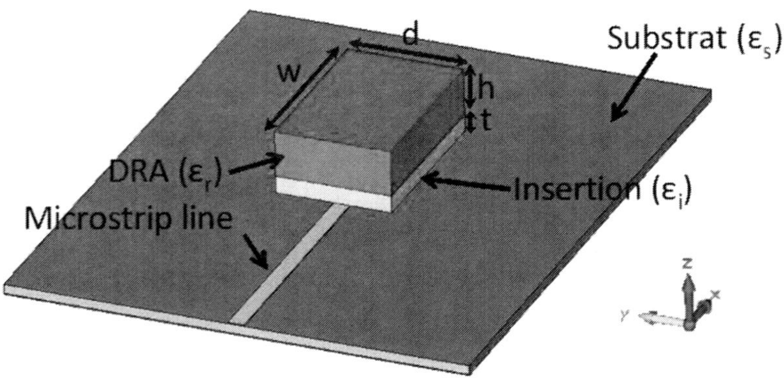

Figure 17. Multisegment DRA

It allows achieving strong coupling when the first insertion has a relatively high dielectric permittivity. This technique is detailed in [12] and [31]. The Table 2 summarizes a parametrical study done in [31] for one layer inserted (Figure 17) with w=7.875 mm, d=2 mm, h=3.175 and ε_r=10. It is mounted on a 0.762 mm height substrate of permittivity ε_s=3. The TE_{111} mode of the DRA is excited with a 50Ω microstrop line.

Table 2. A parametrical study done in [31] for one layer inserted

t (mm)	ε_i	Measured f_0(GHz)	Bandwidth
0	-	15.2	21%
0.25	20	14.7	18%
0.635	20	14.5	18%
1	20	13.9	16%
0.25	40	14.7	20%
0.635	40	13.7	13%
1	40	12.9	5%
0.25	100	14.7	16%
0.635	100	13.1	7%
1	100	10.8	5%

Thus, a thin layer insertion allows improving the coupling of modes inside the DRA while decreasing the resonant frequency thanks to the decrease of the effective dielectric permittivity of the DRA. As the previous technique, the downside is the decrease of the impedance bandwidth.

4.3. Circular sector DRAs

To clearly explain this miniaturization technique, we need to take up the cylindrical DRA example with the equation 1 of the resonant frequencies. In [32], DRA size and resonance frequencies significant reductions have been demonstrated by using cylindrical sector DRA which is shown in Figure 18.

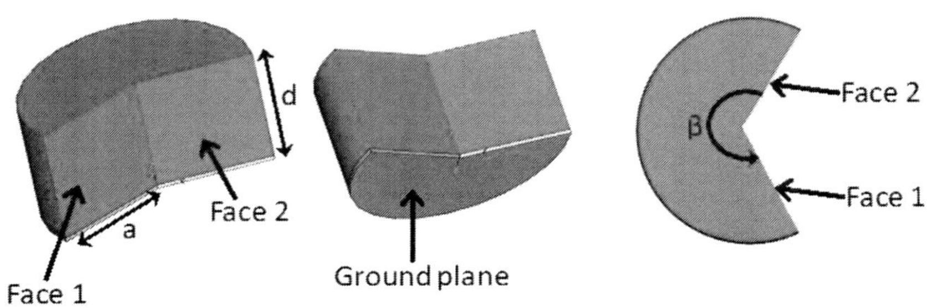

Figure 18. Circular sector DRA

As shows the Figure 18, a cylindrical sector DRA shape consists of a cylindrical DRA of radius *a* and height *d* mounted on a metallic ground plane, with a sector of dielectric material removed. β is the angle between the face 1 and 2, which can be metalized or left open. Thus, a cylindrical sector DRA is formed when $\beta < 2\pi$. For such a DRA, considering the equation 1, the n subscript can be substitute by the ν subscript, which is a positive real number that depends on the boundary conditions on the sector faces as well as the sector angle β. In this case, first excited modes are writing such as HE_{pm+} and the corresponding resonant frequencies as defined by the following equation:

$$f_{\upsilon pm} = \frac{c}{2\pi\sqrt{\varepsilon_r \mu_r}}\sqrt{\left(\frac{X'_{\upsilon p}}{a}\right)^2 + \left(\frac{(2m+1)\pi}{2d}\right)^2} = \frac{c}{2\pi a\sqrt{\varepsilon_r \mu_r}}\sqrt{\left(X'_{\upsilon p}\right)^2 + \left(\frac{(2m+1)\pi . a}{2d}\right)^2} \quad (7)$$

It should be noted that for the $\beta = 2\pi$ case, ν=n.

An important point has to be highlighted: For a given cylindrical sector DRA (radius, height, permittivity and permeability), the resonant frequency is only depending on the X'_p value. The lower the X'_p value will be, the lower the resonant frequency will be. X'_p values are summarized in the Table 3.

Table 3. X'_p values according to the ν and p values

	ν=0	ν=1/4	ν=1/3	ν=1/2	ν=2/3	ν=1	ν=2
p=1	3.832	0.769	0.910	1.166	1.401	1.841	3.054
p=2	7.016	4.225	4.353	4.604	4.851	5.331	6.706

By applying the good boundary conditions on each faces, fundamental modes of the different shapes presented in Table 4 can be determined. This table is reminding the X'_p values with the corresponding resonant frequencies for a DRA such as a=40mm and d=45mm with a dielectric permittivity equals to 10.

Following these results, resonance frequencies depend on whether parts of some faces are coated with a metal or not. A good trade-off between antenna size and low resonant frequency is to choose the third line of the Table 4, when one face is coated by metal. Finally, this study has demonstrated that the metallization of some DRA faces allows creating new resonant modes, which have resonant frequencies lower than the fundamental modes inside a classical cylindrical DRA. It should be noted that these results have been done with cavity resonator model, i.e. the outer surfaces of the cavity are approximated by perfect magnetic walls. They thus show an approximate analysis of the fields inside the resonator. However, this model provides reasonable accuracy for prediction of resonant frequencies.

Table 4. Excited modes with the corresponding DRA shapes

Mode	ν	p	$X'_{\nu p}$	$f_{\nu pm}$	Shape
$HE_{21\delta}$	2	1	3.054	1.27 GHz	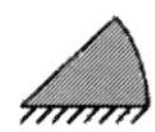
$HE_{11\delta}$	1	1	1.841	872 MHz	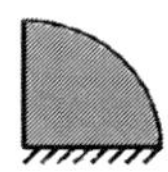
$HE_{1/2\,1\,\delta}$	1/2	1	1.166	687 MHz	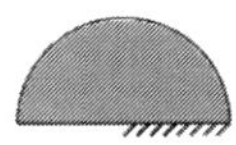
$HE_{1/3\,1\,\delta}$	1/3	1	0.910	629 MHz	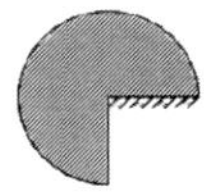
$HE_{1/4\,1\,\delta}$	1/4	1	0.769	602 MHz	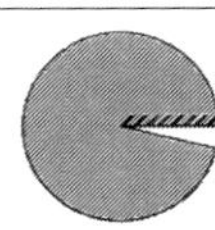

5. NEW APPROACHES FOR WIRELESS APPLICATIONS

Antenna design for mobile communications is often problematic by the necessity to implement multiple and/or ultra wideband applications on the same small terminal.

Some of currently antennas integrated in portable wireless systems have a planar structure based on microstrip patches or PIFAs [33-35]. These kinds of antennas present a low efficiency, especially when small, because of metallic losses. DRAs do not suffer from such losses, which makes them a good alternative for these more conventional antennas. That is why, in recent years, much attention has been given to DRA miniaturization [16,36] in order to integrate them inside mobile handheld.

This part is divided in two sub-sections. The first one will focus on the bandwidth enhancement of a DRA for ultra wideband applications, and the second part will aim multiband applications. The common thread is the miniaturization of DRAs while obtaining good performances. For that, new hybrid modes will be studied with the application of partial electric boundary on sides of DRAs.

5.1. For ultra wideband application

A tendency and well known ultra wideband application is the Digital Video Broadcasting-Handheld (DVB-H). Indeed, the allocated frequency range is divided in two sub-bands, i.e. [470 MHz – 790 MHz] and [790 MHz – 862 MHz]. Considering the entire DVB-H frequency range, it presents 58% of bandwidth and a few kind of antenna can cover it instantaneously while being miniature and having good performances.

As presented in previous parts, for a single-mode excitation, the DRAs bandwidth doesn't exceed 15%. Recently, different shapes and stacked resonators have been proposed to enhance the bandwidth. These techniques are generally difficult to implement without increasing the size of the antenna. A novel DRA design method is proposed in this part with detailed parametric studies [37]. After seeing the proposed DRA and its miniaturization technique, the study will focus on the bandwidth enhancement. In order to design and optimize the proposed antenna, both eigenmode solver and the Finite Integration Temporal method of CST Microwave Studio are used. The measurement and simulation results will be shown, followed by a discussion.

Previous parts have detailed the miniaturization technique by using a circular sector DRAs. It has been shown that resonance frequencies depend on whether parts of some faces are coated with a metal or not. According to the equation 7 and in order to easily integrate the antenna inside a terminal, *d* has to be lower than 25 mm. In this case, the lowest mathematical resonance frequency which can be obtained is 949MHz. The best compromise between the size and the desired frequency band is to use the $HE_{½1}$ mode (Table 4). In order to entirely cover the DVB-H band, the non-coated face of the selected shape is transformed to a cubic part. The final designed DR shape is shown Figure 19.

(a) (b)

Figure 19. Proposed DR shape (a) and metallization of two different sides (b)

As mentioned previously, coating some fáces with a metal allows resonance frequencies decreasing. A further parametric study is performed in order to determine how to coat some faces with a metal and how to choose the resonator position on the ground plane. In the Figure 19, *a* and *b* are defined as the metallization lengths of two different metallic sides.

Figure 20 represents the variation of the first and the second resonance frequencies mode according to *a* and *b* lengths.

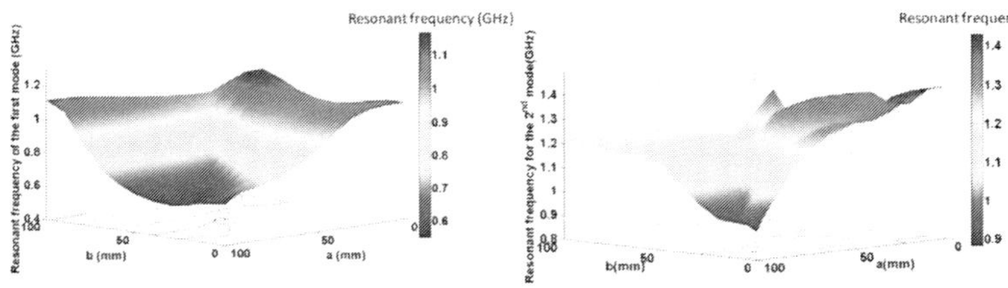

(a) (b)

Figure 20. Variation of the resonance frequency of the 1st (a) and 2nd (b) modes according to a and b lengths

The resonance frequencies decrease significantly when the two metallic sides have the same potential and each one covers half of the face. So, the optimum metallization lengths are *a*=50mm and *b*=50mm and the two lowest resonance frequencies are 568MHz and 1GHz. Through the above results, the chosen structure looks like the Figure 19 with *a*=*b*=50mm. Now, it is necessary first of all to glance over the E-field distribution inside the DR. Figure 21 shows the E-field inside the DR for the first and the second mode respectively at 568 MHz and 1 GHz. The first mode is derivative from the $HE_{\frac{1}{2}1}$ mode.

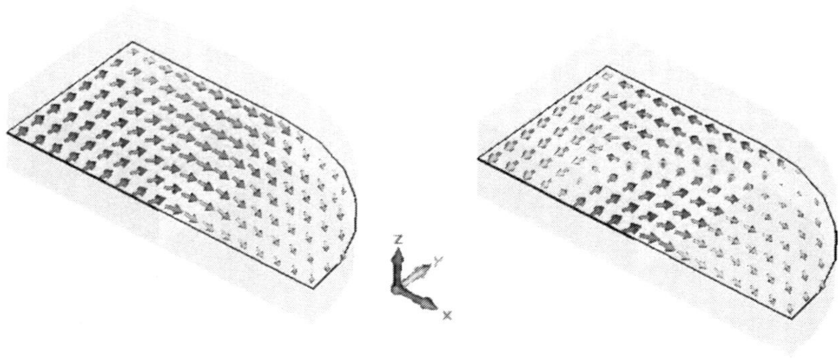

Figure 21. E fields of the two first modes

In the following, a ground plane is inserted on the lower metallic face as shown Figure 22. The ground plane size is 230mm x 130mm, chosen to correspond to a standard DVB-H handheld receiver. The DR of $\lambda_0/7x\lambda_0/13x\lambda_0/28$ dimensions at 470 MHz is mounted on such ground plane.

Considering the E fields distribution, a probe is chosen to feed the DRA. It is placed on the lateral metallic side in order to simultaneously excite the two first modes (Figure 22).

To reach an ultra wideband, the probe position, diameter and length are tuned. The optimal probe diameter and length values are 1.5mm and 49mm.

Following this optimization, input impedance and return loss are shown in the Figure 23.

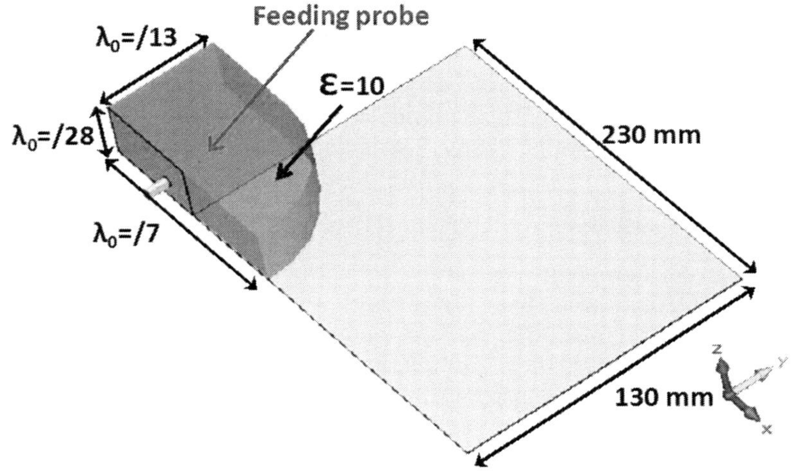

Figure 22. Antenna design fed by a coaxial probe

Figure 23. Input impedance (a) and S11 parameter (b) of the considered DRA

They show that two modes exist at 0.6 MHz and 0.93 MHz, excited by the coaxial probe. As expected, a difference appears between resonance frequency values which are obtained by the eigenmode solver and electromagnetic simulation. This discrepancy is due to the presence of the probe and the ground plane. Furthermore, the boundary conditions applied during the modal analysis are perfect electric or magnetic conditions on the DR walls contrarily to the electromagnetic simulation.

The simulated coefficient reflection shows 70% of bandwidth for a -8 dB impedance bandwidth definition over the frequency range [466 MHz – 935 MHz]. The -8dB impedance bandwidth definition is sufficient to achieve a good efficiency. Thus, the antenna satisfies the DVB-H system specifications mentioned previously. To demonstrate the antenna performances, the antenna has been realized with a ceramic material. It is fed by a probe and is half-mounted on a ground plane. The manufactured antenna is shown Figure 24.

The input impedance and the coefficient reflection are shown Figure 25.

There is a good agreement between simulations and measurements. It should be underlined that these simulations series take into account that the probe is inserted in a 2 mm diameter hole instead of a 1.5 mm hole due to mechanical constraint, therefore an air gap appears around the probe. This air gap results in lowering the DRA effective dielectric constant, which is turn lower the Q-factor accompanied by a shift in the resonance frequency. Furthermore, after material characterizations, the ceramic's dielectric constant turns out equal to 9.5 at 500 MHz. So, the -8dB impedance matching bandwidth is included in [540 MHz – 1.05 GHz].

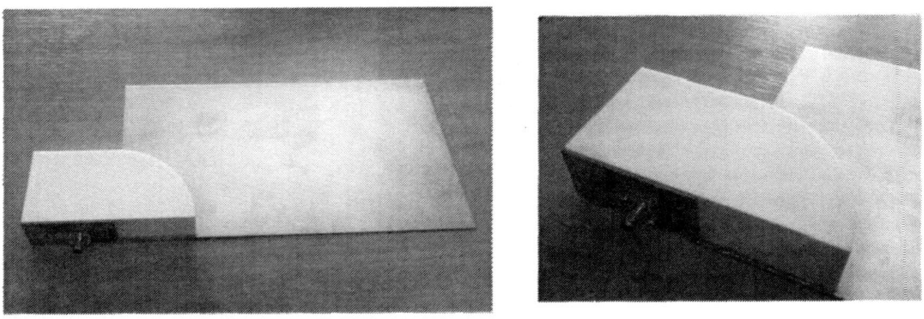

Figure 24. Antenna prototype

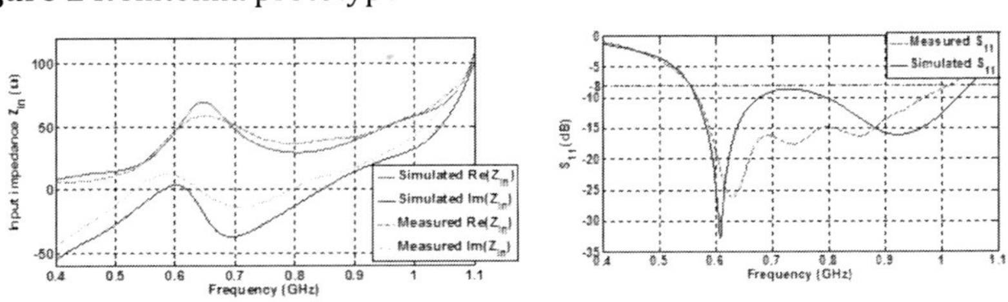

(a) (b)

Figure 25. Measured and simulated input impedances (a) and S_{11} parameters (b)

Radiation patterns of the antenna were characterized in an anechoic chamber. Simulated and measured radiation patterns at 620MHz and 870MHz are shown Table 5.

It can be noticed that the *xz*-plane and *yz*-plane radiation patterns indicate a correct omnidirectional radiation patterns. The antenna operates at two different modes inside the operating band, so the antenna doesn't offer the same radiation pattern at the two resonance frequencies, particularly in the *xy*-plane. The simulated and measured total efficiencies of the proposed antenna are illustrated Figure 26.

This efficiency disagreement is probably due to the coaxial feeding cable, the discrepancy between the simulated and measured reflection coefficient and the dielectric loss tangent.

In spite of this 10% difference, there is a good agreement between measured and simulated efficiencies. The measured efficiency remains higher than 75% above 600MHz.

This part has proposed a novel dielectric resonator antenna. Parametric studies have been realized to decrease resonance frequencies and to increase impedance matching bandwidth. The results have shown that by shaping the dielectric structure and coating some faces with a metal, resonance frequencies have been reduced for a fixed structure size. Good performances have been obtained and the proposed antenna can be used for DVB-H and/or GSM900 applications with a wide bandwidth and a good efficiency.

Table 5. Simulated (blue line) and measured (red line) radiation patterns at 620 MHz and 870 MHz

	$\phi=0°$	$\phi=90°$	$\theta=90°$
f=620 MHz	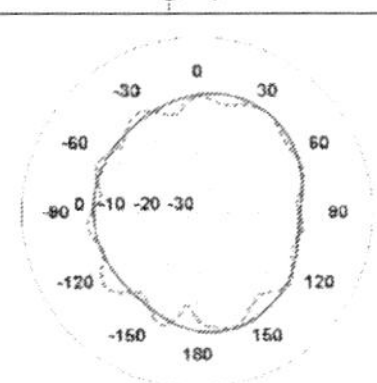	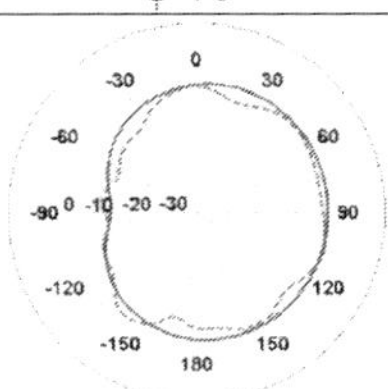	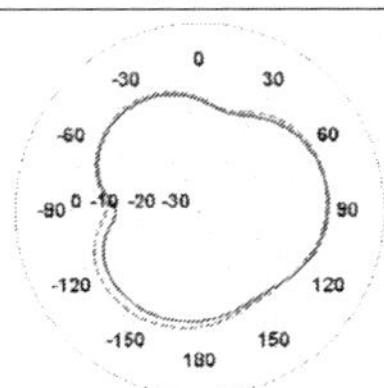
f=870 MHz	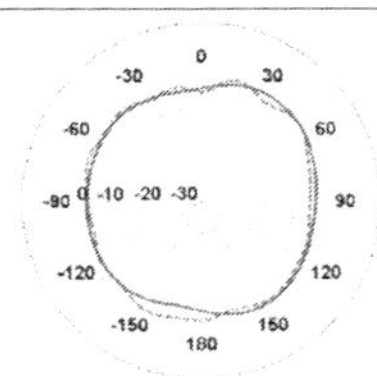	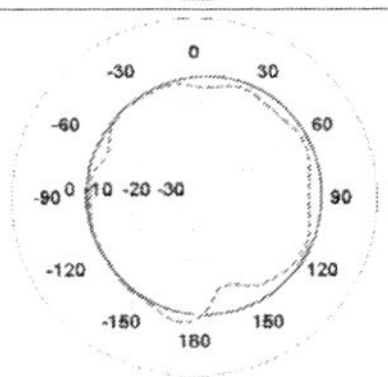	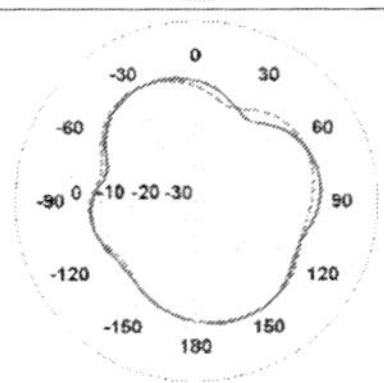

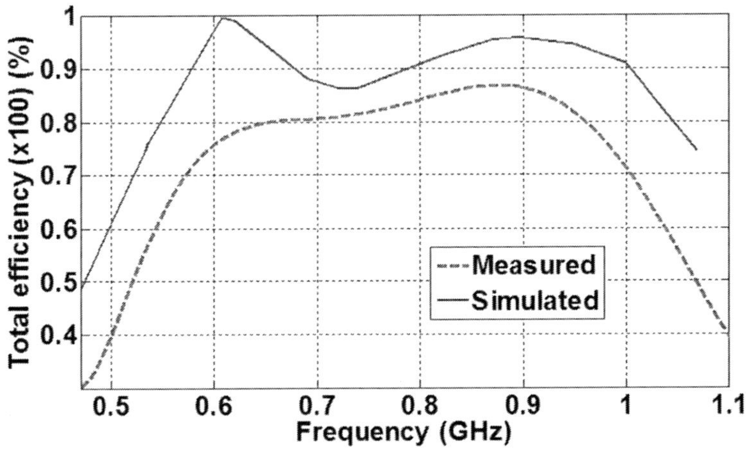

Figure 26. Measured and simulated total efficiencies of the considered DRA

5.2. For multiband application

In the last decade, the huge demand for mobile and portable communication systems has led to an increased need for more compact antenna designs. This aspect is even more critical when several wireless technologies have to be integrated on the same mobile wireless communicator. All the new services and the increased user density are driving the antenna design toward multi-band operation.

Recently, many studies have been devoted to multi-band antennas [38-41], some of them dealing with DRAs [42-44]. A dielectric resonator indeed supports more than one resonant mode at two close frequencies, which allows them to meet the requirements of different applications with a unique device. Some studies furthermore use both the dielectric resonator and the feeding mechanism as radiator elements [43-45]. This explains why the DRAs present a major advantage for multi-standard devices, when compared to other kinds of antennas.

The objective of this part is to show the integration of a small antenna in a multi-band mobile handheld device, working on the nine channels of the second sub-band of the DVB-H, i.e. [790 MHz – 862 MHz], the WiFi band at 2.4 GHz and the WiMax band at 3.5 GHz. Additionally, in order to improve the quality and the reliability of the wireless links, i.e. obtain pattern diversity, two antennas will be integrated in the same device. Setting maximum limits, we have decided to integrate two orthogonally aligned antennas in the allocated space of 30 mm x 41 mm ($\lambda_0/13$ x $\lambda_0/9$ at 800

MHz), on a 230 mm x 130 mm ground plane. Each DRA will therefore have to be very compact to be able to fit in such a limited area, and will operate around 850 MHz, 2.4 GHz and 3.5 GHz, thus covering the nine channels of the DVB-H band, the WiMax and the WiFi band.

Firstly, only one radiating element is described. It will be done thanks to a modal analysis [46] of the dielectric resonator. The DRA design and the choice of the dielectric permittivity will be discussed. As for the previous sub-section, both the Finite Integration Temporal method and the eigenmode solver of CST Microwave Studio were used to carry out this work.

Getting three resonant frequencies for a single dielectric resonator is aiming. Firstly, the resonator has to be integrated in a handheld receiver, which means that it will be placed on a FR4 substrate (ε_r=4.9). Secondly, to be integrated in a handheld device, the allocated space for the antenna system must not exceed 41mm x 30 mm x 4mm as shown Figure 27.

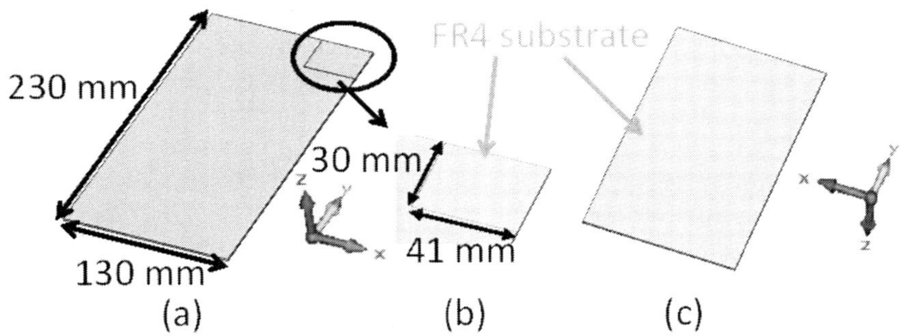

Figure 27. Top view of the defined PCB card (a), allocated size for the antenna system (b) and bottom view of the defined PCB card (c)

The dimensions and dielectric permittivity of each resonator need to be chosen according to these constraints, to ensure the integration of the antenna in the final device.

As a result, the dimensions of one resonator were chosen to be 25 mm x 10 mm x 4 mm, with a very high dielectric constant $_r$ of 37. The resulting geometry is depicted Figure 28.

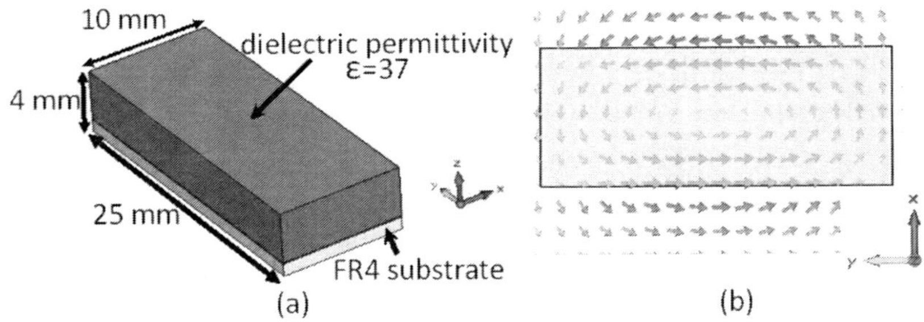

Figure 28. Dimensions and properties of one resonator (a) and E-field distribution of the first natural mode for the resonator in (b) in the z=2mm plane

The first natural mode of this resonator is the $TE_{11\delta}$ (Figure 28) which resonates at 3.99 GHz. This resonance frequency remains too high for the intended applications. It has however been shown [37], that this resonance frequency depends on the metallization of the DRA's faces. It is therefore necessary to envision the feeding mechanism of the resonator, before performing the modal analysis. Indeed, if the antenna is not fed by proximity coupling, the excitation cannot be ignored during the modal analysis. Furthermore, it can be used to adjust the resonance frequency of the resonator, especially in the case of an electrically small DRA. In this study, the feeding will indeed play a preeminent role. The chosen excitation is a line printed on the FR4 substrate and positioned under the dielectric resonator as shown Figure 29. The E-field distribution of the first mode (Figure 29) is completely different from the one obtained without this line. Indeed, as stated before, the E-field and H-field distributions inside the DRA depend on the boundary conditions on its faces. The feeding line introduces partially perfect electric conducting conditions, which disturb the field distribution inside the resonator when compared with the $TE_{11\delta}$ mode. As a result, the resonance frequency of each mode will vary in accordance with the length and width of the feeding line (defined in the Figure 29).

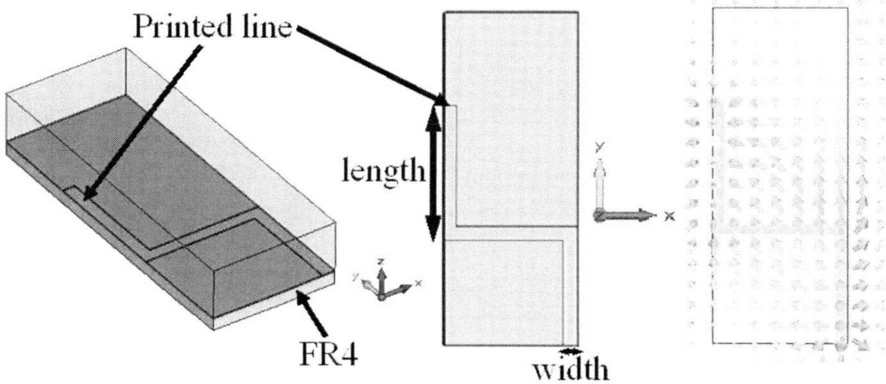

Figure 29. Design of the DRA fed by the printed line and E-field distribution of the first natural mode for the resonator fed by the line

Considering the previous study, the resonator design method is as follows.

- First of all, in spite of the high dielectric permittivity, there is no mode around 800 MHz. In order to allow the antenna to operate on the nine channels of the DVB-H band, the printed line will be designed to resonate around 800 MHz. It will therefore behave like a printed monopole loaded by a dielectric. The length of this line, which will from now be referred as the "monopole", will be set to obtain the first resonance frequency around 800 MHz.

- This being done, the second and third band will be covered by the resonances of the dielectric resonator disturbed by the presence of the monopole. As previously explained, the length, width and shape of this monopole entail a modification of the DRA resonance frequencies. The shape and width of the monopole will therefore be optimized to obtain the desired resonance frequencies, while its length remains set by the first resonance.

- An important point concerns the radiation Q-factor. The high dielectric permittivity involves a high Q-factor [7]. With such an $_r$, the radiation Q-factor is also important, making it difficult to obtain a wide impedance bandwidth for a given mode. So, to have a suitable impedance bandwidth, the antenna will have to be matched between two peaks of the real part of input impedance. By this way, the resonances won't have to be close to the operating bands. The goal being to match the DRA over the WiFi and WiMax bands, the monopole (its shape and width) has to be optimized to obtain resonance frequencies around 2 GHz, 2.8 GHz and 4 GHz.

- A modal analysis has been performed to show the variations of the resonance frequencies of the first three modes, according to the shape and width of the monopole. Figure 30 shows the resonance frequency of the first mode according to the monopole geometry. With the first resonance frequency set to match the antenna at 2.4 GHz and the length of the monopole set to have the first resonance at 800 MHz, the graph of the modal analysis allows an easy determination of the monopole width, which is 1 mm.

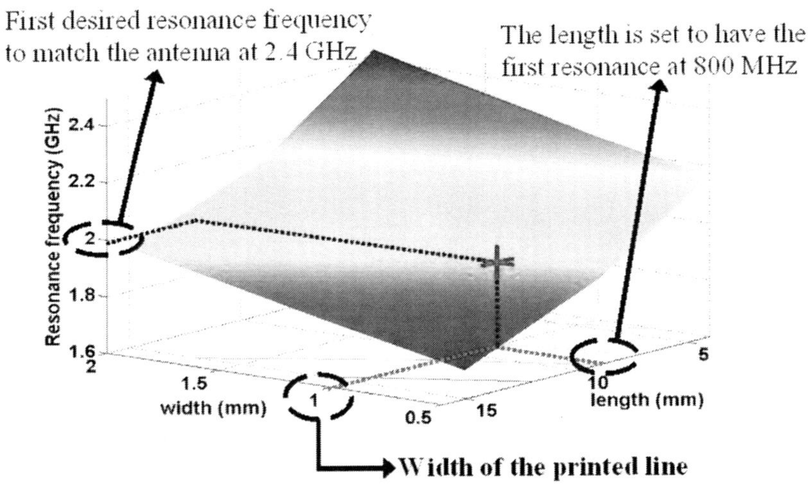

Figure 30. Resonance frequency of the first dielectric resonator mode according to the width and the length of the monopole previously defined

Same studies have been performed for different shapes of the monopole. All these studies allowed the shape, length and width of the monopole to be set, which led to the final design of the dielectric resonator. Table 6 shows the values of the resonance frequencies for the first three modes inside the resonator, which were obtained through the modal analysis.

- Electromagnetic study of the DRA

In order to validate the previous modal study, the dielectric resonator, placed in the area dedicated to the antenna (Figure 27) and fed by a 50Ω discrete port, has been simulated with the FIT method using CST Microwave Studio. It must be noticed that a discrete port is modeled by a lumped element, consisting of a current source with a 50 Ω inner impedance that excites and absorbs power.

Table 6. Values of Resonance Frequency for the Three First Modes

Mode	Resonance frequency
First mode	1.969 GHz
Second mode	2.773 GHz
Third mode	4.135 GHz

The Figure 31 shows the simulated input impedance of the dielectric resonator with its feed. The resonance frequencies are in agreement with the modal analysis. The radiation Q-factor is important, and the input impedance variations confirm that the antenna matching is easier between two resonances. The reflection coefficient is shown Figure 31. The antenna is matched on all of the desired bands, i.e. the nine channels of DVB-H going from 790 MHz to 862 MHz, the WiFi band at 2.4 GHz and the WiMax band at 3.5 GHz. It can be noted that the matching over the first band is obtained due to the resonance of the λ/4 monopole.

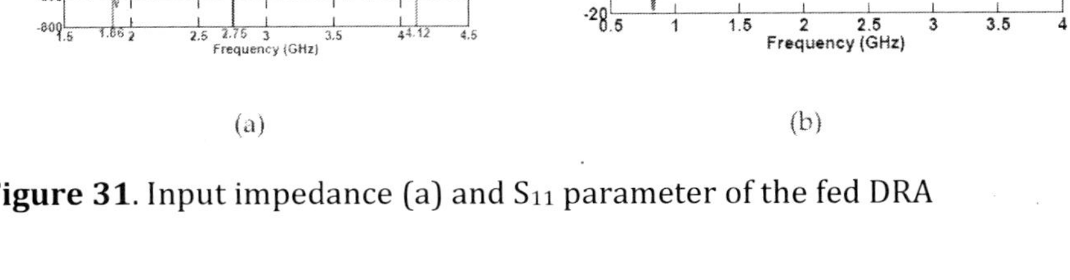

(a) (b)

Figure 31. Input impedance (a) and S_{11} parameter of the fed DRA

The dielectric resonator must now be integrated in its context, i. e. on a 230 mm x 130 mm ground plane, chosen to correspond to a standard DVB-H handheld receiver. As explained before, another specification was to obtain a reconfigurable radiation pattern. Thus, two instances of the previously studied resonator are orthogonally integrated on the ground plane.

- Final structure

Based on the previous parametric study, the final structure has been designed as shown Figure 32. In order to obtain pattern diversity, two dielectric resonators are orthogonally disposed in a 30 mm x 41 mm area, both fed by a printed line acting as a monopole. Each line is fed by a 50Ω coaxial cable. They are studied on a 230 mm x 130 mm ground plane, as defined by the specifications. It will be shown in a following section that the antenna matching is not affected by the ground plane dimensions. In order to ascertain the performances of this antenna, a prototype has been fabricated as shown Figure 32.

The resonators are manufactured with a ceramic material with a dielectric permittivity of 37 and a loss tangent tanδ=0.005 on the 0.5-10 GHz band. During the simulation and measurements, each resonator has been excited by a printed line fed by a coaxial cable.

During the manufacturing process of the antenna, a special care has to be given to minimize the air gap between the excitation and the resonator. In the case of this prototype, the resonators have been pressed onto the PCB to avoid this air gap.

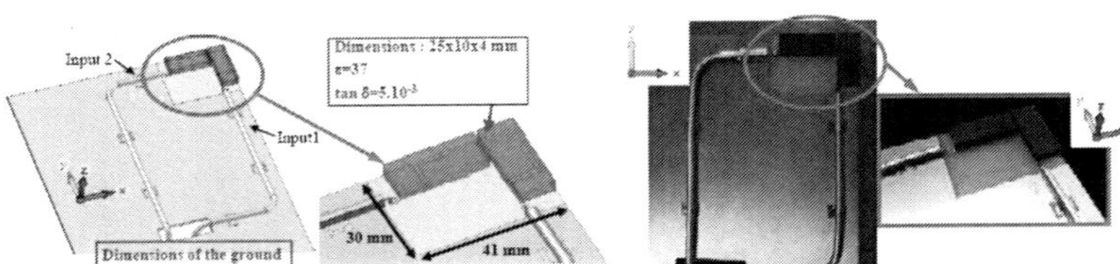

(a) (b)

Figure 32. Final simulated design (a) and the corresponding prototype (b)

- Measured and simulated performances of the antenna

The comparison between the simulated and measured results is now studied. The first ones have been obtained using the transient solver of CST Microwave Studio, while the measurements have been performed inside an anechoic chamber. Both the simulated S11 and S22 parameters are compared with the measured ones Figure 33.

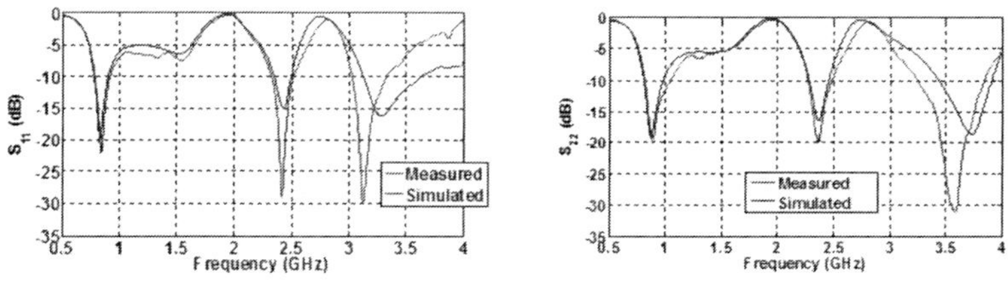

(a) (b)

Figure 33. Measured and simulated S_{11} (a) and S_{22} (b) parameters of the DRA

The measurements and simulations are in very good agreement. Moreover, the antenna is matched over all the desired bands and for both inputs.

Radiation patterns: The radiation patterns have been measured inside an anechoic chamber. Table 7 shows the 3D simulated radiation patterns for both inputs at 830 MHz, 2.4 GHz and 3.5 GHz.

It can be seen that the radiation pattern at a given frequency will depend on the excited port. While promising, this result is not sufficient to conclude that the radiation pattern is reconfigured. More details are explained in [47] (this requires the characterization of the whole system in a reverberation chamber, in order to determine the correlation coefficient).

Thus, this study started with modal analyses, which allowed the shape and dimensions of the antenna's excitation to be defined with a dual objective. Indeed, this line had first to behave like a monopole and cover the nine channels of the DVB-H band (going from 790 MHz to 862 MHz). Secondly, it had to excite the dielectric resonator and set its resonance frequencies so as to match the antenna on the WiFi and WiMax bands.

After performing these preliminary studies, two instances of the conceived dielectric resonator have been orthogonally integrated on a 230 mm x 130 mm ground plane, which is consistent with a tablet. Finally, the antenna system, which only occupies a 30 mm x 41 mm area, is matched on the three desired bands, i.e. the nine channels of the DVB-H band, the WiFi band and the WiMax band with pattern diversity.

This part has presented the design method, the realization and the measurement of a two compact DRAs, one for ultra wideband application and the second for multiband applications.

	f=830 MHz	f=2.4 GHz	f=3.5 GHz
1st input	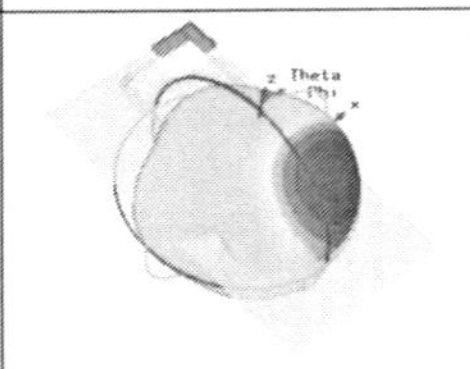	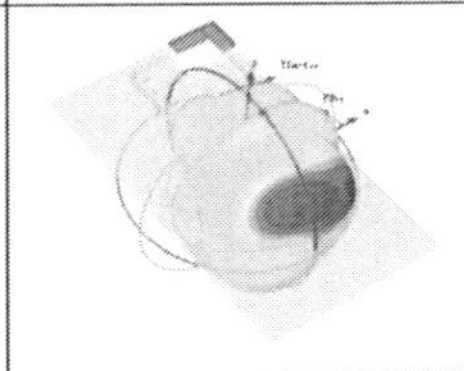	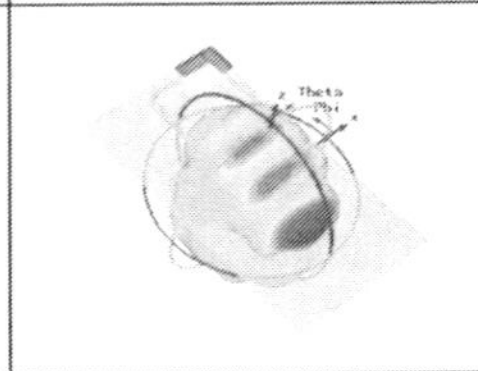
2nd input	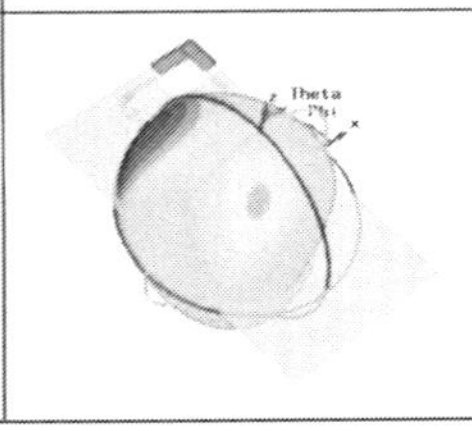	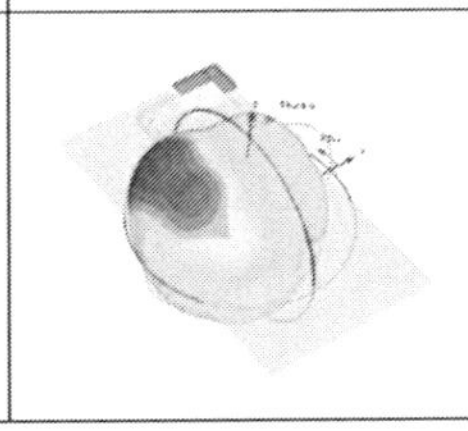	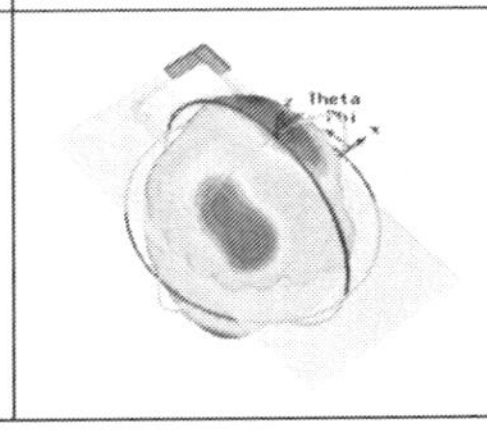

Table 7. 3D Radiation patterns at 830 MHz, 2.4 GHz and 3.5 GHz for the two inputs

6. CONCLUSION

To conclude, an affordable chapter has been presented allowing the reader to find an overview of main DRA shapes, properties and approaches while appreciating the influence and the impact of the dielectric material properties. Indeed, a broad spectrum of dielectric materials can be used depending on the intended application. In addition to the advantages common to all DRAs described at the beginning of this chapter, a dedicated part has focused on other advantages of compact DRAs, which are desirable for many emerging wireless and mobile communication systems. Finally, a specific part had presented relevant data for postgraduate researchers, antenna design engineers in general and particularly the ones engaged in the innovative design of mobile and wireless systems by focusing on the hybrid modes creation to enhance the bandwidth or develop multiband antennas.

REFERENCE

1. R. D. Richtmyer, "Dielectric Resonator", J. Appl. Phys., vol. 10, pp. 391-398, Jun. 1939
2. D. Kajfez and P. Guillon, Eds., Dielectric Resonators. Norwood, MA: Artech House, 1986
3. Cohn, S.B.;, "Microwave Bandpass Filters Containing High-Q Dielectric Resonators," Microwave Theory and Techniques, IEEE Transactions on, vol.16, no.4, pp.218227Apr 1968
4. S. J. Fiedziuszko, "Microwave Dielectric Resonators", Microwave Journal, vol. 29, September 1986, pp 189-200
5. S.A. Long, M.W. McAllister and L.C. Shen, "The Resonant Dielectric Cavity Antenna", IEEE Transactions on Antennas and Propagation, Vol.n°3, March 1983406412
6. A. Petosa, A. Ittipiboon, Y.M.M. Antar and D. Roscoe, "Recent Advances in Dielectric Resonator Antenna Technology", IEEE Antennas and Propagation Magazine, Vol. 40, n°3, 06/1998, pp. 35-48
7. K.M Luk and K.W Leung, "Dielectric Resonator Antennas", Electronic & Electrical Engineering Research Studies
8. R.K Mongia and A. Ittipiboon, "Theoretical And Experimental Investigations on Rectangular Dielectric Resonator Antenna", IEEE Transactions on Antennas and Propagation, Vol. 45, n°9, September 1997, pp. 1348-1356
9. D. Drossos, Z. Wu and L.E. Davis, "Theoretical and experimental investigation of cylindrical Dielectric Resonator Antennas", Microwave and Optical Technology Letters, Vol. 13, No. 3, pp. 119-123, October 1996
10. R. K. Mongia and P. Bhartia, "Dielectric Resonator Antennas- A review and General Design Relations for resonant Frequency and Bandwidth", International Journal of Microwave and Millimeter-wave Computer-Aided Engineering, Vol. 4, No. 3, pp. 230-247, Mar. 1994
11. A.A. Kishk, B. Ahn and D. Kajfez, "Broadband stacked dielectric resonator antennas", IEE Electronics Letters, Vol. 25, n°18, Aug. 1989, pp. 1232-1233

12. A. Petosa, “Dielectric Resonator Antenna Handbook”, Artech House, Boston/London, 2007

13. R. Chair, A. A. Kishk, K. F. Lee, “Wideband Stair-Shaped Dielectric Resonator Antennas,” IET Microwaves, Antennas & Propagation, Vol. 1, Issue 2, pp. 299-305, April 2007

14. Wei Huang and Ahmed Kishk,”Compact Wideband Multi-Layer Cylindrical Dielectric Resonator Antenna,” IET Microwave Antenna and Propagation, Vol.59981005October 2007

15. R. Chair, A. A. Kishk, K. F. Lee, “Low Profile Wideband Embedded Dielectric Resonator Antenna,” IET Microwaves, Antennas & Propagation, Vol. 1, Issue 2, pp. 294- 298, April 2007

16. Y. Gao, B. L. Ooi, W. B. Ewe, A. P. Popov, “A compact wideband hybrid dielectric resonator antenna”, IEEE Microw.Wirel. Compon. Lett., 2006, 16, (4), pp. 227-229

17. K. A. A. Wei Huang, “Use of electric and magnetic conductors to reduce the DRA size”, Int. Workshop on Antenna Technology: Small and Smart Antennas Metamaterials and Applications, IWAT ‘07, 2007

18. J.M Ide, S.P Kingsley, S.G O’Keefe, S.A Saario, "A novel wide band antenna for WLAN applications," Antennas and Propagation Society International Symposium,2005IEEE, 4Ano., 243246vol. 4A, 3-8 July 2005

19. L. Huitema, M. Koubeissi, C. Decroze, T. Monediere, "Handheld Dielectric Resonator Antenna for Ultra Wideband Applications", 2010 IEEE International Workshop on Antenna Technology: iWAT2010: « Small Antennas and Novel Metamaterials» March 1-3, 2010, pp. 1-4, Portugal

20. L. Huitema, M. Koubeissi, C. Decroze, T. Monediere, "Compact and multiband dielectric resonator antenna with reconfigurable radiation pattern," Antennas and Propagation (EuCAP), 2010 Proceedings of the Fourth European Conference on, vol., no., pp.1-4, 12-16 April 2010

21. A. Sangiovanni, J. Y. Dauvignac and Ch. Pichot, “Stacked dielectric resonator antenna for multifrequency operation”, Microw. and Opt. Techn. Lett., vol. 18, pp. 303-306, July 1998

22. Z. Fan and Y.M.M. Antar, "Slot-Coupled DR Antenna for Dual-Frequency Operation," IEEE Trans. Antennas and Propagation, Vol. 45, No. 2, Feb. 97,pp. 306-308

23. G.P. Junker, A.A. Kishk, A.W. Glisson and D. Kajfez, "Effect of an air gap around the coaxial probe exciting a cylindrical dielectric resonator antennas", Electronics Letters, Vol. 30, No. 3, pp. 177-178, 3rd February 1994

24. G.P. Junker, A.A. Kishk, A.W. Glisson and D. Kajfez, "Effect of air gap on cylindrical dielectric resonator antennas operating in TM01 mode", Electronics Letters, Vol. 30, No. 2, pp. 97-98, 20th January 1994

25. Kwok-Wa Leung; Kwai-Man Luk; Lai, K.Y.A.; Deyun Lin;, "Theory and experiment of an aperture-coupled hemispherical dielectric resonator antenna," Antennas and Propagation, IEEE Transactions on, vol.43, no.11, pp.1192-1198, Nov 1995

26. A.A. Kishk, A. Ittipiboon, Y. Antar, M. Cuhaci "Slot Excitation of the dielectric disk radiator", IEEE Transactions on Antennas and propagation, vol. 43, N°2, pp.198-201, Feb 1993

27. A.W. Glisson, D. Kajfez and J. James, "Evaluation of modes in dielectric resonators using a surface integral equation formulation," IEEE Trans. Microwave Theory Tech., Vol. MTT-31, pp. 1023-1029, 1983

28. D. Kajfez, A. W. Glisson and J. James, "Computed modal field distributions for isolated dielectric resonators," IEEE Trans. Microwave Theory Tech., Vol.MTT-32, pp. 1609-1616, 1984

29. Y. Kobayashi and S. Tanaka, "Resonant Modes of a dielectric rod resonator short circuited at both ends by parallel conducting plates", IEEE Trans. Microwave Theory and Tech., Vol. 28, pp. 1077-1085, Oct. 1980

30. Van Bladel, J., "On the Resonances of a Dielectric Resonator of Very High Permittivity," Microwave Theory and Techniques, IEEE Transactions on, vol.23, no.2, pp.199208Feb 1975

31. Petosa, A.; Simons, N.; Siushansian, R.; Ittipiboon, A.; Cuhaci, M.;, "Design and analysis of multisegment dielectric resonator antennas," Antennas and Propagation, IEEE Transactions on, vol.48, no.5, pp.738-742, May 2000

32. M.T.K Tam and R.D Murch, "Compact circular sector and annular sector dielectric resonator antennas", IEEE Transactions on antennas and Propagation, Vol.n° 5, 1999837842

33. C. W. Ling, C. Y. Lee, C. Y. Tang, and S. J. Chung "Analysis and Application of an On-Package Planar Inverted-F Antenna", IEEE

Transactions on Antennas and Propagation, Vol. 55, n°6, June 2007, pp. 1774- 1780.

34. M. J. Ammann and L. E. Doyle, "A loaded inverted-F antenna for mobile handset," Microwave Opt. Technol. Lett., vol. 28, pp. 226-228, 2001.

35. M. Ali andG. J. Hayes, "Analysis of integrated inverted-F antennas for Bluetooth application," in Proc. IEEE-APS Conf. Antennas and Propagation for Wireless Communications, Waltham, MA, 2000, pp. 21-24.

36. K. A. A. Wei Huang, "Use of electric and magnetic conductors to reduce the DRA size", Int. Workshop on Antenna Technology: Small and Smart Antennas Metamaterials and Applications, IWAT '07, 2007

37. L. Huitema, M. Koubeissi, C. Decroze, T. Monediere, "Ultrawideband Dielectric Resonator Antenna for DVB-H and GSM Applications" IEEE Antennas and Wireless Propagation letter, vol. 8, pp. 1021-1027, 2009

38. Y.-Y. Wang and S.-J. Chung, "A new dual-band antenna for WLAN applications," in Proc.IEEE AP-S Int. Symp., Jun. 20-25, 2004326112614

39. 521903 EOF1905 EOFJan, J. Y. and L. C. Tseng, "Small planar monopole antenna with a shorted parasitic inverted-L wire for wireless communications in the 2.4-, 5.2-, and 5.8-GHz bands," IEEE Trans. Antennas Propag., Vol. 52, 1903-1905, 2004

40. Z. D. Liu andP. S. Hall, "Dual-band antenna for handheld portable telephones," Electron. Lett., vol. 32, pp. 609-610, 1996.

41. J. Y. Jan andL. C. Tseng, "Planar monopole antennas for 2.4/5.2 GHz dual-band application," in Proc. IEEE-APS Int. Symp. Dig., Columbus, OH, 2003, pp. 158-161

42. K. Hady, A. A. Kishk and D. Kajfez, "Dual-Band Compact DRA With Circular and Monopole-Like Linear Polarizations as a Concept for GPS and WLAN Applications", IEEE Trans. on ant. and prpoag., Vol. 57, No. 9, 2591-2598, September 2009.

43. Rao, Q.; Denidni, T.A.; Sebak, A.R.; Johnston, R.H.;, "Compact Independent Dual-Band Hybrid Resonator Antenna With Multifunctional Beams," Antennas and Wireless Propagation Letters, IEEE, vol.5, no.1, pp.239-242, Dec. 2006

44. Rotaru, M.; Sykulski, J.K.;, "Numerical investigation on compact multimode dielectric resonator antennas of very high permittivity," Science, Measurement & Technology, IET, vol.3, no.3, pp.217-228, May 2009

45. Qinjiang Rao; Denidni, T.A.; Sebak, A.R.;, "A hybrid resonator antenna suitable for wireless communication applications at 1.9 and 2.45 GHz," Antennas and Wireless Propagation Letters, IEEE, vol.4, no., pp.2005

46. R.A Kranenburg, S.A Long, "Microstrip transmission line excitation of dielectric resonator antennas," Electronics Letters, vol.24, no.18, pp.1156-1157, 1 Sep 1988

47. L. Huitema, M. Koubeissi, M. Mouhamadou, E. Arnaud, C. Decroze And T. Monediere, "Compact and Multiband Dielectric Resonator Antenna with Pattern Diversity for Multi Standard Mobile Handheld Devices", IEEE Transaction on antennas and propagation, vol. 59, pp.4201-4208, 2011

2

CHAPTER

POLYMERIC DIELECTRIC MATERIALS

Zulkifli Ahmad[1]

[1] *University Sains Malaysia, Malaysia*

1. INTRODUCTION

1.1. Background and brief history

The definition for dielectric constant relates to the permittivity of the material (symbol use here ε). The permittivity expresses the ability of a material to polarise in response to an applied field. It is the ratio of the permittivity of the dielectric to the permittivity of a vacuum. Physically it means the greater the polarisation developed by a material in an applied field of given strength, the greater the dielectric constant will be. Traditionally dielectric materials are made from inorganic substances eg. mica and silicon dioxide. However polymers are gaining wider use as dielectric materials. This is due to the easier processing, flexibility, able to tailor made for specific uses and better resistance to chemical attack. As early as mid-60's polymers eg polyvinyl fluoride [1] and aromatic-containing polymers [2] are used as dielectric materials in capacitors. Further improvement in organic film fabrication was established as revealed in US Patent 4153925. Polymers can be fabricated fairly easily into thin film by solution casting and spin coating, immersion in organic substrate, electron or UV radiation and glow discharge methods. This is mainly due to lower thermal properties such as glass transition and melting temperature which contribute to a lower temperature processing windows.

Their solubility is controllable without offsetting their intrinsic properties. In the case of inorganic material and ceramic, they have much higher thermal properties hence temperature requirement leads to an extreme end of processing temperatures. On the other hand polymers cannot stand too high a temperature. Their coefficient of thermal expansion is relatively larger than ceramic materials and susceptible to atmospheric and hydrolytic degradation. Table 1 shows the values of dielectric properties of several polymers with comparisons with several inorganic materials.

Inorganic/ceramics materials have higher dielectric constant than polymers. Water has a relatively high dielectric constant. This is quite cumbersome as any traces of moisture trapped or absorb will dramatically alter the desired dielectric properties. Inorganic materials generally have higher dielectric constant compared to polymeric materials. Intrinsically they contains ions and polar groups which contribute to their high dielectric constant. Air having a dielectric constant of 1.02 is taken as reference dielectric.

Table 1. Dielectric constant of several polymers and inorganic materials. (Adapted from Ref 3)

Materials	Dielectric constant, ε	Materials	Dielectric constant, ε
TiO_2	100	Fluorinated polyimide	2.5 – 2.9
H_2O	78	Methylsilsesquioxane	2.6 – 2.8
neoprene	9.8	Polyarelene ether	2.8 – 2.9
PVDF	6.0	Polyethylene	2.3 – 2.7
SiO2	3.9 – 4.5	Polystyrene	2.5 – 2.9
Fluorosilicate glass	3.2 – 4.0	Teflon AF	2.1
Polyimide	2.8 – 3.2	Air	1.02

1.2. Application of polymeric dielectric materials.

Both dielectrics with low and high dielectric constant are essential in electronic industries. Low dielectric constant is required basically as insulators. They are known as passivation materials. Their applications ranged in isolating signal-carrying conductors from each other, fast signal propagation, interlayer dielectric to reduce the resistance-capacitance (RC) time delays, crosstalk and power dissipation in the high density and high speed integration [4]. They are of necessity in very dense multi-layered IC's, wherein coupling between very close metal lines need to be suppressed to prevent degradation in device performance. This role involve packaging and encapsulation. In electronic packaging, they separate interlayers and provide isolated pathways for electronic devices connection in multilayer printed circuit boards. As the trends are towards miniaturization in microprocessor fabrication, any decrease in relative permittivity will reduces the deleterious effect of stray and coupling capacitances. Dielectric naterials are also employed to encapsulate the balls which bridged the die and substrate. This encapsulation is specifically called underfill which helps to protect any circuitary failures as well as reducing thermal mismatch between the bridging layers.(Figure 1) In LED encapsulation low dielectric materials is used for insulation at the lead frame housing.

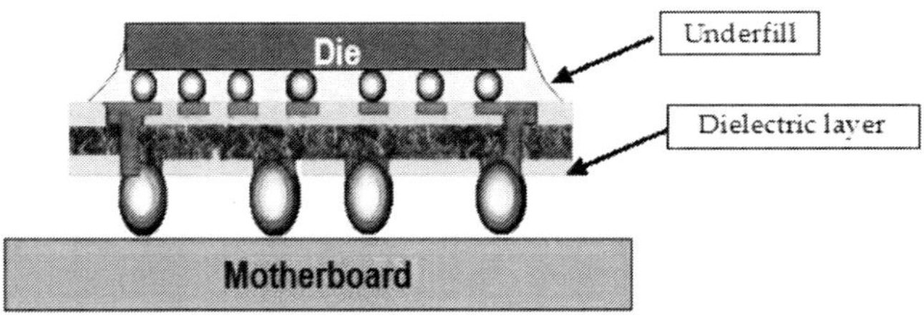

Figure 1. Application of dielectric polymers in IC packaging

As an active components, designing is geared towards high ε value and are used as polarizable media for capacitors, in apparatus used for the propagation or reflection of electromagnetic waves, and for a variety of artifacts, such as rectifiers and semiconductor devices, piezoelectric transducers, dielectric amplifiers, and memory elements. Despite being insulators, hence non-polar, these materials can be made polar by introducing small amount of impurities. In this state, the material is able to store large amount of charges at small applied electrical field. This is the case with polyvinylidene fluoride when introduced with impurities chlorotrifluoroethylene.[5] Indeed several works have been performed on polymers like polyimide with added Al_2O_3, $BaTiO_3$ and ZrO_3 'impurities' [6,7,8] which showed an improved dielectric constant. Once there is large charge storage, it can be readily released on demand. In a rectifier, a capacitor is used to smooth off the pulsating direct current.

2. THEORY OF DIELECTRIC PROPERTIES IN POLYMER

2.1. Mechanism of interaction with electric field

Quantitative treatment of a dielectric in an electric field can be summarized using Clausius–Mossotti equation (1).

$$P = \frac{\varepsilon_r - 1}{\varepsilon_r + 2} \cdot \frac{M}{\rho} = \frac{N_A \alpha}{3\varepsilon_0}$$

(1)

P is the molar polarisability, *ϒr* ϒr is the relative permittivity, *ϒo* ϒo is the permittivity in vacuum, M is molecular weight of a repeat unit, ϟϟ is density, □ □ is polarisability, N_a is the Avogadro constant. This equation shows that dielectric constant is dependent on polarisability and free volume of the constituents element present in the materials. Polarisability refer to the proportionality constant for the formation of dipole under the influence of electric field. Thus its value is typical for each different type of atom or molecule.[9] The relation between polarizability with the permittivity of the dielectric material can be shown as in Equation (2):

$$\varepsilon_r = 1 + \frac{N\alpha}{\varepsilon_0}$$

(2)

It shows that relative permittivity *ϒr* ϒr is the ratio of total permittivity of one mole of material with that in vacuum. The dependency of free volume of relative permittivity thus originate from the volume involved in one mole of the material. Again the molar volume is characteristic of each different type of atom or molecule. Molar polarization therefore is obtained if the molar volume is introduced into these derivations leading to Clausius–Mossotti equation.

Physically, polarisability is induced when there is electric field applied onto the materials. In the absence of electric field, the electrons are distributed evenly around the nuclei. When the electric field is applied the electron cloud is displaced from the nuclei in the direction opposite to the applied field. This result in separation of positive and negative charges and the molecules behave like an electric dipole. There are three mode of polarizations [10]:

i. Electronic polarization – slight displacement of electrons with respect to the nucleus.

ii. Atomic polarization – distortion of atomic position in a molecule or lattice

iii. Orientational polarization – For polar molecules, there is a tendency for permanent dipole to align by the electric field to give a net polarization in that direction

When a static electric field is applied on to these materials the dipoles become permanently polarized giving a dielectric constant as ε_{static}. However if the field changes as when alternating electric current is applied, polarization will also oscillate with the changing electric field. All three modes of polarization contributing to the overall dielectric constant will be dependent on the frequency of the oscillating electric field. Obviously the electronic polarization is instantaneous as it is able to follow in phase with the changing electric field compared to atomic polarization which in turn better able to follow the oscillating electric field compared to the orientational polarization. Certain structures and elements display a higher polarisibility than the others. Aromatic rings, sulphur, iodine and bromine are considered as highly polarisable. The present of these groups induced an increase in dielectric constant. The π bond in the aromatic rings is loosely attached compared to the sigma bond. Therefore it is easily polarized. For large size atoms like bromine and iodine, the electron cloud is so large and further apart from the influence of electrostatic attraction of the positive nucleus. It is expected to display a high polarisibility. This is as oppose to fluorine which has small atomic radius and concentrated negative charge. It is able to hold the electron cloud much tightly resulted in a low polarisability. This will induce a lower dielectric constant.

Free volume is also an important factor in determining the dielectric constant. Free volume is defined as the volume which is not occupied by the polymeric material. The free volume associated with one mole of repeat units of the polymer may be estimated by subtracting the occupied molar volume of a repeat unit, *V*o, from the total molar volume, M/ρ, where M is the molar mass of the repeat unit. [10] The fractional free volume V_n is given by Equation (3):

$$V_n = \frac{M / \rho - V_0}{M / \rho} \tag{3}$$

The addition of pendant groups, flexible bridging units, and bulky groups which limit chain packing density have been utilised to enhance free volume. [21] The presence of free volume in the form of pores will similarly result in a decrease in dielectric constant as it being occupied by air whose relative permittivity is about one. A higher fractional free volume means that the density of the material will be lower resulting in a lower polarisible group per unit volume. Replacement of hydrogen with fluorine result in lowering of dielectric constant since fluorine occupies higher volume. Thus beside being low polarisability, introduction of fluorine induce a significant decrease of dielectric constant through an increase in free volume.

2.2. Effect of chain polarity

Polymers can be polar or non-polar. This feature affect significantly the dielectric properties. Examples of polar polymers include PMMA, PVC, PA (Nylon), PC while non-polar polymers include PTFE (and many other fluoropolymers), PE, PP and PS. Under alternating electric field, polar polymers require sometime to aligned the dipoles. At very low frequencies the dipoles have sufficient time to align with the field before it changes direction. At very high frequencies the dipoles do not have time to align before the field changes direction. At intermediate frequencies the dipoles move but have not completed their movement before the field changes direction and they must realign with the changed field. The electronic polarization and to some extent atomic polarization, is instantaneous weather at high or low frequency for both polar and non polar polymers. Therefore, polar polymers at low frequencies (eg 60 Hz) generally have dielectric constants of between 3 and 9 and at high frequencies (eg 100 Hz) generally have dielectric constants of between 3 and 5. For non-polar polymer the dielectric constant is independent of the alternating current frequency because the electron polarization is effectively instantaneous hence they always have dielectric constants of less than 3. The chain geometry determines whether a polymer is polar or non-polar. If the polymer is held in a fix confirmation, the resulting dipole will depend whether their dipole moments reinforce or cancell each other. In the case of extended configuration of PTFE, the high dipole moment of $-CF_2-$ units at each alternating carbon backbone cancelled each other since their vector are in opposite directions. Its dielectric constant therefore is low (2.1). On the other hand, PVC has its dipole moment directing parallel to each other resulting in reinforcement of dipole. Its dielectric constant is 4.5. This is illustrated as in Figure 2.

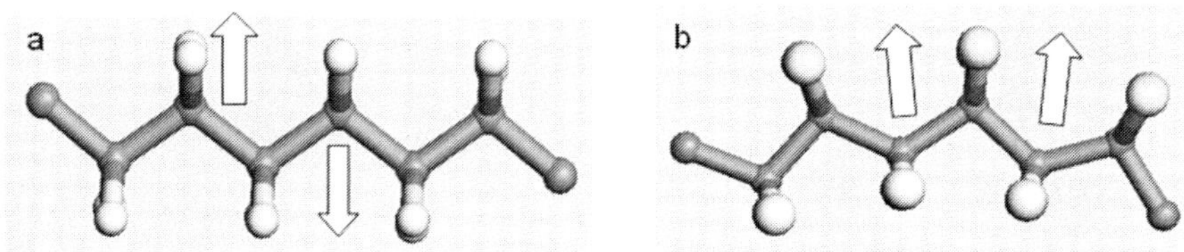

Figure 2. PTFE (a) and PVC (b) with arrow showing the net dipole moment.

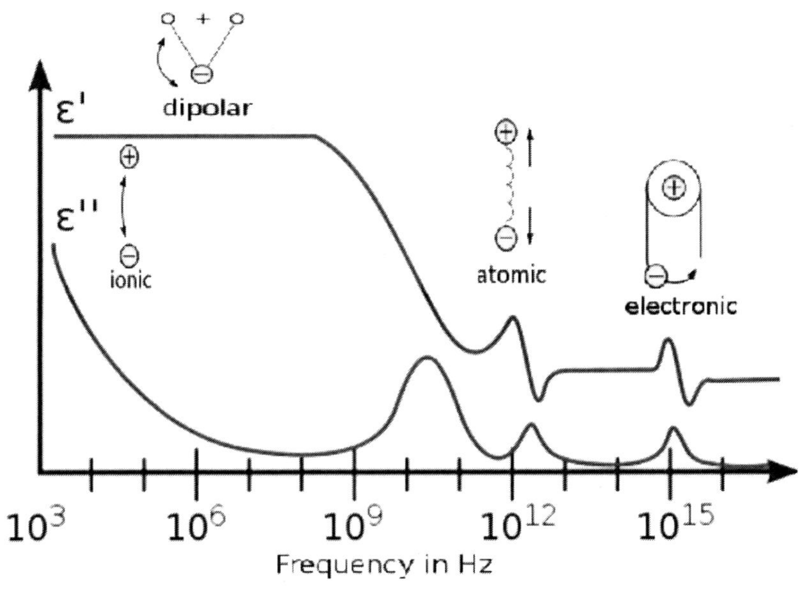

Figure 3. Dielectric constant and loss dispersion of dielectric materials against frequency (adapted from Wikipedia)

The designing of dielectric material so as to achived the desired dielectric properties should take careful consideration of net polarity of the structure. This has been exemplified by the opposite effect in indiscriminately subsitituting fluorine atom into a polyimide chain resulting in an increase in otherwise low dielectric constant material.[11] There is no dipole polarization contribution for non-polar polymers as found in polar polymers. This different mode of mechanism lead to the resonance spectra in the case of electronic polarization which occur at frequency beyond 10^{12} Hz. At below this frequency, the relaxation spectra prevail relating to the behavior of dipole polarization. This observation can best be summarized as in the following Figure 3:

2.3. Relaxation and dielectric loss

Relative permittivity can be express in complex form as in Equation (4) below:

$$\varepsilon^* = \varepsilon' - j\varepsilon''$$

(4)

It consist of the real part which is dielectric constant and the imaginary part which is the dielectric loss. The ratio between the dielectric loss with the dielectric constant is quantified as tan δ ie:

$$\tan \delta = \frac{\varepsilon''}{\varepsilon'}$$

Dielectric loss result from the inability of polarization process in a molecules to follow the rate of change of the oscillating applied electric field. This arise from the relaxation time (τ) in a polymer which is the time taken for the dipoles to return to its original random orientation. It does not occur instatntaneously but the polarization diminished exponentially. If the relaxation time is smaller or comparable to the rate of oscillating electric field, then there would be no or minimum loss. However when the rate of electric field oscillate well faster than the relaxation time, the polarization cannot follow the oscillating frequency resulting in the energy absorption and dissipated as heat. Dipole polarisability is frequency dependent and can be shown as in Equation (5)

$$\alpha_d = \frac{\alpha_o}{1 + j\omega t}$$

(5)

where α_d is the dipole polarisability and α_o is the low frequency (static) polarisability. It normally occur in the microwave region. Figure 3 above shows the variation in real dielectric constant with the imaginary dielectric loss. There is a sudden drop in dipole polarization region (< 10^{12} Hz) for dielectric constant ε accompanied with the maximum dielectric loss ε". This maximum represent the complete failure for the dipole to follow the oscillating electric field beyond which the dipole remain freeze with no effective contribution to the dielectric constant. The mechanism for electronic and atomic polarization occur at higher frequency (shorter wavelength eg. infra-red region). This region involved excitation of electrons which is characterized by the quantized energy level hence is known as resonance behaviour. The dielectric constant display a maximum before a symmetrical drop about a certain frequency. These maximum and minimum represent the optimum polarization in phase with the oscillating frequency. The frequency at which the turning point occur is term the natural frequency ω_o. At this point the frequency of applied electric field is at resonant with the natural frequency hence there is a maximum absorption. Consequently this lead to maximum dielectric loss ε'.

2.4. Effect of temperature

Temperature affect dielectric properties. As the temperature is increased the intermolecular forces between polymer chains is broken which enhances thermal agitation. The polar group will be more free to orient allowing it to keep up with the changing electric field. At lower temperature, the segmental motion of the chain is practically freezed and this will reduce the dielectric constant. At sufficiently higher temperature, the dielectric constant is again reduced due to strong thermal motion which disturb the orientation of the dipoles. At this latter stage the polarization effectively contribute minimal dielectric constant. Beside the kinetic energy acquired, free space in the polymer matrix is of necessity so as to induce segmental movement. Throughout the measured frequency and temperature, electronic and atomic polarization are spontaneous. The dipole polarization, on the other hand, would significantly be affected during heat treatment by effectively reducing the relaxation time (τ) since the polymer chain τ would reduced as the temperature is increased hence the polymer segment would be better able to follow in phase with the oscillating electric field. Significant chain and segmental motions occur in polymers and they are identified as follows [39]:

i. α relaxation: Micro-Brownian motion of the whole chain. Formally this motion is designated as glass transition.

ii. β relaxation: Rotation of polar groups about C-C bond eg. CH_2Cl and $-COOC_2H_5$, conformational flip of cyclic unit.

iii. γ relaxation: libration of phenyl ring and limited C-H segmental chain movement.

The dielectric loss will show maxima at respective relaxation mechanisms as the temperature is increased. The loss in dielectric can be schematically represented as in the following Figure 4

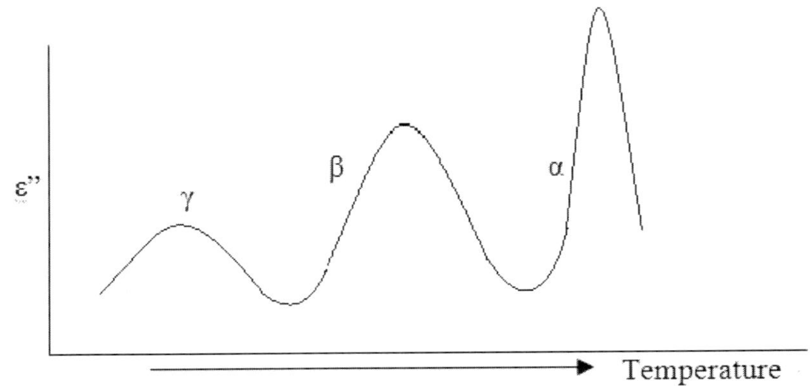

Figure 4. Schematic dielectric loss curve for polymer as temperature is increased.

The γ relaxation occur at lower temperature as it involved small entities of phenyl rings and C-H units whose motion are readily perturbed at low thermal energy. This is followed by β relaxation and finally α relaxation corresponding to the longer scale segmental motion. The broadness for each peaks signify dispersion in relaxation time as the result of different local environment of polarisable groups.

3. STRUCTURE-PROPERTIES RELATIONSHIP.

3.1. Dielectric relaxation.

The earliest model of relaxation behavior is originally derived from Debye relaxation model [12] In this model, real and imaginary part of dielectric constant can be represented as in Figure 5

where

$$\varepsilon' = \varepsilon_{\infty} + \frac{\varepsilon_{o} - \varepsilon_{\infty}}{1 + \omega^2 \tau^2}$$

$$\varepsilon'' = \frac{\varepsilon_{o} - \varepsilon_{\infty}}{1 + \omega^2 \tau^2} \omega \tau$$

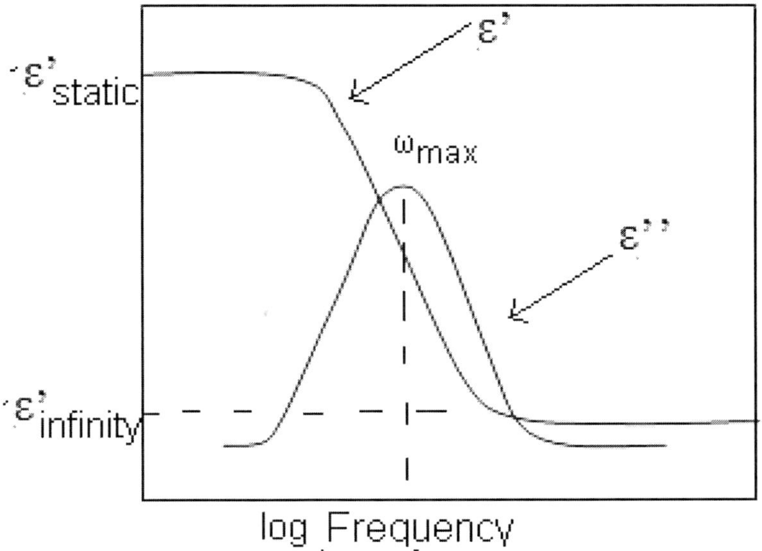

Figure 5. Debye dielectric dispersion curve.

This model relates the dielectric properties with the relaxation time. The relationship between ε'and ε'' can be formulated by eliminating the parameter ωτ to give Equation (6):

$$\left(\varepsilon' - \frac{\varepsilon_s + \varepsilon_\infty}{2}\right)^2 + \varepsilon''^2 = \left(\frac{\varepsilon_s - \varepsilon_\infty}{2}\right)^2 \tag{6}$$

This is a form of a semispherical plot which is popularly known as Cole-Cole plot. See Figure 6.

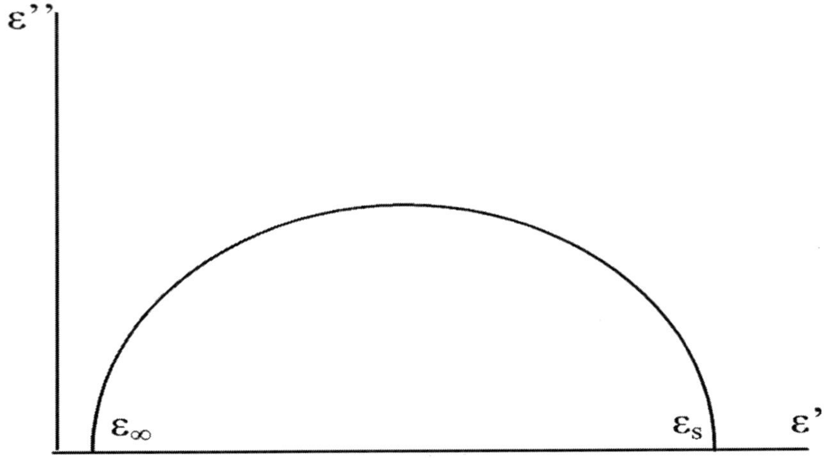

Figure 6. Cole-Cole Plot showing the relationship between dielectric constant and dielectric loss.

The plot shows that at dielectric constant of infinite frequency, ε_∞ and static dielectric constant, ε_s there will be no loss. Maximum loss occur at the midpoint between the two dielectric values. The larger the different between the static and infinite dielectric constant, the higher will be the loss. This model fit very well with polar small molecular liquids. However, polymeric materials are bigger in size, higher viscosity with entanglement between chains. This contribute to visco-elastic properties which requires some modifications to the original model. It can be noted that the above relationship involved only one specific relaxation time. This is contrary in polymeric system whose relaxation time is dependent on mobility of dipoles which behave differently in varying local environments. This result in distribution in relaxtion time. Modification include Cole and Cole semiemperical equation [13] Davidson and Cole [14] Williams and Watt [15] and Navriliak and Nagami [16]. The last modification lead to the new equation (7):

$$\varepsilon' = \varepsilon_\infty + \frac{\varepsilon_{o-}\,\varepsilon_\infty}{(1+\left(\omega^2\tau^2\right)^\alpha).^\beta} \quad (7)$$

where α and β is in the range 0 and 1. No physical meaning as yet is assignable to these parameters.[17] This modification result in a broader peak and smaller loss value with asymmetrical in features. The behaviour of dielectric constant and loss at variable frequencies and temperatures is exemplified in the following Figure 7 for polyvinylchloride.

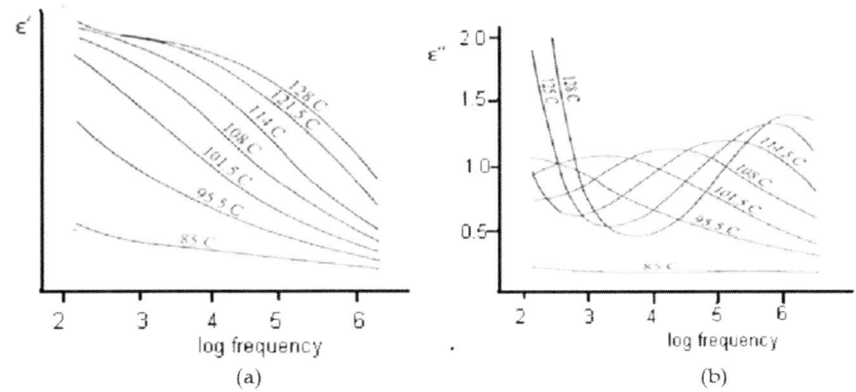

Figure 7. Plot of dielectric constant (a) and dielectric loss with the change in frequency and temperature for polyvinylchloride. (From Ref 18)

Figure 7a shows the variation of ε' and ε" at the region of glass transition (85 ^{0}C) of polyvinylchloride. At the onset of glass transition the PVC showed a relatively low dielectric constant of 4.1 to 3.2 within the measured frequency range. With the increased in temperature, chain mobility begin to increase thus reducing the relaxation time. The dipole polarization of the polymer chain is better able to align in phase with the changing frequency and this account for the increase in dielectric constant as the temperature is increased. However this alignment with the applied oscillating field gradually failed as the frequency is increased. The optimum rate of decreased of dielectric constant occur at higher frequency as the temperature is increased. This correspond to the maximum dielectric loss in Fig 7b. Based on Cole-Cole plot, when there is a big difference in static and infinite dielectric constant, as under high thermal treatment, then the dielectric loss will be correspondingly large. It can be noted that at temperature 128 ^{o}C, there is a large dielectric loss occurring at higher frequency compared to those of lower temperature. Glass transition of polymer is a vital consideration that need to take into account during use of polymers as this affect the dielectric properties substantially. Substitution of fluorine into polyimide, for example, only affect the electronic polarization since PI is mostly used at temperature lower then its Tg (<260 ^{o}C). At this temperature, no effective polar orientation occurr which reduce any possibility of intrusion effect from this mechanism into the dielectric properties. The following Table 2 present the dielectric constant and loss of commercially used polymers.

Table 2. Dielectric parameters for some polymers at various frequencies.

Material	Dielectric constant (ε′)	Loss tangent (tan δ)	Frequency (Hz)
ABS (plastic)	2.0 – 3.5	0.005 – 0.0190	
Butyl rubber	2.35 2.35	0.001 0.0009	1 MHz 3 GHz
Gutta percha	2.6		
HDPE	1.0 – 5.0	0.00004 – 0.001	
Kapton (Type 100) (Type 200)	3.9 2.9		
Neoprene rubber	6.26 4.0	0.038 0.34	1 MHz 3 GHz
Nylon	3.2 - 5		
Polyamide	2.5 – 2.6		
Polycarbonate	2.8 – 3.4	0.00066 – 0.01	
Polypropylene	2.2		
Polystyrene	2.5 – 2.6	0.0001 0.00033	100 MHz 3 GHz
PVC	3		
Silicone (RTV)	3.6		
Teflon (PTFE)	2.0 – 2.1	0.0005 0.00028	100 Hz 3 GHz

3.2. Effect of cross-link between chains

Polymers are often cross-linked to improve their properties. The cross linking or curing process can be conveniently monitored based on relaxation time changes with the progress of reaction. This is exemplified during curing of diglycidylether bisphenol A (DGEBA) with diethyltetraamine (DETA)[19]. During the cross-linking process, the chains are covalently bonded to each other which induce chain rigidity.

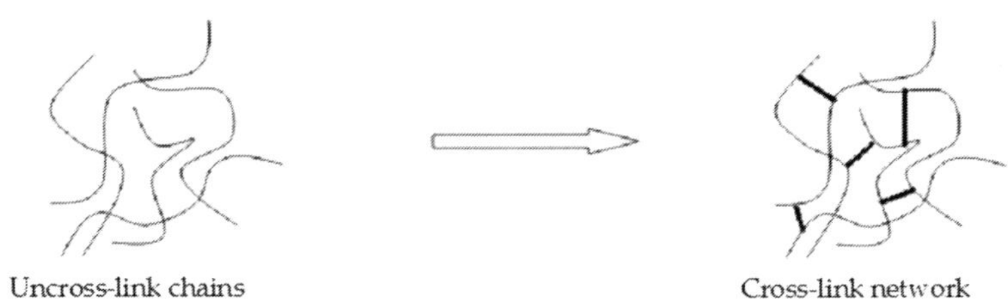

Scheme 1. Affect of crosslink network on rigidity of polymer chains

This rigidity is proportional to cross-link density henceforth affecting the change in relaxation time. This can be illustrated in the following Figure 8:

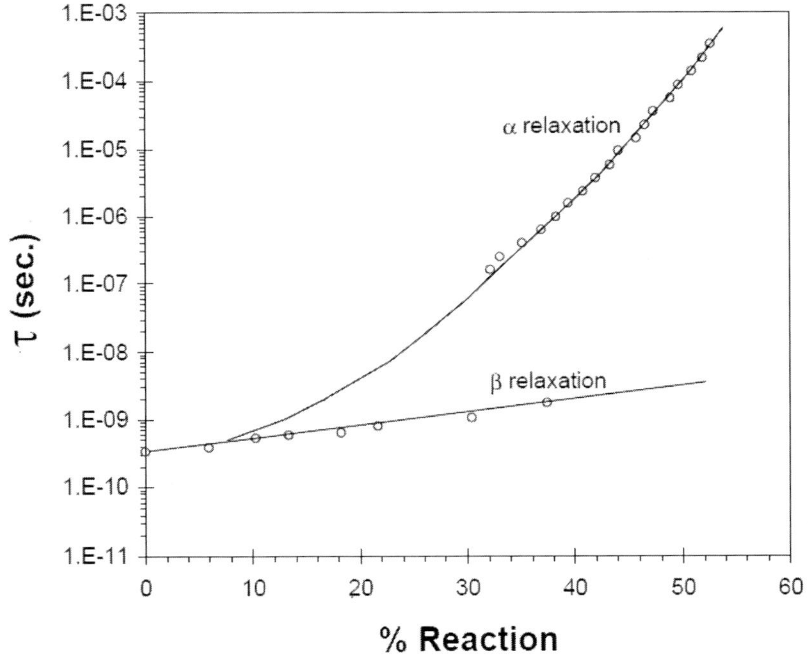

Figure 8. Effect of degree of reaction on the α and β relaxation time of DGEBA-DETA system

Figure 8 shows that both the α and β relaxation time increase with the increase in amount of cross-linking reaction. With the increase in cross-link density, the polymer chains are mostly bounded to each other much tighter hence inducing a longer time to return to their original equilibrium configuration. The rate of increase in α relaxation is higher as it approaches glassy state compared to β relaxation as the former relates to the segmental chain motion of larger in scale compared to the latter. The dielectric constant ε and loss is illustrated in the following Figure 9. In Figure 9a there is a significant drop in dielectric constant which correspond to the maximum frequency for dielectric loss in Figure 9b. This transition represent the frequency at which the dipole polarization is completely out of phase with the applied oscillating electric field. The maximum frequency ω_{max} of dielectric loss was extracted and applied into the equation $\tau = 1/\omega_{max}$ to yield the α relaxation time. It can be observed that as the level of curing is increased the maximum dielectric loss shift towards lower frequency while the change in dielectric constant at α_{static} with $\alpha_{infinity}$ become diminished. This behaviour represent the gradual transition from the rubbery state to glassy state of the polymer with the increase of cross-link density.

3.2.1. Polarizability and free volume

Polarizability and free volume are two important factors that influence the dielectric properties as formulated in the Clausius–Mossotti equation. These effects can be exemplified by introducing fluorine into a polymer chains.[20] Fluorination of polyimide film was performed through gaseous phase in a vacuum chamber. The impregnated fluorine content was determined using XPS analysis and the dielectric constant is shown as in following Table 3:

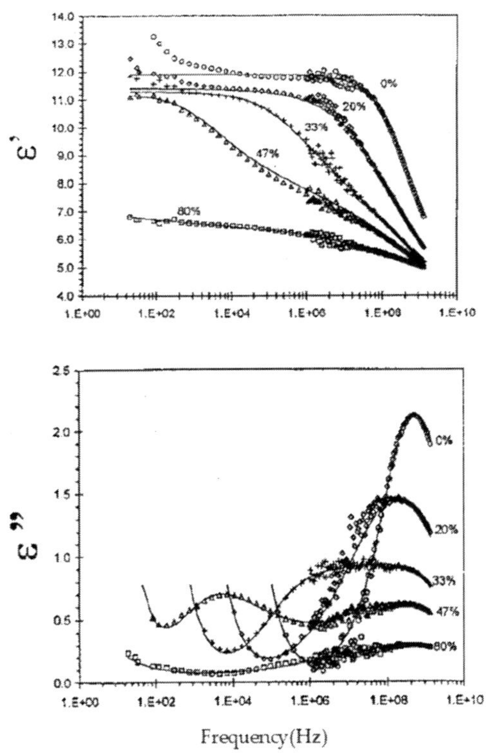

Figure 9. Effect of cross-link density on dielectric constant (ε' above) and dielectric loss (ε'' below) for DGEBA-DETA system

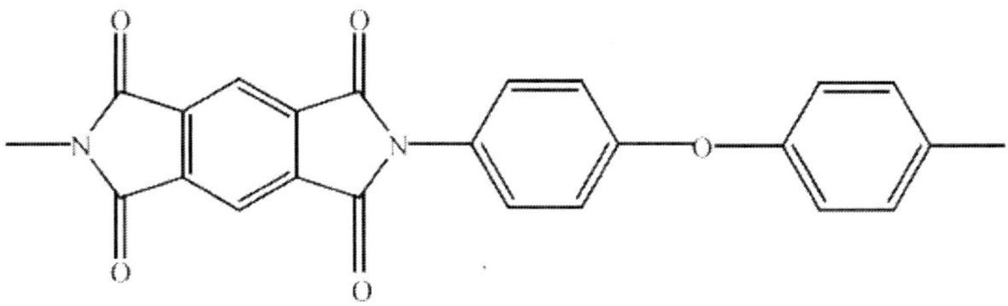

Scheme 2. Repeat unit of the polyimde (from Ref 20)

Table 3. The effect of Fluorine content on the dielectric constant of a polyimide

Sample	F_{1s}/C_{1s}(%)	Dielectric constant, ε	
		10^2(Hz)	10^6 (Hz)
F0	0	2.93	2.90
F2	57.8	2.64	2.60
F3	67.7	2.42	2.41
F4	78.6	2.28	2.27
F5	87.4	2.37	2.26

Similar result was obtained in a series of polyimides synthesised from starting monomers bearing varying percentage of fluorine content. [21] The decreased in dielectric constant as the fluorine content is increased can be explained as due to the low polarizability of fluorine. The electrons of fluorine being very tightly held and close to the nucleus. The polarizability of the fluorinated polyimides is decreased as the number of fluorine atoms is increased, due to the lower electronic polarizability of a C–F bond relative to that of a C–H bond that has been displaced. [22,23] The free volume concomitantly increases due to the relatively large volume of fluorine compared with hydrogen, which reduces the number of polarizable groups per unit volume.

The effect of free volume can be seen when introducing adamantane into a polyimide chain. [24] Adamantane is a bulky group which induce an increase in the free volume. The dielectric constant achieved was 2.7 at 1 KHz. This value is well below the commercial Kapton H film (25.4 μm) with a dielectric constant of approximately 3.5 at 1 kHz and 3.3 at 10 MHz. Besides, hydrophobicity was reduced thus preventing absorption of moisture. Low dielectric loss is important for a good capacitors and insulation. The strategy of introducing bulky substituents is further exemplified in a commercial Avatrel™ dielectric polymer made up of polynorbonene for passivation applications. It has a dielectric constant of 2.55, a loss tangent less than 0.002. These electrical properties held constant up to above 1 GHz. The bulky structures in these polymers are illustrated in the following Figure 10:

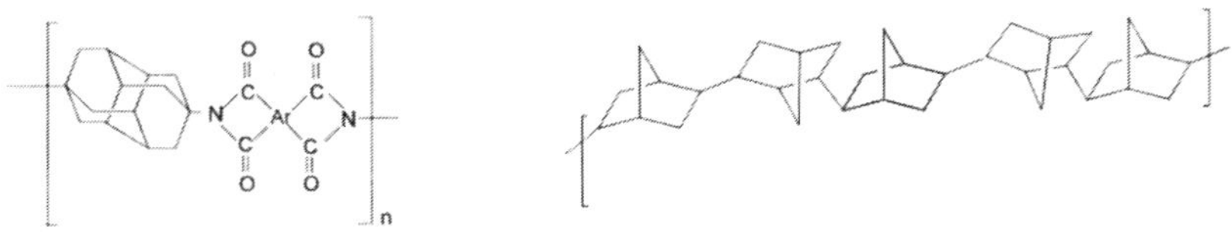

(a) (b)

Figure 10. Adamantane structure incorporated into polyimide chain (a, from Ref 12) and the generic structure for polynorbonene (b).

3.3. Dielectric breakdown

Electrical breakdown occurs when the dielectric strength which is the maximum electric field applicable on dielectric material is exceeded. It underwent catastrophic failure leading to short circuit or blown fuse. This occurs when at a given applied voltage the heat generated due to the losses is greater than the heat dissipated and if the voltage is applied long enough period then the dielectric is unable to reach a state of internal thermal equilibrium. The favourable condition for the occurrence of breakdown is large thickness of the dielectric, high temperature of both the dielectric and the surrounding, continous application of high voltage and large dielectric loss (high tan δ). The last factor is the most important to occur at high frequency. The high humidity in air can similarly affect dielectric breakdown through electrolytic process.

4. DESIGNING OF POLYMER DIELECTRIC MATERIALS

Based on the preceding discussions, designing of polymer dielectric materials can be made using several approaches. The following examples review two approaches undertaken of late namely free volume and copolymerization.

4.1. Free volume

Based on Maxwell-Garnet theory, the presence of second phase of lower dielectric constant in a composite will affect a significant decrease in dielectric constant.[25] This concept was applied in generating foam structure with the introduction of air-filled pores. At least two methods were utilised. One is to synthesised block copolymer of different thermal lability [26] and the other is performing solution etching of soluble component in a composite matrix. [27] The former method involved the use of block copolymer composed of high temperature and high Tg polymer and a second component of lower thermal property which can preferentially undergoes thermal decomposition. One of such a triblock polymer is shown below:

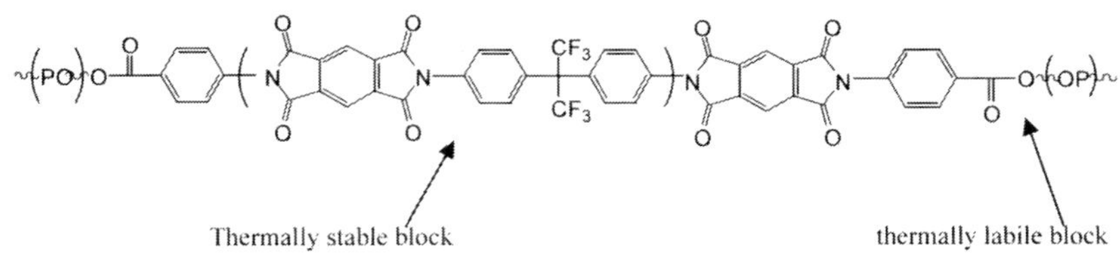

Scheme 3. Triblock polyimide structure illustrating the thermally labile and stable segments.

This triblock composed of thermally stable polyimide and thermally labile phosphate ester block. This copolymer is subjected to thermal treatment such that the temperature is sufficient to degrade the thermally labile block and leaving the thermally stable block intact. A small size scale of microphase saparation is then generated with spherical pore morphology, monodispersed in size and discontinuous. These nanopores are filled with air (ε = 1.0) which ε is responsible for the reduction in dielectric constant. Thermally labile oligomers include polymethylstyrene, polypropylene oxide and polymethylmethacrylate. Nanofoam with dielectric constant of 2.3 was achievable with system made-up from PMDA/4BDAF/PPO triblock of void volume 16%. Figure 11 illustrate the relationship between the void content with the dielectric constant for PMDA/3FDA/PPO triblock system.[28]

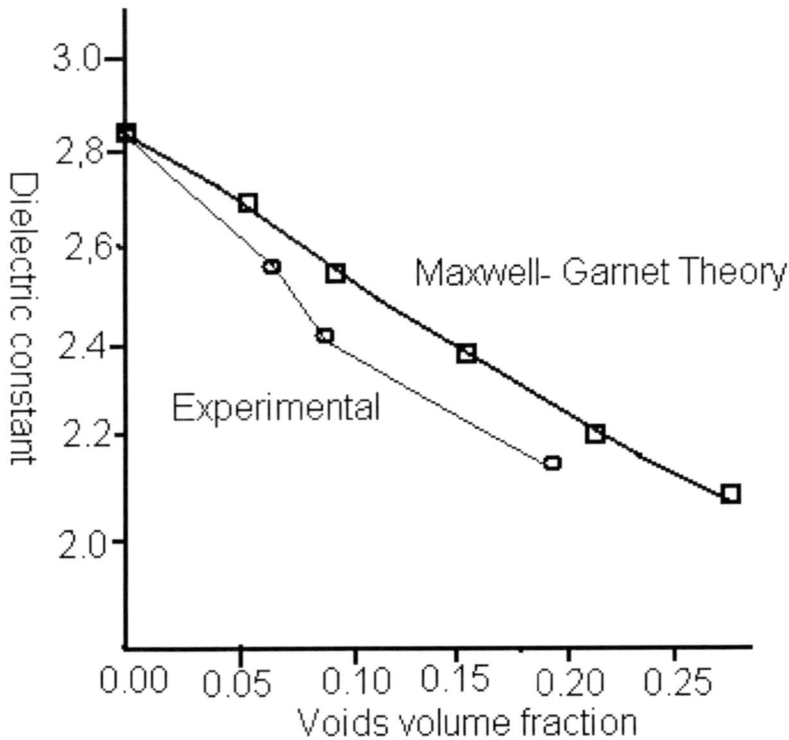

Figure 11. Relation between the dielectric constant with the void fractional volume in PMDA/3FDA/PPO triblock system.

In solution etching method, porosity were achieved by solution etching of soluble component in a nanocomposites leaving the chemically stable matrix intact. This was attempted using polyamic acid, a polyimide prepolymer, as the matrix while inorganic TEOS was incorporated through sol-gel method. Once the inorganic phase was homogeneously distributed in the polymer matrix, the composite was thermally cured followed by hydrofluoride etching. This will dissolved away the acid labile inorganic phase with the generation of nanosize closed cell pore of uniform density. The steps involved during its fabrication is illustrated as in the following Scheme (4):

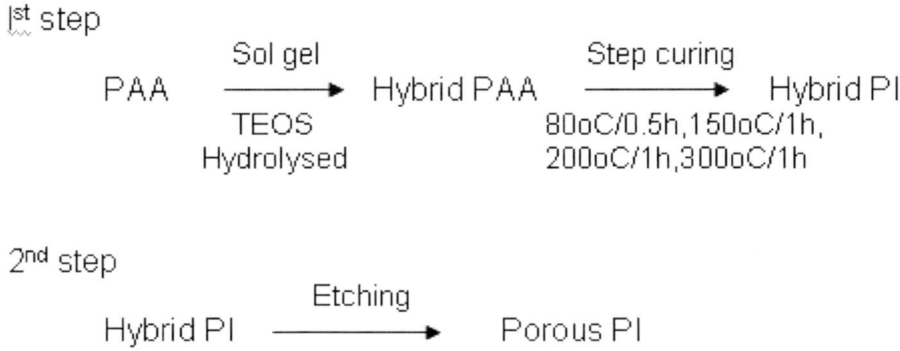

Scheme 4. Preparation of porous Polyimide using sol-gel method

The level of porosity is dependent on the TEOS content incorporated into the polymer matrix. Table 4 shows the dependence in dielectric constant on fluorine content and level of porosity based on TEOS content added during the materials fabrication

	F (Wt %)	0% TEO	10% TEOS	% TEOS
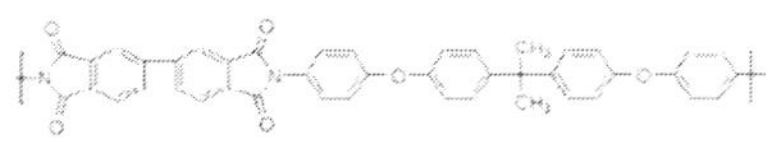 BAPP-BPDA	0	2.71	2.84	3.41
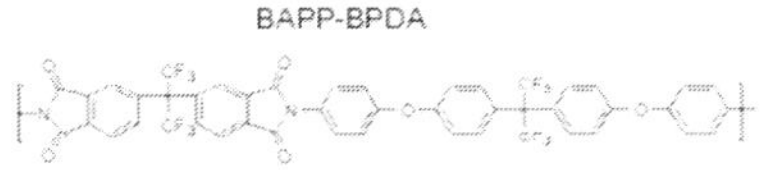 6FDA-BDAF	15	2.45	2.71	3.25
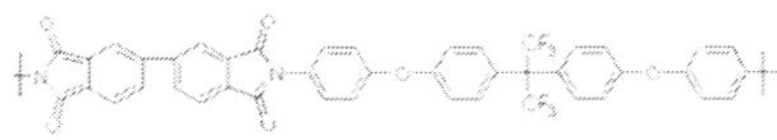 BDAF- BPDA	17	2.61	2.69	3.18
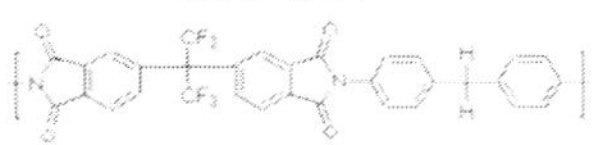 6FDA-MDA	33	2.50	2.62	2.98

Table 4. Dielectric constant of a series of polyimides at varying TEOS content

The results above display a general trend of decreasing dielectric constant as the TEOS concentration used during sol gel technique were increased. This was ascribed to an increased in void structures which reduced the dielectric property as the result of the presence of air. There was a linear decreased in dielectric constant as the weight percent of fluorine content in the structures were increased. Further the rate of decrease is almost constant between different TEOS content. Of the four synthesized polyimides, BAPP-BPDA showed the highest dielectric constant since this sample contains no fluorine. The SEM picture for the fracture surface morphology is shown in the following Figure 12.

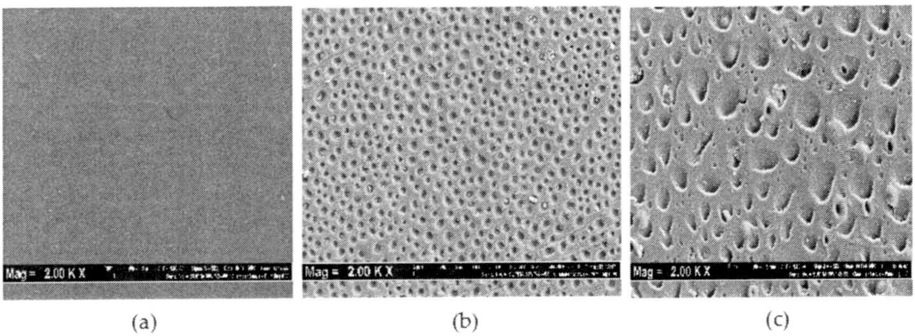

Figure 12. SEM scan of fracture surface of pure (a) PI/SiO_2 10% (b) and PI/SiO_2 20 % porosity (c)

Simpsons *et al* [21] concluded that the presence of fluorine increases the free volume, lower electronic polarization and can either increase or decrease the dielectric constant depending on whether the substitution of the atoms are symmetric or asymmetric.

4.2. Copolymerisation

Copolymerisation of two or more polymers together is a strategy to produce a new materials of tailored dielectric properties. In copolymersation, two or more different monomer units were covalently bound thus producing a synergesic effect of respective constituents. Copolymerisation of polyimide with polysiloxane is popularly performed due to the complementary chemical and mechanical properties between the two. The polyimide has superior thermal and mechanical properties but too intractable to normal processing methods. For example the modulus of polyimides are in the range of 10^9 to 10^{12} Pa but their Tg are above 260 °C. On the other hand the polysiloxane is flexible and easily processable beside having a stable thermal degradation (> 400 °C). Copolymers of these materials produce an optimized dielectric material of practical application for several electronic packagings. Attempt was made with the following structures. [29]

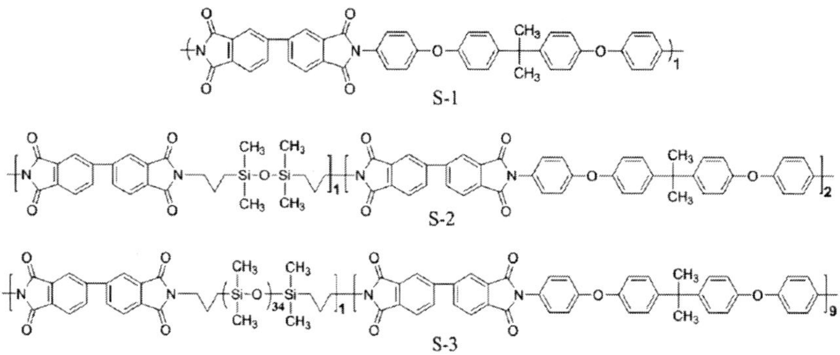

Scheme 5. Series of PI-polysiloxane copolymers

Their dielectric constant are shown in the following Table 5:

Table 5. Effect of silicone content in silicon-polyimide copolymers on dielectric constant.

Sample	Dielectric constant at 1 kHz 293 °C	n (Si-O repeat unit)
S1	2.90	0
S2	2.57	1
S3	2.43	34

The table above shows there is a decreasing trend in dielectric constant with the increase in siloxane units. Silicon is comparatively larger than a carbon atom and the Si - O bond is more flexible than the C - C bond. Thus, the bulky silicone units would be less mobile. Its presence affects the bulk movement of the whole polyimide network which reduces the efficiency of the dipole in reacting to polarity change during treatment with an alternating frequency. Furthermore, the molar polarization significantly decreases as the result of an increase in free volume. Several recent studies have demonstrated a similar trend of a decreasing dielectric constant, with an increasing siloxane content into polyimide structures [30,31]

5. COMPOSITES

The traditionally used inorganic material as a dielectric possesses several superior qualities such as excellent thermal, dielectric and magnetic properties. However they are brittle and consume high energy for processing. [32] On the other hand polymers are more flexible, strong resistivity and offer a tractable prosessibility. The disadvantages of polymeric materials are that they possesses lower thermal and dielectric properties. Combining the two materials in the form of nanocomposites offer an alternative in fabricating material of synergesic properties which displayed a tremendous improvement in dielectric properties with high flexibility and ease of processing. Their combination could readily geared towards miniaturization of electronic devices fabrication.

5.1. Polyimide-ceramic composites

Of late several attempts were made towards this strategy. Incorporation of alumina (Al_2O_3), barium titanate ($BaTiO_3$), titania (TiO_2) and zirconia (ZrO_2) into PI matrix were attempted.. [33,34] Several methods were employed in preparing these nanocomposites. It has been established that method of preparation affect the dielectric properties of these materials. A nanocomposite of PI/Al_2O_3 was prepared by mechanical stirring of prepolymer polyamic acid with the inorganic filler followed by thermal curing. [35] The nanocomposite showed an improved dielectric constant compared to a neat polymer material from about 3.0 to 3.4 at 1 MHz. This values increases correspondingly with the amount of filler loading. A further increase in dielectric constant was achieved when mixing was performed using ultrasonication. It has been shown from SEM result that this improvement was due to a better mixing during the latter treatment. Under these processes, the crystal structure of the inorganic fillers remains intact as shown by XRD data. The effect of good miscibility in improving the dielectric constant was proven when using a 3-Aminopropyltrimethoxysilane-treated (APS) ultrasonication. The APS served as an interface layer between the two immiscible organic PI with inorganic filler which reduced any agglomeration between the different phases. This is brought about possibly through the formation of hydrogen bond between the amine moeity of APS with the polar group of polyimide while the inorganic part of the methoxysilane of APS form secondary interaction with the inorganic fillers. Figure 13 reveals the SEM images of PI/ Al_2O_3 composites doped by the treated Al_2O_3 powder.

PA0 demonstrated a neat and clean morphology. The Al_2O_3 particles were homogeneously dispersed into PI matrix in all PA10, PA20 and PA30. The inset images revealed the average size of Al_2O_3 was around 2μm - 4μm.

There was no obvious aggregation observed suggesting the improved compatibility between PI matrix and Al_2O_3 attributed to the APS coupling agent. The bahaviour of PI-nanocomposites for $BaTiO_3$, TiO_2 and ZrO_2 displayed similar trend with that of PI-Al_2O_3 nanocomposites. They can be summarized as in the following Figures 14:

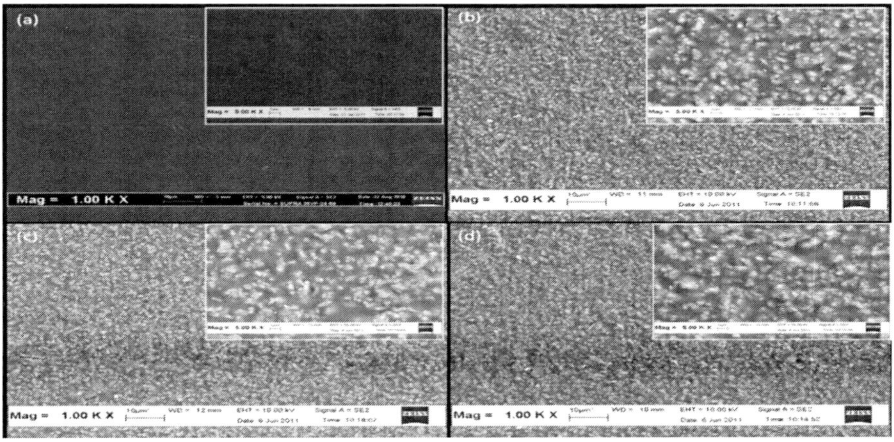

Figure 13. The morphology of PA0, PA10, PA20 and PA30 (a, b, c and d), respectively.

All composite systems displayed a decreased in dielectric constant with the increase in frequencies. The dielectric constants increased as the inorganic filler content were increased. This can be attributed to the increase in polarizability group with the incorporation of the inorganic fillers which replace significant part of the PI in the matrix. As the result the polarizable units per unit volume and the space charge polarizability which occurred at the interfaces between PI matrix and inorganic particles were increased. Fig 13(a) shows the dielectric constant of PI/$BaTiO_3$ composite films demonstrating the highest value of dielectric constant followed by PI/TiO_2, PI/Al_2O_3, PI/ZrO_2 and neat PI films. Apparently this property is dependent on the dielectric constant of the respective fillers. $BaTiO_3$ was known to display highest value of dielectric constant [36] followed by TiO_2, ZrO_2 and Al_2O_3 in their neat form. $BaTiO_3$ possesed perovskite structure which is capable to polarize in the absence of electric field. This feature remains in the composite as the crystal structure remains intact as established in XRD data. The low dielectric constant for PI/ZrO_2 was attributed to the poor compatibility between phases resulted in the presence of voids and even led to cracks. The presence of voids naturally induce a low dielectric constant.

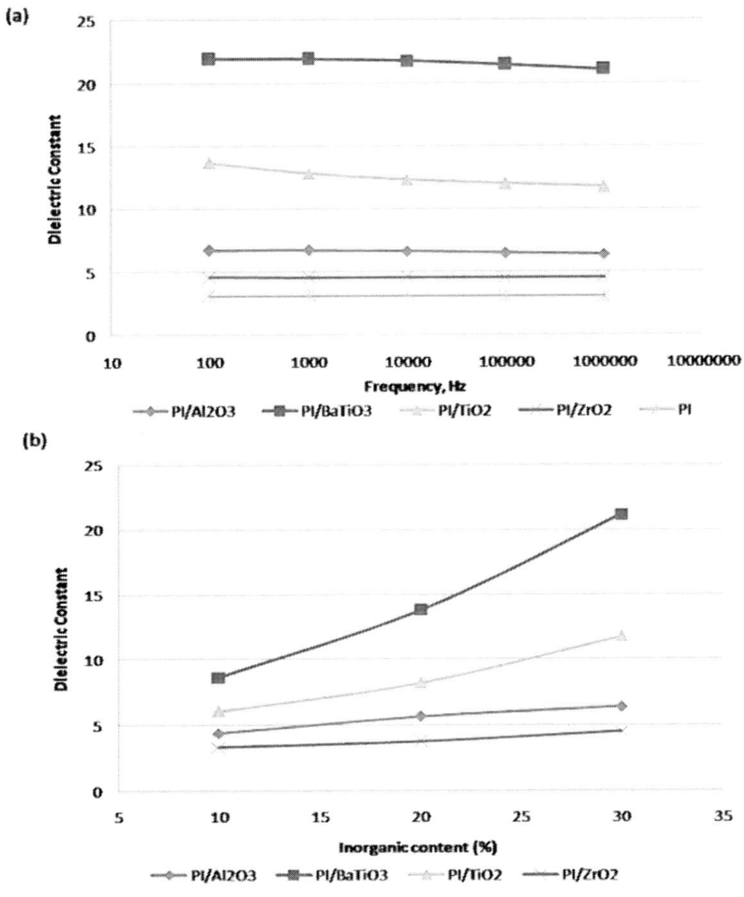

Figure 14. Dielectric constant of PI/inorganic (a) with 30 wt% inorganic content at varying frequency and (b) at 1MHz for several type of inorganic fillers.

5.2. Composite models

Several models were proposed in predicting the dielectric constant of the composites which include Maxwell-Wagner model, Logarithmic Mixing Law and Bruggeman Model. [37] These models allow designing of composite materials based on respective dielectric constant of the polymer, inorganic filler, composition ratio as well as the filler sizes. The slight discrepancy of these models which do not fit to most composite systems are mainly due to inconsistency in treatment for the interphase interaction hence further modification is required. An interphase interaction factor, K, was introduced during fitting into this models.[38] A typical plot of composite dielectric constant with respect to the volume fraction of the fillers is illustrated in the following Figure 15:

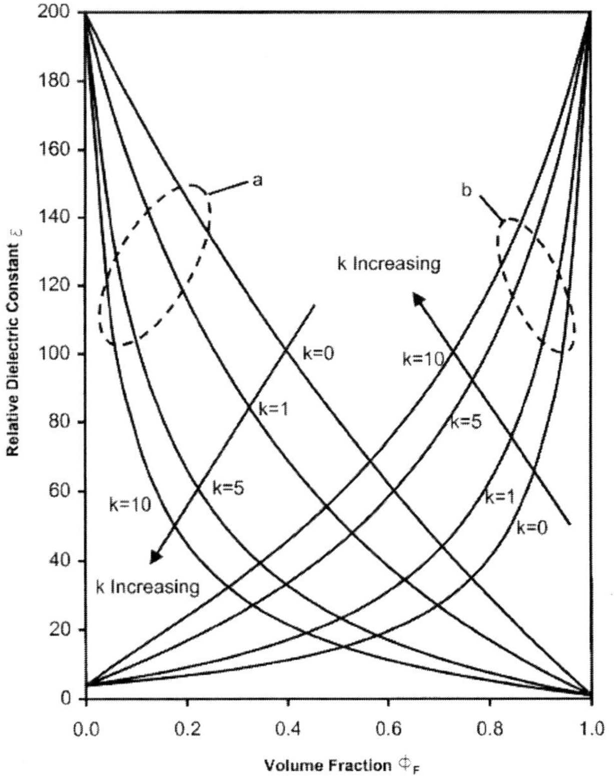

Figure 15. The prediction of the effective dielectric constant as a function of filler volume fraction for different K values. (a) The case of $\varepsilon_{polymer} > \varepsilon_{filler}$. (b) The case of $\varepsilon_{polymer} > \varepsilon_{filler}$ (Adapted from Ref 38)

At K = 0, there is no interaction between phases while a high K values showed a strong interaction. This interaction also dependent on the filler sizes. For a given volume fraction filler, a smaller particle size has a larger fraction of interphase volume in the region between the filler and the matrix granting more polarization to operate. Thus they lead to a relative increase in dielectric constant.

A major concern with polymer/ceramic composites is the heterogeneity in phase which lead to formation of cracks and voids. This effect is known as Maxwell-Wagner effect which reduce the dielectric constant. A more serious type of heterogeneity is that the composite comprised of conductive inorganic fillers which could lead to a mistaken interpretation of dipole polarization occurring at very low frequency region.

6. CONCLUSION

Polymers offer an alternative to the traditionally inorganic and ceramic material as dielectric amterials. This is due to their highly flexible, tractable processing, good chemical stability and readily tunable properties. The main drawback is they have lower thermal stability which limit their wider applications. Generally their dielectric constant is lower than non-polymeric materials. The mechanism which contribute to the dielectric properties are the interaction of electric field with electronic, atomic and dipole poalarization. These are dependent on polarizabity of constituents structure and the free volume as formulated in Clausius-Mossotti equation. The electronic and to some extent the atomic polarization are instantaneous throughout the measureable range of frequencies. However in dipole polarization there is relaxation time allowing an exponential decay of motion to return to equilibrium state. This different behavior contribute significantly to the values of dielectric constant and dielectric loss. These properties can be utilized to detect for any local or segmental motion during change in frequency and temperature treatment. Dielectric properties can be design by introducing polarizable groups into polymer chains, increasing free volume by inducing porosity as well as copolymerization. Increasing dielectric constant can be effectively made by producing nanocomposites with inorganic fillers possessing high dielectric constant.

ACKNOWLEDGEMENT

The author wish to express a grateful acknowledgement to Universiti Sains Malaysia for financial support through Short Term and Research University (Individual) Grants awarded for the accomplishment of related works quoted in this paper.

REFERENCE

1. Milton H P, Electric Capacitor, US Patent 3311801 assigned to Intron International, NY, Mar 1967
2. Alfred S C and Charles E W Jr, Electric Capacitor and method of Making the same, US Patent 3252830 assigned to General Electric Company, NY may 1966
3. R. Yuxing, C. L. Davud, Properties, of. Microstructures, Ultralow. Low-Temperature-Processable-Dielectric, Polyimide. Porous, Journal. Films, Electronic. of, Materials, 2008377
4. R. R. Tummala, E. J. Rymaszewski, Packaging. Microelectronics, Nostrand. Handbook Van, New. Reinhold, York, 1989Chap. I; Numata S, Fujisaki K. Makino D and Kinjo N. Proceedings of the 2nd Technical Conference on Polyimides, Society of Plastic Engineers, Inc., October 1985, Ellenville, New York, 164
5. V. Ranjan, L. Yu, M. B. Nardelli, J. Bernholc, Equilibria. Phase, High. in, Density. P. V. D. F. Energy-Based, Phys. Polymers, Rev. Lett. 2007
6. S. H. Xie, B. K. Zhu, J. B. Li, X. Z. Wei, Z. K. Xu, Preparation, of. properties, nitride. polyimide/aluminum, composites, Polymer Testing, 2004
7. L. Liu, B. Liang, W. Wang, Q. Lei, Preparation of polyimide/inorganic nanoparticle hybrid films by sol-gel method. Journal of Composite Materials, 2006
8. H. Li, G. Liu, B. Liu, W. Chen, S. Chen, properties. Dielectric, polyimide. of, O. Al, synthesized. hybrids, in-situ. by, polymerization, Materials Letters, 2007
9. C. S. Indulkar, S. Thiruvengadam, Introduction. An, Electronic. to, Materials. Engineering, 2008S Chand and Company, New Delhi, India.
10. T. Blythe, D. Bloor, Proeprties. Electrical, Polymers. of, Ed. 2nd, 2005Cambridge University Press.
11. G. Hougham, G. Tesoro, A. Viehbeck, J. D. Chapple-Sokol, Effects. Polarization, Fluorine. of, the. on, Permittivity. Relative, Polyimides. in, Macromolecules, 1994
12. P. Debye, molecules. Polar, 1929Chemical Cataloque Company, reprinted in New York by Dover Publications.

13. R. R. Cole, K. S. Cole, Dispersion, in. Absorption, I. Dielectrics, Alternating Current Characteristics,1941J Chem Phys, 9, 341.

14. D. W. Davidson, R. H. Cole, relaxation. Dielectric, glycerine. of, 1950J. Chem Phys., 18, 1417

15. G. Williams, D. C. Watt, dielectric. Non-symmetrical, behaviour. relaxation, from. a. arising, empirical. simple, function. decay, 1970Trans Faraday Soc 66, 80

16. S. Navriliak, S. J. Havriliak, Dielectric, Relaxation. Mechanical, Materials. in-Analysis, Interpretation, to. Application, Polymers, 1997Munich, Hanser.

17. A. A. Lukichev, method. Graphical, the. for, relaxation. Debye-like, analysis. spectra, of. Journal, Solids. Non-Crystalline, 20122012358447

18. Y. Ishada, Colloid, Z.,1960

19. M. Oplicki, J. M. Kenny, Chem. Makromol, Symp. Makromol, 1993

20. S. J. Park, K. S. Cho, S. H. A. Kim, on. study, characteristics. dielectric, fluorinated. of, thin. polyimide, Journal. film, Colloid. of, Science. Interface, 20042004272384390

21. J. O. Simpson, Clair. A. K. St, insight. Fundamental, developing. on, dielectric. low, polyimides. constant, Solid. Thin, Volumes. Films, 1997

22. G. Nansé, E. Papirer, P. Fioiux, F. Mocuet, A. Tressaud, Carbon, 1997

23. P. S. Alegaonkar, A. B. Mandale, S. R. Sainkar, V. N. Bhoraskar, Instrum. Nucl, Phys. Methods, Sect. B. Res, Interact. Beam, Atoms. Mater, 2002

24. Y. T. Chern, C. M. Huang, Synthesis, of. a. Characterisation, polyimides. new, from. derived, Polymer. 4 9 -diaminodiamantane, 1998

25. D. J. Bergman, D. Stroud, physical. The, of. properties, inhomogeneous. macroscopically, Solid. media, Phys. State, 1992

26. J. Hedrick, J. Labadie, T. P. Russel, D. Hofer, V. Wakharker, temperature. High, foam. polymer, 1993Polymer, 34(22) 1747

27. Nicholas Ang Soon Ming, Muhammad Bisyrul Hafi Othman, Hazizan Md Akil, Zulkifli Ahmad Dielectric Constant Dependence on Fluorine Content and Porosity of Polyimides, Journal of Applied Polymer Science, Volume,2011

28. J. L. Hedrick, K. K. Carter, J. W. Labadie, R. D. Miller, W. Volksen, C. J. Hawker, D. Y. Yoon, T. P. Russell, J. E. Mc Grath, R. M. Briber, Polyimide. Nanoporous, Polym. Adv, Sci, 1999199940143

29. Muhammad Bisyrul Hafi Othman, Mohamad Riduwan Ramli, Looi Yien Tyng, Zulkifli Ahmad, Hazizan Md. Akil, Dielectric constant and refractive index of poly (siloxane-imide) block copolymer, Materials and Design,201120113231733182

30. L. Tao, H. Yang, J. Liu, L. Fan, S. Yang, Synthesis and characterization of highly optical transparent and low dielectric constant fluorinated polyimides. Polymer 2009

31. A. Ghosh, S. Banerjee, mechanical. Thermal, properties. dielectric, novel. of, copoly. fluorinated, siloxane. s. (imide, J Appl Polym Sci 20082008109232940

32. S. F. Wang, T. R. Wang, K. C. Cheng, Y. P. Hsaio, of. Characteristics, barium. polyimide, composite. titanate, Ceramics. films, . International, 2009

33. Y. Bai, Z. Y. Cheng, V. Bharti, H. S. Xu, Q. M. Zhang, High-dielectric-constant ceramic-powder polymer composites. Applied Physics Letters, 2000

34. D. H. Kuo, C. C. Chang, T. Y. Su, W. K. Wang, B. Y. Lin, Dielectric behaviours of multi-doped BaTiO3/epoxy composites. Journal of the European Ceramic Society, 200111711177

35. Alias. Asliza, Ahmad. Zulkifli, Badri. Ahmad, Preparation. Ismail, polyimide. of, O. Al, films. composite, solid. improved, Materials. dielectrics, Science, B. Engineering, 2011

36. B. Viswanathan, 2006Structutre and properties of solid state materials, Oxford Alpha Science Internal LTd, 4377

37. Scaife B K P, Principles of Dielectric, Oxford Science Publication, USA198; NE Hill, Dielectric Properties and Molecular Behaviour, Van Nostrand-Rheinhold, England 1969

38. T. V. Hung, G. S. Frank, model. Towards, engineering. based, optoelectronic. of, materials. packaging, constant. dielectric, modeling, Journal. Microelectronics, 2002200233409415

39. R. J. Young, to. Introduction, Polymers, 1989Chapman & Hall Ltd.

3
CHAPTER

COMPOSITE MATERIALS DAMAGE MODELING BASED ON DIELECTRIC PROPERTIES

Rassel Raihan[1*], Fazle Rabbi[2], Vamsee Vadlamudi[2], Kenneth Reifsnider[1]

[1]*University of Texas at Arlington, Arlington, USA*

[2]*University of South Carolina, Columbia, USA*

ABSTRACT

Composite materials, by nature, are universally dielectric. The distribution of the phases, including voids and cracks, has a major influence on the dielectric properties of the composite materials. The dielectric relaxation behavior measured by Broadband Dielectric Spectroscopy (BbDS) is often caused by interfacial polarization, which is known as Maxwell-Wagner-Sillars polarization that develops because of the heterogeneity of the composite materials. A prominent mechanism in the low frequency range is driven by charge accumulation at the interphases between different constituent phases. In our previous work, we observed in-situ changes in dielectric behavior during static tensile testing, and also studied the effects of applied mechanical and ambient environments on composite material damage states based on the evaluation of dielectric spectral analysis parameters. In the present work, a two dimensional conformal computational model was developed using a COMSOL™ multi-physics module to interpret the effective dielectric behavior of the resulting composite as a function of applied frequency spectra, especially the effects of volume fraction, the distribution of the defects inside of the material volume, and the influence of the permittivity and Ohmic conductivity of the host materials and defects.

Keywords: Polymer Matrix Composite Materials, Dielectric Properties, Degradation of Composite Materials, Broadband Dielectric Spectroscopy (BbDS)

1. INTRODUCTION

The applications of composite materials are now widespread because of their various advantages over conventional isotropic materials. These heterogeneous material system's properties can be tailored based on the needs of the application and design. Aerospace and automotive industries are using composite materials to reduce weight to increase fuel efficiency, and also for energy storage and structural stability. The automobile Company Volvo has developed structural composite materials which can store and discharge electrical energy while also being used as a car body structure, fabricated from carbon fibers and a polymer resin [1] [2] .

It is necessary to understand the material state changes caused by applied mechanical, thermal, and electrical fields to design and synthesize an effective material system. These complex material systems degrade progressively under combined applied field conditions. To evaluate such material state changes there are many computational tools and methods but most of them do not give a direct and quantitative assessment of the damage state. Numerous experimental techniques and methods have also been developed to measure such material state changes but most of them do not give a direct and quantitative assessment of the damage state.

During the service life of composite materials many degradation processes occur and generally this degradation initiates and evolves by microdamage development, especially matrix microcracking and crack growth, delamination, fiber fracture, fiber-matrix debonding, and microbuckling [1] . In our previous work, Figure 1, we have shown that the analysis of the dielectric data gives us information about the types of material state changes throughout the mechanical life of composite materials [2] [3] .

Broadband Dielectric Spectroscopy (BbDS) measures the interaction of EMF with a material system over a wide range of frequencies which is shown in Figure 2. The response to that broad frequency range typically contains information about molecular and collective dipolar fluctuations, and charge transport and polarization effects that occur at inner and outer boundaries as they affect the form of different dielectric properties of the composite material under study.

Various researchers have used finite element methods (FEM) to model the effective dielectric properties of periodic and random composites containing inclusions of various shapes [4] [5] . In this present work, we used COMSOL Multiphysics™ for conformal modeling, and to reduce the complicacy of the model we only considered the interfacial polarization which is caused by the permittivity and conductivity difference between two constituents. We assumed that the composite materials were homogeneous, and represented defects/cracks as inclusions as shown in Figure 3.

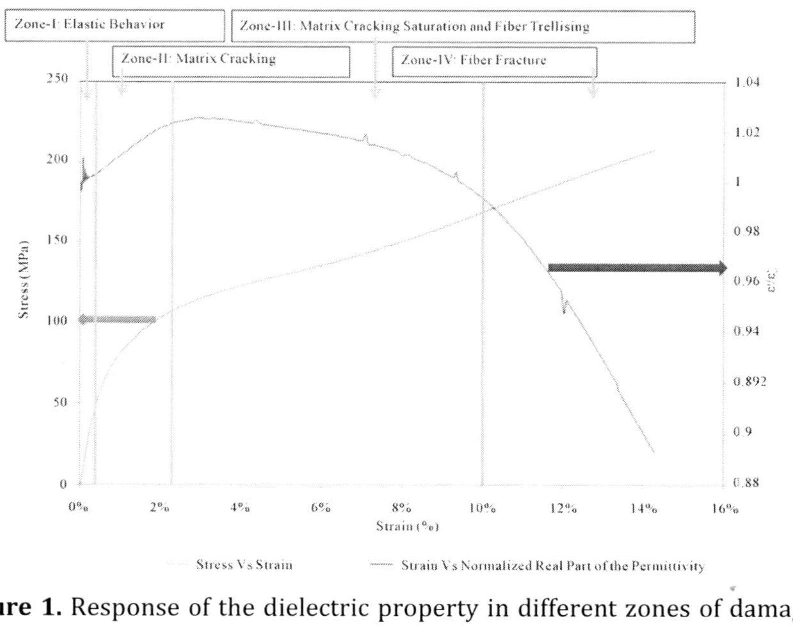

Figure 1. Response of the dielectric property in different zones of damage progression [2] .

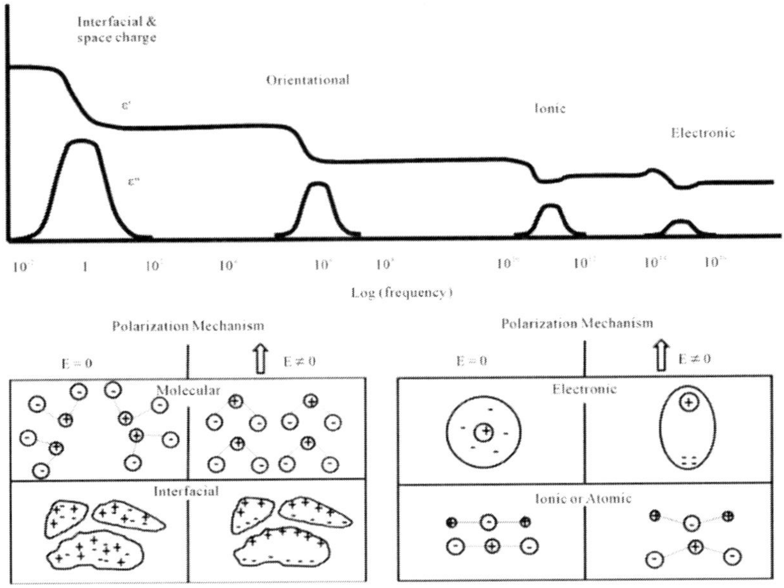

Figure 2. Dielectric responses of material constituents at broad band frequency ranges and different polarization mechanism.

2. BASIC EQUATIONS

In classical dielectrics the relation between the applied electric field E and the dielectric displacement D is linear and can be expressed as [6] ,

$$\boldsymbol{D} = \varepsilon_o \varepsilon_r \boldsymbol{E} \tag{1}$$

where, ε_o is the permittivity of the free space and ε_r is the relative permittivity of the dielectric material.

If ρ is the charge density, from Maxwell's equations we know that the dielectric displacement follows the following relationship

$$\nabla \cdot \boldsymbol{D} = \rho \tag{2}$$

For current density J we can state the following from the continuity equation

$$\nabla \cdot \boldsymbol{J} = -\frac{\mathrm{d}\rho}{\mathrm{d}t} \tag{3}$$

From also Ohm's law we know

$$\boldsymbol{J} = \sigma \boldsymbol{E} \tag{4}$$

Here σ is the conductivity of the material.

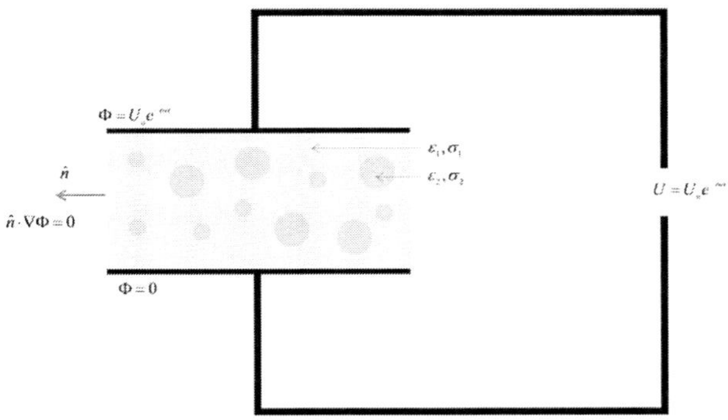

Figure 3. Schematic of the calculation of effective dielectric properties of composites.

So from equations (2) and (3) we obtain

$$\nabla \cdot \left(\boldsymbol{J} + \frac{\mathrm{d}\boldsymbol{D}}{\mathrm{d}t} \right) = 0 \tag{5}$$

Now using 1, 4 and 5 we can write the following

$$\nabla \cdot \left(\sigma \boldsymbol{E} + \frac{\mathrm{d}\left(\varepsilon_o \varepsilon_r \boldsymbol{E}\right)}{\mathrm{d}t} \right) = 0 \tag{6}$$

In case of a sinusoidal applied electric field E of angular frequency ω

$$\nabla \cdot \left(\sigma + i\omega\varepsilon_o\varepsilon_r \right) \boldsymbol{E} = 0 \tag{7}$$

We know

$$\boldsymbol{E} = -\nabla\Phi \tag{8}$$

From equation (7) and (8) we get

$$\nabla\cdot\left[\left(\sigma+i\omega\varepsilon_o\varepsilon_r\right)\nabla\Phi\right]=0 \tag{9}$$

From equation (9), we can tell that in a heterogeneous material the product of the physical properties (some form of the conductivity and permittivity) and the slope of the potential must be a constant as we cross material boundaries. For the quasi-static case with harmonic input fields, the gradient of that product vanishes. The interacting field is a result of the charge difference at the interface, and unless the conductivity and permittivity of adjacent material phases are identical, there is a disruption of charge transfer at the material boundary which results in internal polarization.

To solve equation (9), we set the potential on the top electrode to be,

$$\Phi=U=U_o\mathrm{e}^{-i\omega t} \tag{10}$$

And on the bottom electrode,

$$\Phi=0 \tag{11}$$

Boundary conditions on the interfaces are,

$$\Phi_1=\Phi_2 \tag{12}$$

$$\varepsilon_1\hat{\boldsymbol{n}}\cdot\nabla\Phi_1=\varepsilon_2\hat{\boldsymbol{n}}\cdot\nabla\Phi_2 \tag{13}$$

where $\hat{\boldsymbol{n}}$ is the unit vector normal to the interface surface. To eliminate fringe effects on the side planes, we set

$$\hat{\boldsymbol{n}}\cdot\nabla\Phi=0 \tag{14}$$

here $\hat{\boldsymbol{n}}$ is the unit vector normal to the side plane.

3. RESULTS AND DISCUSSION

3.1. Two Phase Model

For this model, an undamaged composite material is considered to be a homogeneous material and the cracks (here as circular inclusions) are considered to be the second phase inside of that homogeneous material system. Permittivity and ohmic conductivity of the host material were taken to be $\varepsilon_1 = 5$ and $\sigma_1 = 10^{-13}$ S/m and for the inclusion permittivity and ohmic conductivity, $\varepsilon_2 = 2$ and $\sigma_2 = 10^{-15}$ S/m, were chosen which are values close to those of the ambient air permittivity and conductivity [7] . Because of the difference in the permittivities and conductivities of the phases, the accumulation of charge at the interphase boundaries causes an undulation of the space distribution of the potential which is shown in the Figure 4 and Figure 5. Figure 4 shows the potential distribution around the inclusion and Figure 5 shows the potential distribution along the horizontal center line. It can be seen that around the boundary of the phases there is a potential nonlinearity (in the figure the nonlinearity is shown inside two ellipses) that is caused by the charge accumulation at the interface between the host material and inclusion. Figure 6 shows that the space charge accumulation is higher, which is caused by the dissimilarity of the material properties around the inclusion boundary in the presence of the applied electric field.

Computer simulations were performed for different volume fractions of the inclusions. Figure 7 shows that the space charge density increases with an increase of the inclusion volume fraction. In the frequency range above 1 Hz the space charge density is constant but below 1 Hz a nonlinear increase is observed in the space charge density around the inclusion interface as shown in Figure 8.

Figure 9 shows the change of the real and imaginary parts of the global permittivities with the increase of volume fraction of the inclusion as a function of frequency. At a high frequency the period of potential oscillations is not sufficient for charge accumulation but at low frequency the charge has enough time to accumulate around the interface which leads to interfacial polarization (Maxwell-Wagner-Sillar polarization); that is why there is an increase in the real part of the permittivity (shown in Figure 10) and dielectric loss at the lower frequencies.

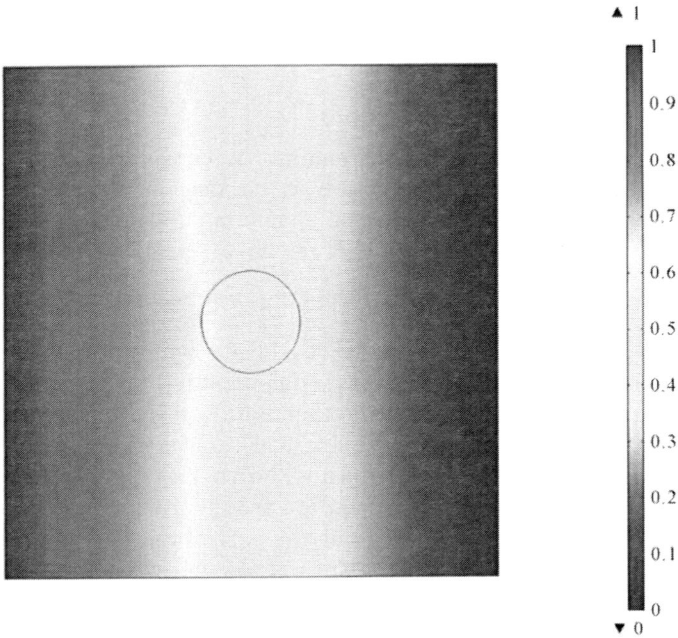

Figure 4. Potential distributions around the inclusion.

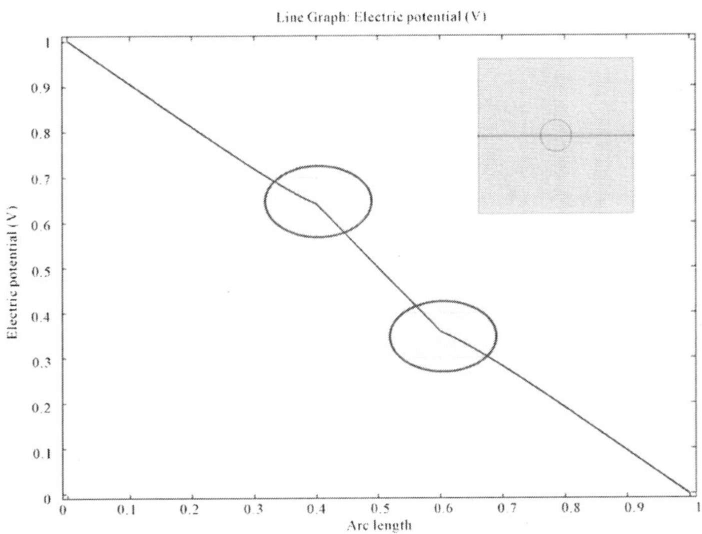

Figure 5. Potential distributions along the line.

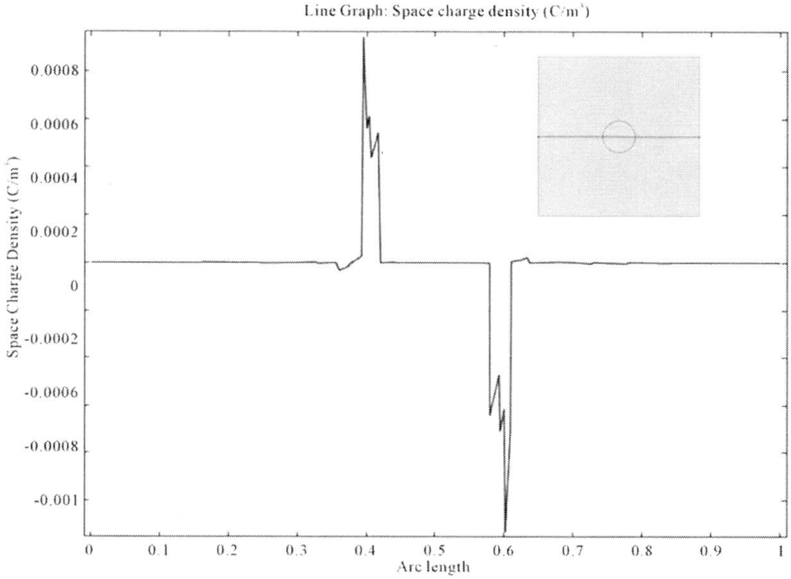

Figure 6. Space charge densities along the line.

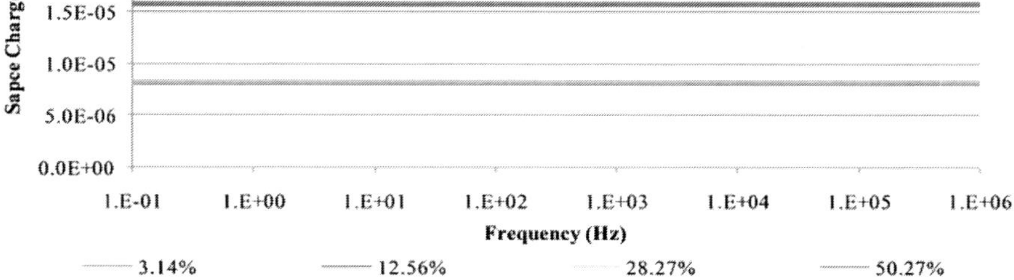

Figure 7. Space charge densities around the inclusion interface of different volume fraction in different frequency.

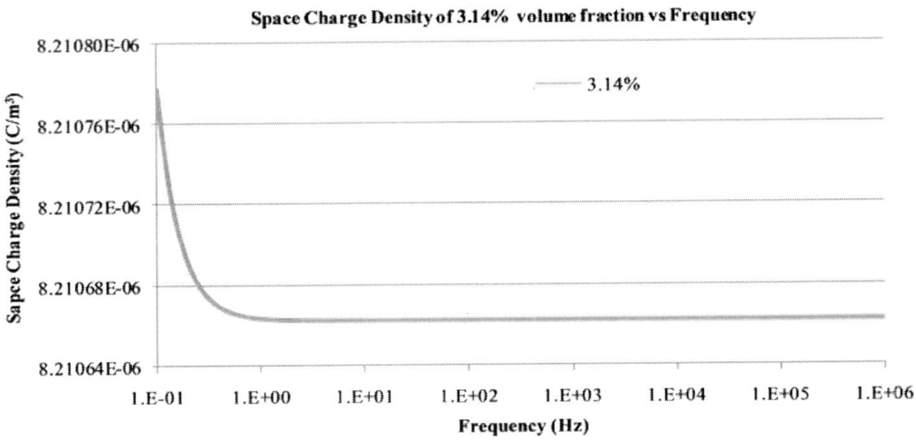

Figure 8. Space charge density change of 3.14% volume fraction of inclusion with frequency.

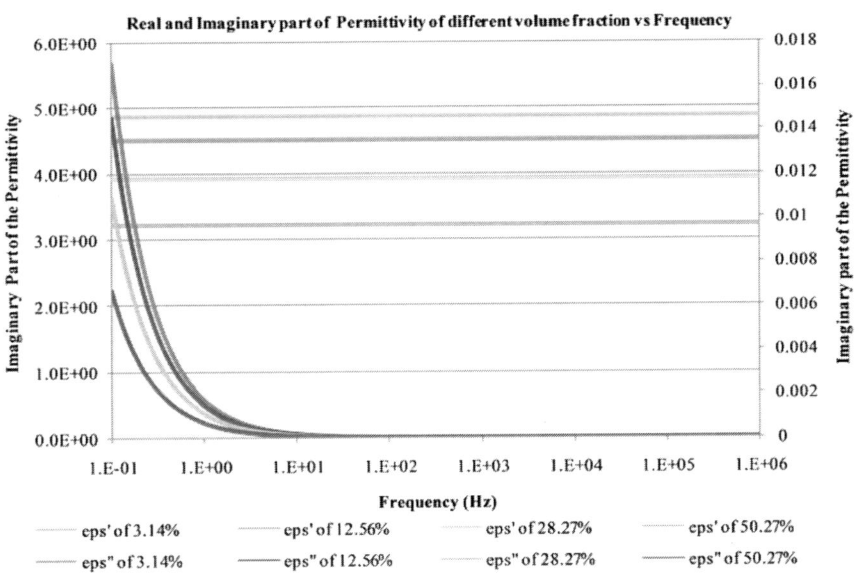

Figure 9. Real and imaginary part of the permittivity with different volume fraction of the inclusion.

For different volume fractions of inclusion, the real part of the permittivity was calculated from the computer simulation for a frequency of 10 Hz.

Figure 11 is the comparison of real part of the permittivity change with increasing volume fraction of the inclusion phase. The relation between the real part of the permittivity and volume fraction is almost linear, but when the real part of the permittivity is plotted with surface area fraction (surface area fraction is the ratio of inclusion surface to the material surface) of the inclusion there is clearly a nonlinear relationship predicted. For low volume fractions the effect of surface area fraction of the inclusion is more dominant than the volume fraction, but for the higher volume fractions it is opposite.

Figure 12 illustrates the comparison between computer simulation results with increasing inclusion volume/ surface-area fraction and experimental results. Figure 12(a) and Figure 12(b) show slight increases in the real part of the impedance for low volume/surface-area fractions. Figure 12(c) shows the experimental results for the real part of the impedance change with the strain. The real part of the impedance increases below the low strain (below 5%) for the off axis sample where matrix microcracking is dominant and distributed throughout the material system.

3.2. Three Phase Model

Composite materials are filled with various additive materials to achieve the desired mechanical, thermal and electrical properties. Typical filler materials used for the present modeling are carbon or glass fibers. The use of these fibers as filler materials introduces a water sensitive component into the polymer composites. Glass fibers are well known for their water affinity on their surfaces. Currently, epoxies are widely used matrix materials in

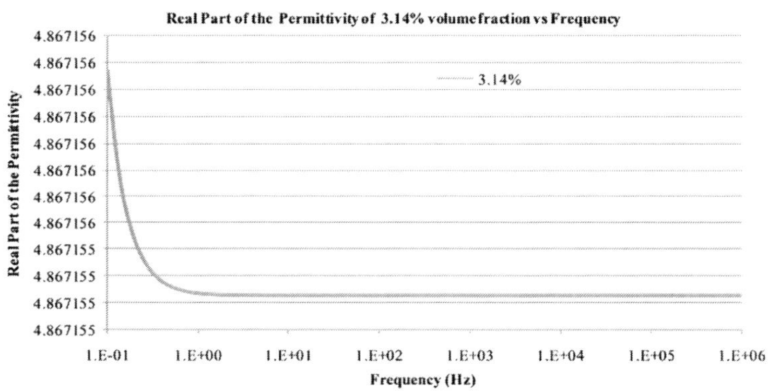

Figure 10. Real part of the permittivity of the material with 3.14% volume fraction of inclusion.

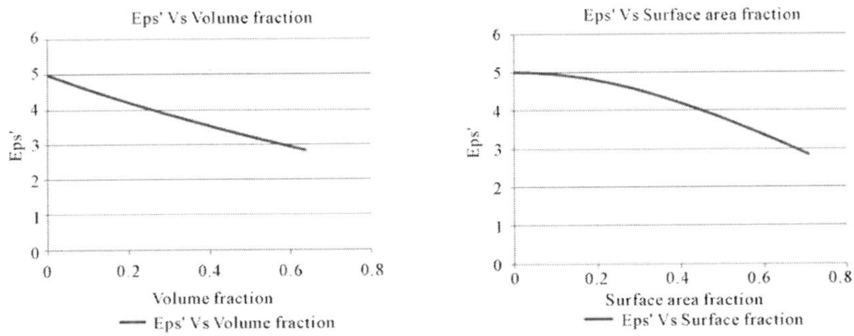

Figure 11. Eps' comparison with volume fraction and surfacearea fraction of the inclusion.

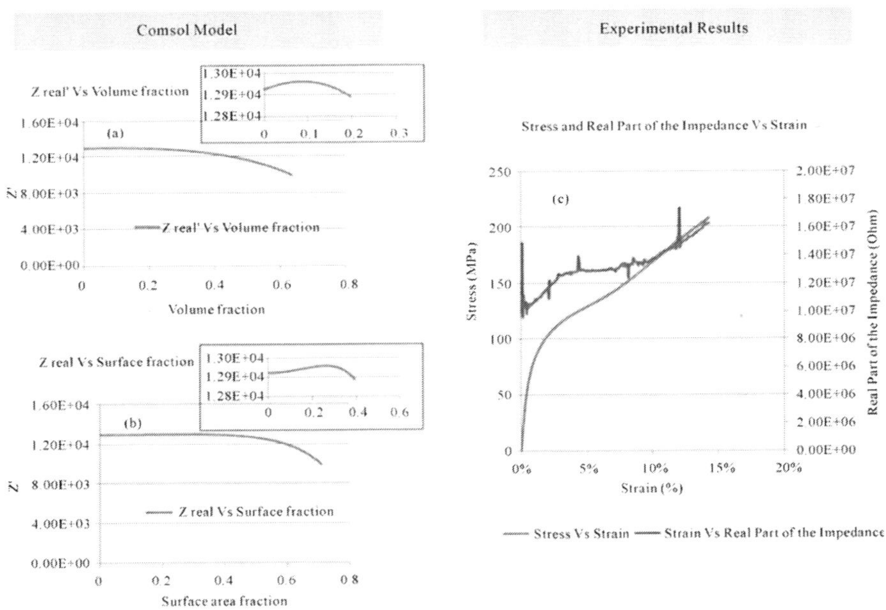

Figure 12. Comparison of Comsol™ simulation result and experimental results.

composite industries, which also have the potential of being sensitive to moist conditions or humid environments. Soles and Yee [8] found that a network of nanopores that is inherent in the epoxy structure helps free water to traverse the epoxy; they found the average size of nanopore diameters to vary from 5 to 6.1 Å and account for 3% - 7% of the total volume of the epoxy material. The approximate diameter of a kinetic water molecule is just 3.0 Å, so via the nanopores network the moisture can easily traverse into the epoxy. They also found that the volume fraction of nanopores does not affect the diffusion coefficient of water and argued that polar groups coincident with the nanopores are the rate-limiting factor in the diffusion process, which could explain why the diffusion coefficient is essentially independent of the nanopore content. In their Figure 13 they explain how the water transport happens in epoxy networks.

There are many theories about the state of water molecules in polymers. Adamson [9] suggested that moisture can transfer in epoxy resins in the form of either liquid or vapor. It is proposed by Tencer [10] that it is also possible that vapor water molecules undergo a phase transformation and condense to the liquid phase. This condensed moisture was stated to be either in the form of discrete droplets on the surface or in the form of a uniform monolayer [11] .

Water has a higher dielectric permittivity and conductivity than the glass fiber and matrix, so it has strong effects on the dielectric properties, i.e. relative permittivity and dielectric loss, of the material system. In the literature it is well established that water absorption increases the dielectric constant of the dielectric material [12] -[16] . This dielectric loss is observed in the low frequency range. Water diffuses through the interface and also weakens the interfacial strength of filler and matrix.

When composite materials go through degradation processes, microcracks typically form and these micro- cracks can also be filled with moist air, and condensed or adsorbed water layers can form on the surface of those defects. In our two phase model we saw an interfacial polarization (Maxwell-Wagner-Sillars polarization) that is present in the low frequency region of the frequency spectra. If a water layer is present on the surface of the defect it will become electrically conductive. Since the host material and defect have low electrical conductivity and permittivity is not significantly high, this will give rise to interfacial polarization.

Figure 14 illustrates the tri-layer computational model used for the study where yellow, gray and blue parts represent respectively the host material, defect, and a conductive layer. The permittivity and ohmic conductivity of the host material were taken to be $\varepsilon_1 = 5$ and $\sigma_1 = 10^{-13}$ S/m and for the inclusion the permittivity and ohmic conductivity had values of $\varepsilon_2 = 2$ and $\sigma_2 = 10^{-15}$ S/m. For the conductive layer a different permittivity ε_3 and conductivity σ_3 (higher than host and defect properties) was used to see the effect on the effective dielectric properties of the material system.

Figure 15 shows the potential distributions around the inclusion of the tri-layer model which is different than the potential difference shown in Figure 4 for the two phase inclusion model. Because of the conductive layer around the inclusion there is a large undulation of the space distribution of the electric potential.

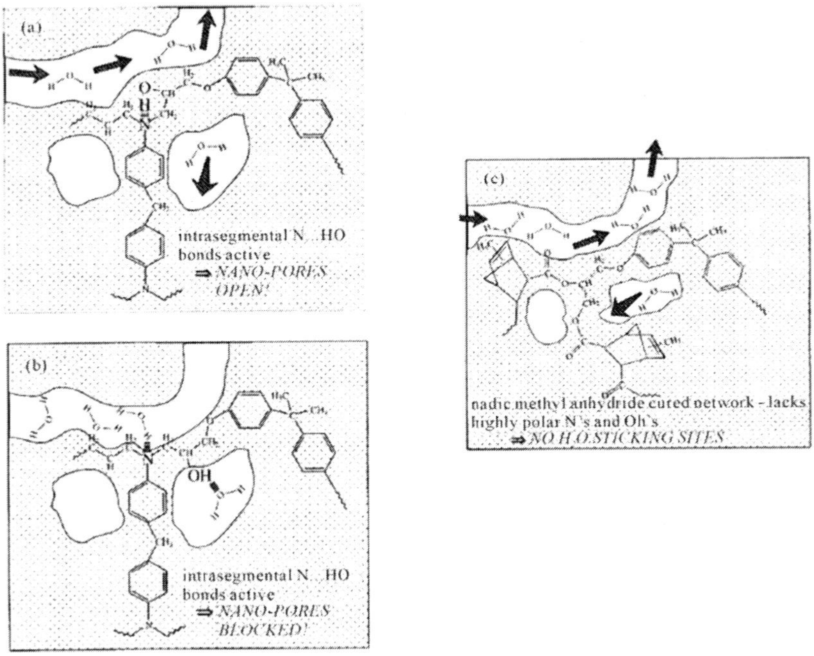

Figure 13. A plausible picture of moisture diffusion through the nanopores of an amine- containing epoxy (Figure from reference [7]).

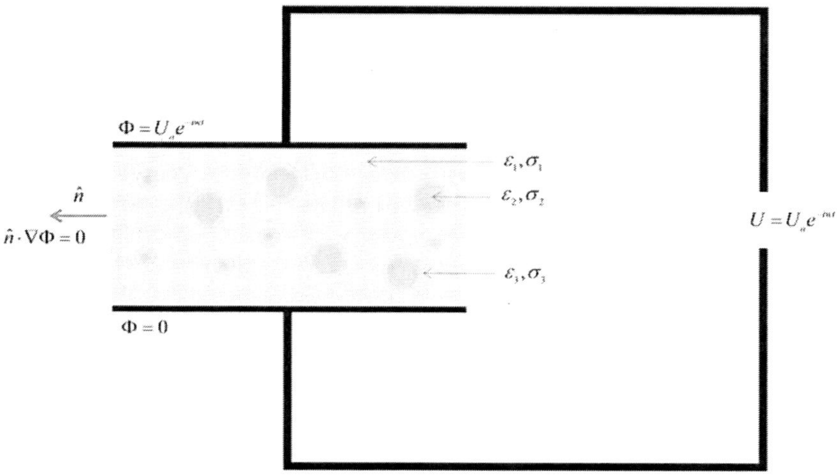

Figure 14. Tri-layer model.

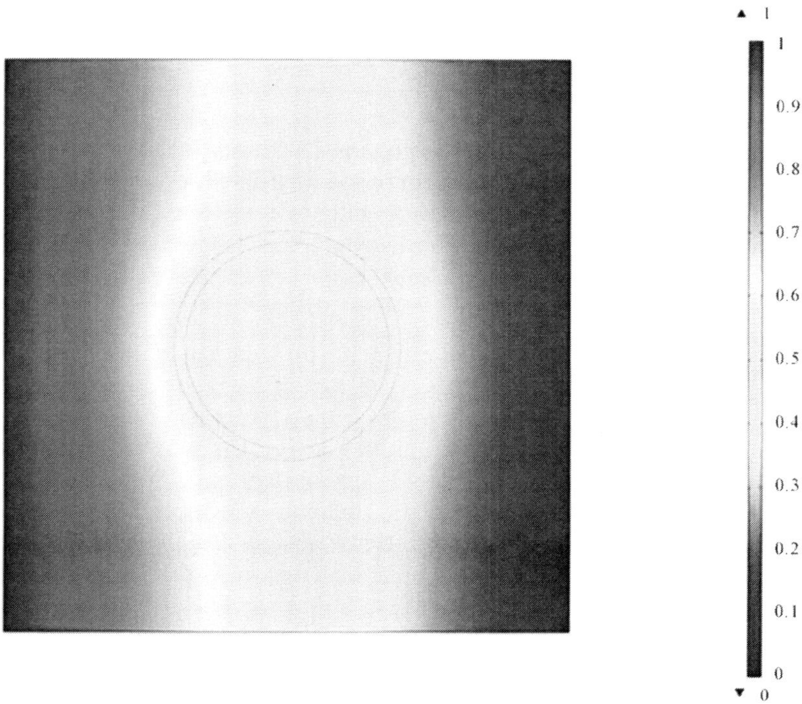

Figure 15. Potential distributions around the inclusion with a conductive layer.

For the tri-layer model, the total volume fraction is the sum of the volume fraction of the defect and the volume fraction of the conductive layer. For all of the cases of tri-layer modelling, the conductive layer thickness was specified as 0.5 micro meter. We observed that for two phase models, the real part of the permittivity was almost linear but in Figure 16 we can see that for the three phase case there is an increase in real part of the permittivity for lower volume fractions and then a decreasing trend. As there is a conductive layer in between the defect and the host matrix, the interfacial polarization plays a vital role for this type of behavior. The subsequent decrease of the real part of the permittivity for higher volume fractions is caused by the dominance of the volume of the defects which is higher than the interfacial polarization contributed by the conductive layer.

It is also clear from Figure 17 that for the same conductivity and permittivity values of the conductive layer, for low frequency the real part of the permittivity of the material system is higher than the value for higher frequency.

Figure 18 shows the dependence of dielectric constant on the conductivity of the conductive layer. The real part of the permittivity at different volume fractions for the same permittivity and frequency behave differently for different surface layer conductivities.

As shown in Figure 19, for the same volume fraction of the inclusion but variable frequency of the input field, there is a step-like increase of the real part of the permittivity over a narrow frequency range, and the dielectric loss also has the peak in that region, where the Maxell-Wagner-Sillars polarization dominates.

Figure 20 shows the relation of the real part of the permittivity with the frequency for all volume fractions of the inclusion. That simulation was also done for just the matrix material (this is the host material, as we considered it homogeneous). Since there were no other phases presents for that case, there was no chance of charge accumulation and there is no predicted change of the dielectric constant for the matrix-only case. We can see for higher volume fractions the dielectric relaxation strength (difference between the real parts of the permittivity at low frequency and high frequency) also increases. At higher frequency the real part of the permittivity drops as a function of volume fraction because charge accumulation does not occur at the interface at those frequencies.

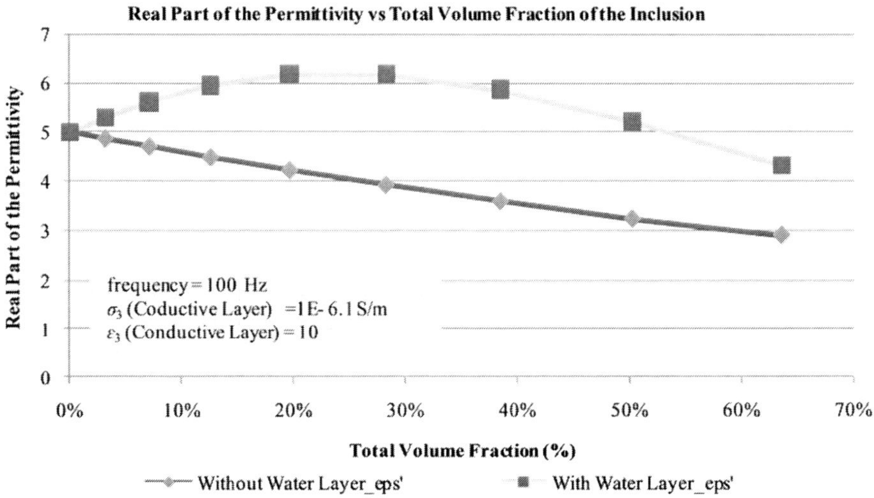

Figure 16. Variation of real part of the permittivity with increasing volume fraction. For the tri-layer model, total volumefraction is the sum of the volume fraction of defect.

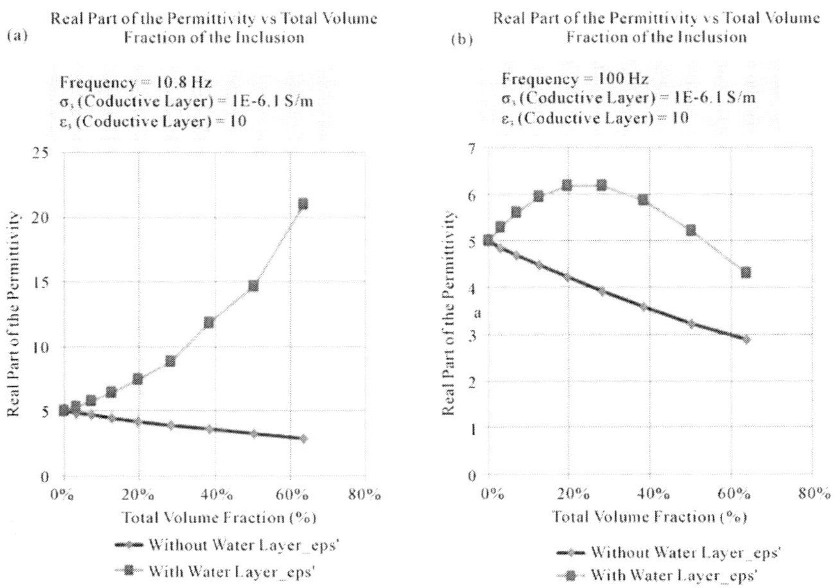

Figure 17. Frequency dependency of the real part of permittivity for different volume fraction.

Dielectric loss (the imaginary part of the permittivity) also varies with volume fraction and it is illustrated in Figure 21. For high volume fraction dielectrics, the loss changes somewhat and the peak of the loss also increases.

The corresponding Cole-Cole plot, Figure 22, also shows the shift in relaxation for different volume fractions.

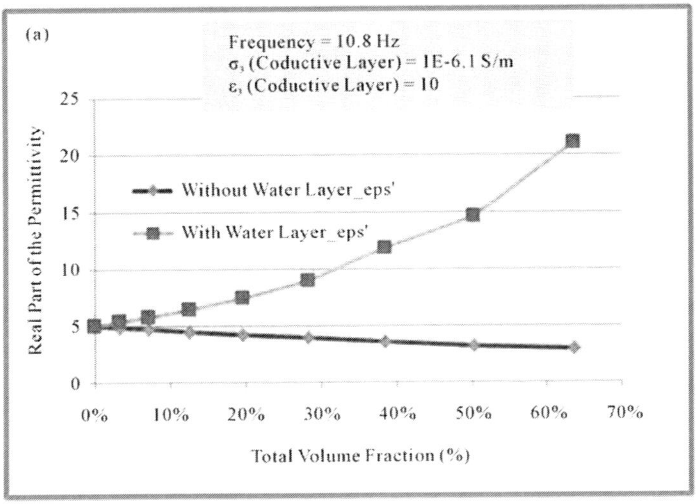

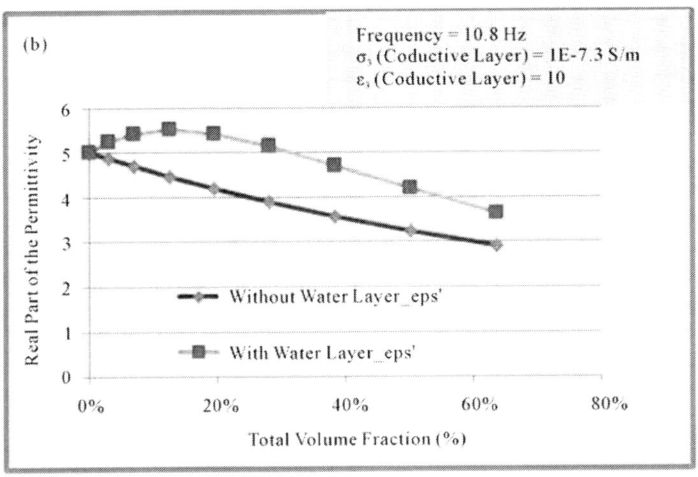

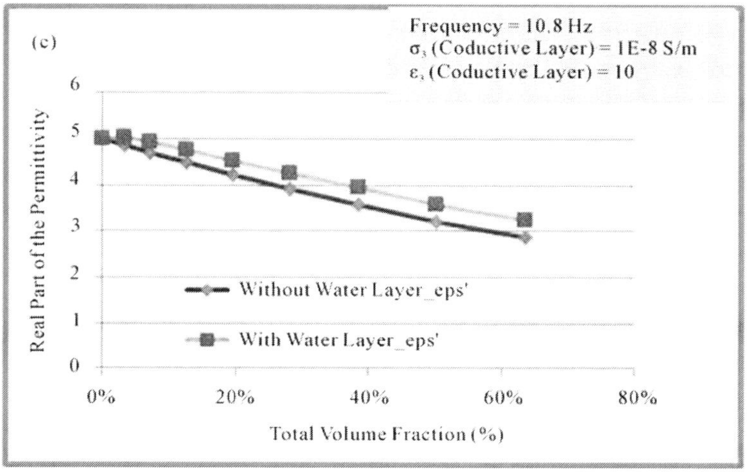

Figure 18. Dielectric properties dependence on the conductivity.

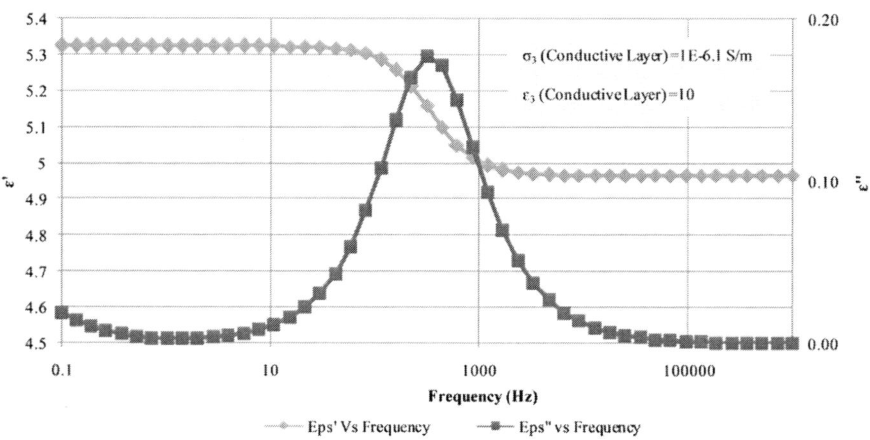

Figure 19. Real and imaginary part of the permittivity Vs frequency for 3.14% total volume fraction of the inclusion.

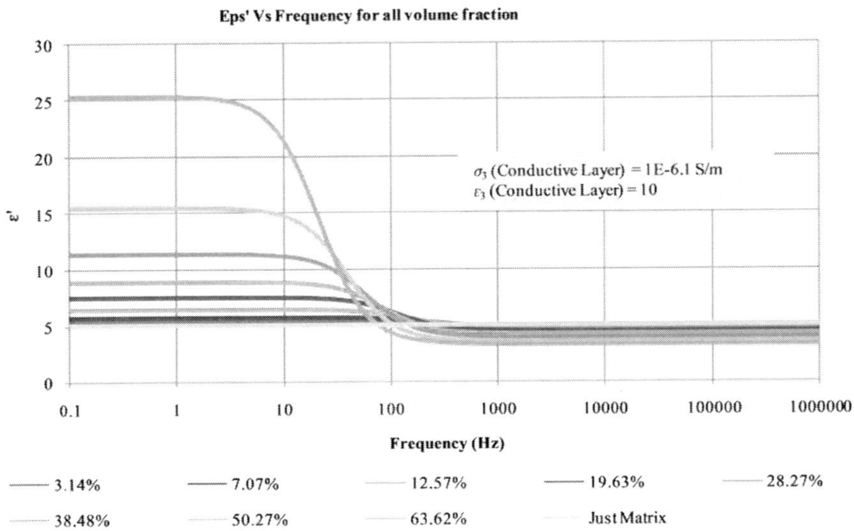

Figure 20. Real part of the permittivity in frequency spectra of all volume fractions.

3.3. Distributed Damage Model

A distributed damage model was created to see the effect of the distribution of the damage. A dielectric study was performed for a certain volume fraction of inclusion, and then that inclusion was divided into 10 inclusions while keeping the total volume fraction the same.

Figures 23-26 show the change of dielectric properties of a single damage volume and distributed damage volumes with the same amount of volume fraction without any conductive layer around the defects. The dielectric loss increased for the distributed damage because of the presence of more interfacial polarization.

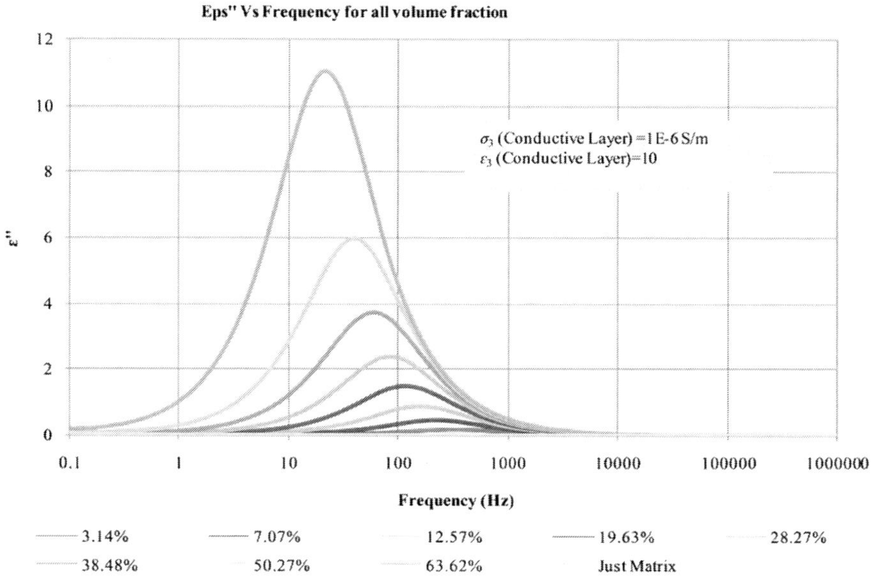

Figure 21. Imaginary part of the permittivity in frequency spectra of all volume fractions.

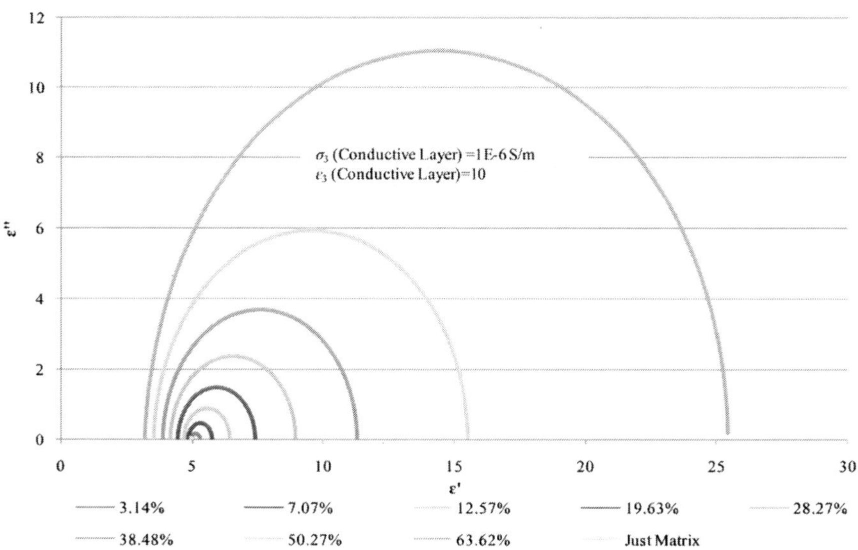

Figure 22. Cole-Cole plot of different volume fraction.

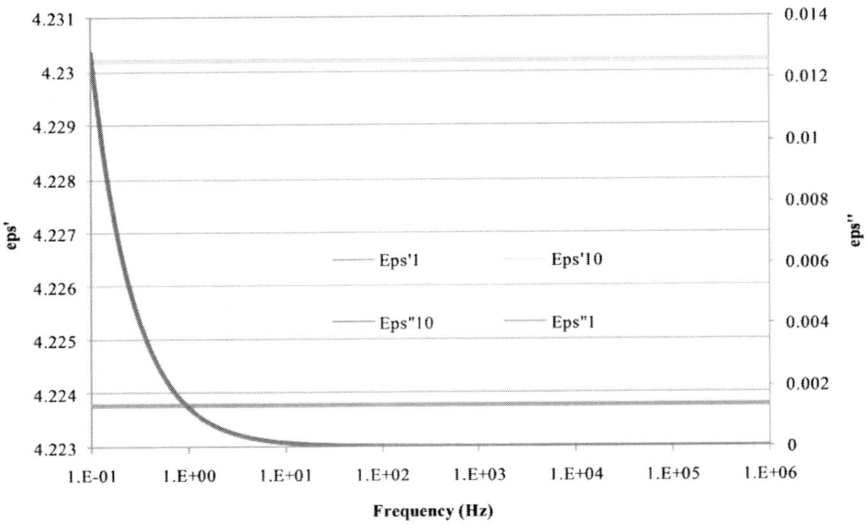

Figure 23. Dielectric Properties without conductive layer for same volume fraction but different number of inclusion.

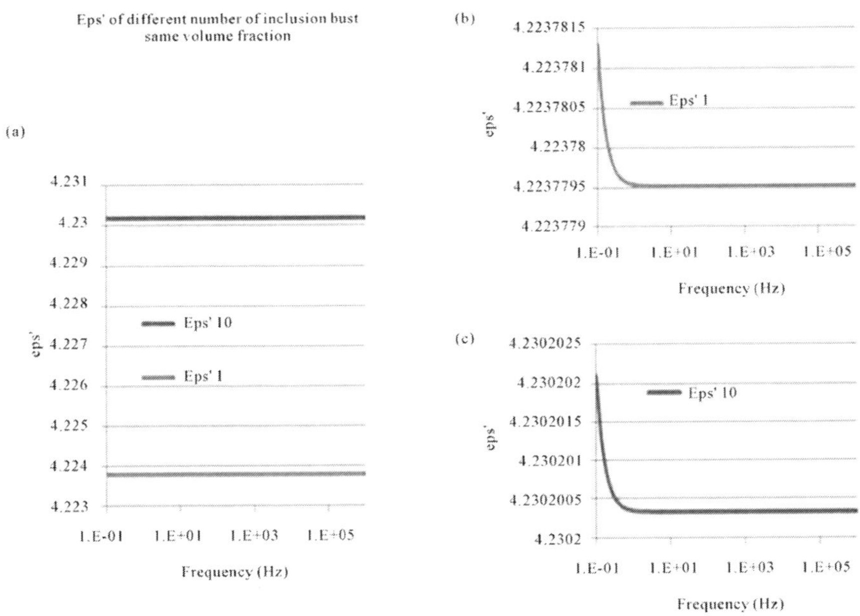

Figure 24. Real Part of the permittivity for different number of inclusion but same volume fraction.

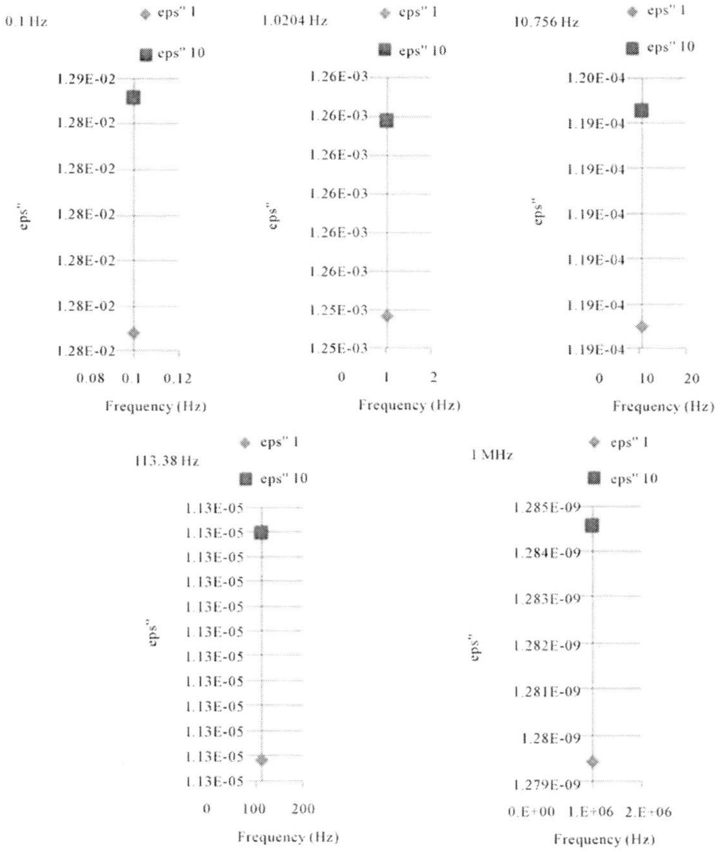

Figure 25. Dielectric losses at different frequency for different number of inclusion but same volume fraction without any conductive layer.

Figures 27-29 show the change of dielectric properties of a single damage phase and distributed damage with the same amount of volume fraction with a conductive layer around the defect. The dielectric loss increased for the distributed damage because of more interfacial polarization and it is more evident than the prior case because the conductive layer around the defect leads to increased interfacial polarization.

The difference between the static permittivity and the limiting high frequency dielectric permittivity is called the Dielectric relaxation strength (DRS), $\Delta\varepsilon$, as given in Equation (15).

$$\Delta\varepsilon = \varepsilon_s - \varepsilon_\infty \tag{15}$$

where, ε_s is the static permittivity and ε_∞ is the limiting high frequency dielectric constant. To calculate DRS from the experimental data we subtract the value of real part of the permittivity at 1 MHz from the value of real part of the permittivity at 0.1 Hz.

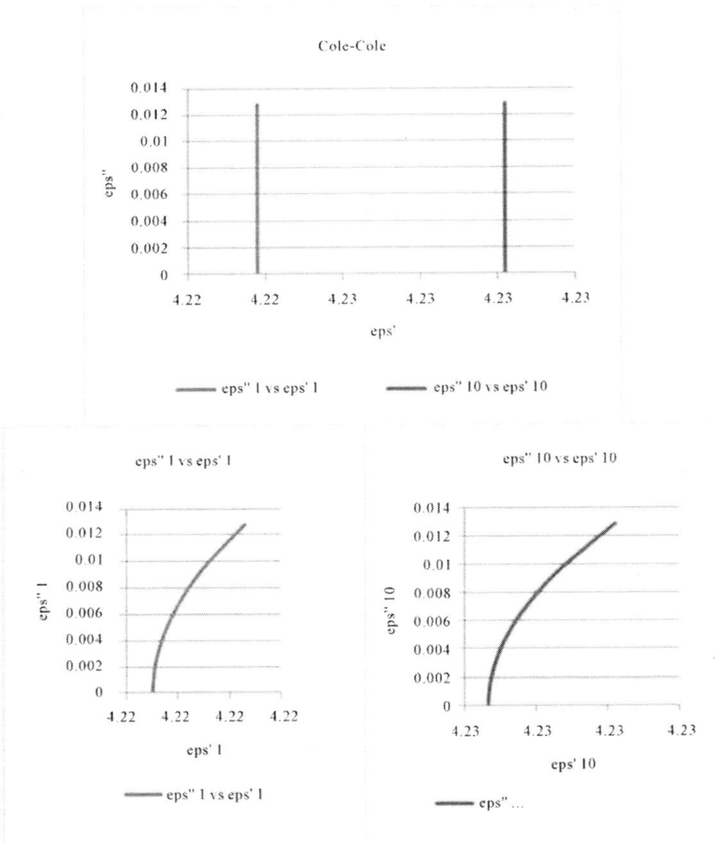

Figure 26. Cole-Cole plot of different number of inclusion without conductive layer.

Figure 30(a) shows the increase of DRS with the increase of damage state defects which is in agreement with what we saw in the computational data

(Figure 20 and Figure 27). Figure 30(b) shows the increase in dielectric loss with the increase of damage, and this loss increased more in the lower frequency region when the defects had conductive solution layers on their surface, which in also in agreement with the computational model (Figure 21 and Figure 28).

4. CONCLUSIONS

In this paper, we have demonstrated a computational model to predict global dielectric property changes caused by increasing defects inside of a materials system. We show that the dielectric character of the defects, their volume, and the morphology of the defect surfaces play an important role in the overall dielectric properties of the materials system during degradation. The data presented here also demonstrate the possibility of using dielectric properties to model and interpret the progressive damage of heterogonous materials systems.

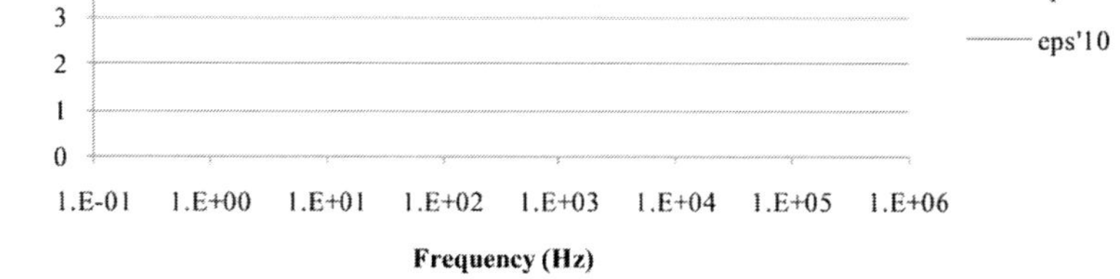

Figure 27. Real part of the Permittivity of different number inclusion but same volume fraction with conductive layer.

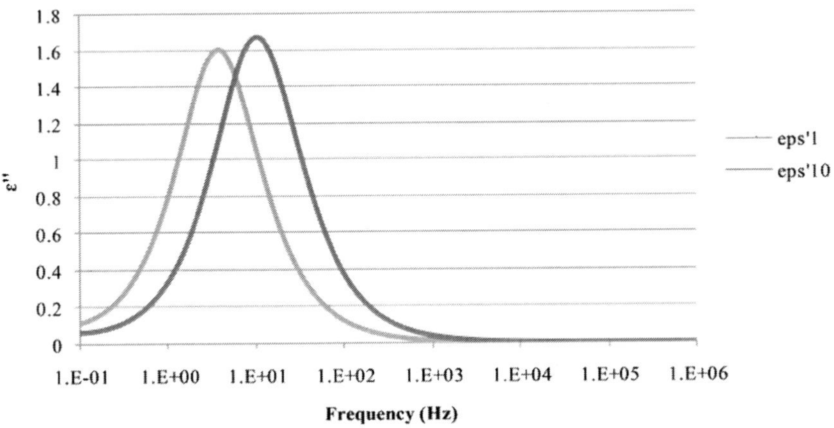

Figure 28. Imaginary part of the Permittivity of different number inclusion but same volume fraction with conductive layer.

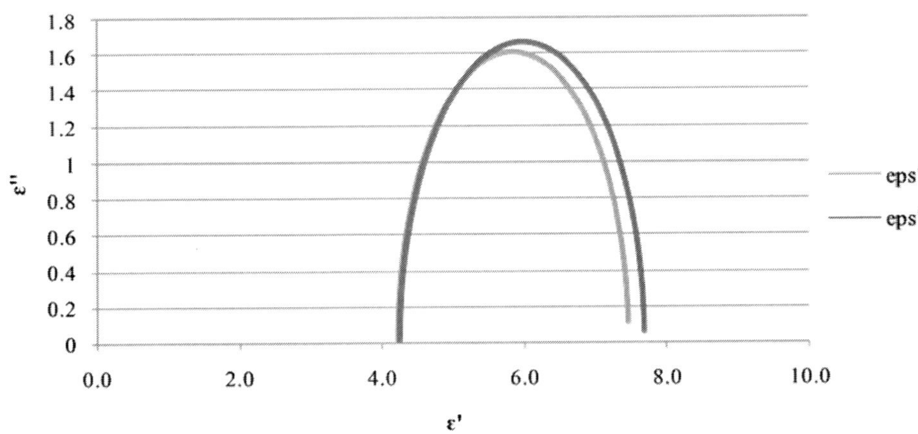

Figure 29. Cole-Cole plot of different number of inclusion with conductive layer.

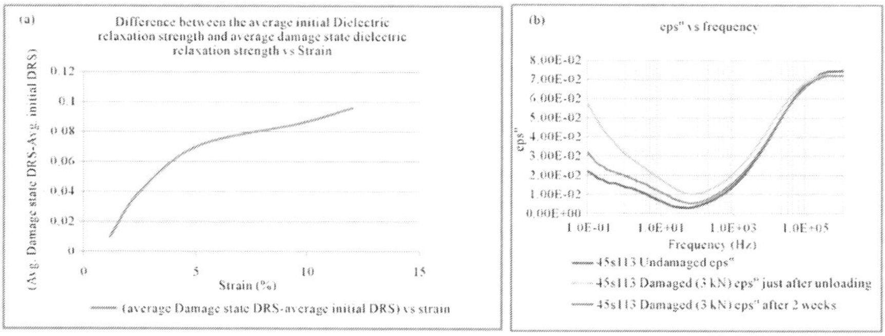

Figure 30. Experimental results of dielectric properties change of damaged composite materials.

In general, we have shown that the dielectric properties of heterogeneous systems are influenced by various physical factors: electrical and structural interactions between particles, heterogeneity of morphological and electrical properties of the constituent phases, frequency dependence of electrical phase parameters, intra-parti- cle structure, particle shape, size, orientation and, volume and surface fraction of the constituent phases. This dependence complicates the determination of the electrical parameters of heterogeneous materials from the observed global dielectric relaxation spectra, but also presents us with an opportunity to recover important information not only about the electrical and structural properties of constituents but also about the interactions between constituents, including the parent materials and damage phases. Further theoretical and experimental investigation is required to fully understand the changes in dielectric spectra associated with many of the specific damage accumulation events and local details in heterogeneous material systems.

From the results presented in this paper, it can be concluded that analysis of the dielectric data gives us information about the type of material state changes throughout the mechanical life of a composite material. It should be emphasized that these changes in the dielectric properties are distinct and measurable changes in material state, and that they are caused by a non-conservative, non-equilibrium material response to the applied fields. Opportunities for further understanding include the identification of the material and physical limitations of this method of characterization, e.g., specimen size, material property ranges, and specimen shapes that are most and least suited to the approach. A robust study of the interpretation of dielectric data associated with specific damage modes and details is also needed.

CITE THIS PAPER

RasselRaihan,FazleRabbi,VamseeVadlamudi,KennethReifsnider, (2015) Composite Materials Damage Modeling Based on Dielectric Properties. *Materials Sciences and Applications*,**06**,1033-1053. doi: 10.4236/msa.2015.611103

REFERENCES

1. Reifsnider, K.L. and Case, S.W. (2002) Damage Tolerance and Durability of Material Systems. John Wiley and Sons, New York.
2. 2. Raihan, R., Adkins, J.M., Baker, J., Rabbi, F. and Reifsnider, K. (2014) Relationship of Dielectric Property Change to Composite Material State Degradation. Composites Science and Technology, 105, 160-165.
3. Raihan, R., Reifsnider, K., Cacuci, D. and Liu, Q. (2015) Dielectric Signatures and Interpretive Analysis for Changes of State in Composite Materials. ZAMM-Journal of Applied Mathematics and Mechanics/Zeitschrift für Angewandte Mathematik und Mechanik.
4. Tuncer, E., Serdyuk, Y.V. and Gubanski, S.M. (2001) Dielectric Mixtures—Electrical Properties and Modeling.
5. Brosseau, C., Beroual, A. and Boudida, A. (2000) How Do Shape Anisotropy and Spatial Orientation of the Constituents Affect the Permittivity of Dielectrich Eterostructures? Journal of Applied Physics, 88, 7278-7288.
6. Baker, J., Adkins, J.M., Rabbi, F., Liu, Q., Reifsnider, K. and Raihan, R. (2014) Meso-Design of Heterogeneous Dielectric Material Systems: Structure Property Relationships. Journal of Advanced Dielectrics, 4, 1450008.
7. Pawar, S.D., Murugavel, P. and Lal, D.M. (2009) Effect of Relative Humidity and Sea Level Pressure on Electrical Conductivity of Air over Indian Ocean. Journal of Geophysical Research: Atmospheres, 114(D2).
8. Soles, C. and Yee, A. (2000) A Discussion of the Molecular Mechanisms of Moisture Transport in Epoxy Resins. Journal of Polymer Science, Part B: Polymer Physics, 38, 792-802.

9. Adamson, M.J. (1980) Thermal Expansion and Swelling of Cured Epoxy Resin Used in Graphite/Epoxy Composite Materials. Journal of Material Science, 15, 1736-1745.
10. Tencer, M. (1994) Moisture Ingress into Nonhermetic Enclosures and Packages—A Quasisteady State Model for Diffusion and Attenuation of Ambient Humidity Variations. IEEE 44th Electronic Components Technology Conference, Washington DC.
11. Shirangi, M.H. and Michel, B. (2010) Mechanism of Moisture Diffusion, Hygroscopic Swelling, and Adhesion Degradation in Epoxy Molding Compounds. Moisture Sensitivity of Plastic Packages of IC Devices, Springer, 29-69.
12. Banhegyi, G. and Karasz, F.E. (1986) The Effect of Adsorbed Water on the Dielectric Properties of CaCO3 Filled Polyethylene Composites. Journal of Polymer Science Part B: Polymer Physics, 24, 209-228.
13. Banhegyi, G., Hedvig, P. and Karasz, F.E. (1988) DC Dielectric Study of Polyethylene/CaCO3 Composites. Colloid and Polymer Science, 266, 701-715.
14. Cotinaud, M., Bonniau, P. and Bunsell, A.R. (1982) The Effect of Water Absorption on the Electrical Properties of Glass-Fibre Reinforced Epoxy Composites. Journal of Materials Science, 17, 867-877.
15. Reid, J.D., Lawrence, W.H. and Buck, R.P. (1986) Dielectric Properties of an Epoxy Resin and Its Composite I. Moisture Effects on Dipole Relaxation. Journal of Applied Polymer Science, 31, 1771-1784.
16. Paquin, L., St-Onge, H. and Wertheimer, M.R. (1982) The Complex Permittivity of Polyethylene/Mica Composites. IEEE Transactions on Electrical Insulation, 5, 399-404.

4
CHAPTER

MICROWAVE DIELECTRICS BASED ON COMPLEX OXIDE SYSTEMS

A.G. Belous

[1] *V.I. Vernadskii Institute of General and Inorganic Chemistry of the Ukrainian NAS, Kyiv, Ukraine*

1. INTRODUCTION

The development of modern telecommunication systems calls for the creation of novel materials with a high level of electrophysical parameters in the MW range. These materials must have in the MW range a high permittivity ($\varepsilon \geq 20$), a low dielectric loss ($tg\ \delta \leq 10^{-3} - 10^{-4}$) and a high thermostability of electrophysical properties (temperature coefficient of permittivity ($TC\varepsilon$) or resonant frequency (τ_f) $\sim 10^{-6}K^{-1}$). Such materials can be used in the development of resonant elements of radio-frequency filters, solid state oscillators, substrates for hybrid MW circuits, allow the size of communication systems to be greatly reduced and improve their parameters. Moreover, the use of them reduces the manufacturing and operating costs for modern communication systems.

The choice of the permittivity value of MW materials is largely determined by the frequency range of the operation of communication systems, the type of exciting wave and the requirement of the optimal size of dielectric element. The value of ε determines the size of radio components. The influence of microminiaturization is based on the fact that the electromagnetic wavelength in dielectric decreases in inverse proportion to $\sqrt{\varepsilon}$. Therefore, in the decimeter wave band, high-Q thermostable materials with high permittivity value ($\varepsilon \geq$ 80-600) are required, whereas in the centimeter and millimeter wave bands, thermostable materials with $\varepsilon \sim$ 15-30 but with extremely high Q ($Q\times f \geq 80000$, where $Q = 1/\ tg\ \delta$ and f is frequency in GHz) are needed.

It should be noted that low dielectric loss in the MW range $10^9 - 10^{11}$ Hz is characteristic only of optical and infrared polarization mechanisms. Other polarization mechanisms give rise, as a rule, to considerable dielectric loss [1].

In the case of optical polarization, dielectrics are characterized by a low negative temperature coefficient of permittivity ($TC\varepsilon \sim 10^{-5}K^{-1}$). However, the dielectric contribution of optical polarization is usually small. Therefore, large permittivity values together with high temperature stability of dielectric parameters and low dielectric loss can be observed only in the dielectrics where the main contribution to polarization is made by infrared polarization mechanism [1, 2]. This mechanism is bound up with cation and anion sublattice displacement in electric field, which is only possible in ionic crystals. The contribution of infrared polarization mechanism to permittivity may be $\Delta\ \varepsilon_{ir} = 1 - 10^4$ in the MW range. The temperature instability of ε increases, as a rule, with increasing ε. The large magnitude of infrared polarization is usually due to the presence of a soft mode in crystal, whose frequency varies by the critical law $\omega_T = A\sqrt{T - Q}$; this leads in accordance with the Liddein–Sax–Teller (LST) relation:

$$\frac{\varepsilon_{MW}}{\varepsilon_{Opt}} = \prod_i \frac{\omega_L^2}{\omega_T^2} \tag{1}$$

to the Curie – Weiss law for permittivity:

$$\varepsilon_{MW} = \varepsilon_L + \frac{C}{T-Q} \tag{2}$$

where ω_L and ω_T are the frequencies of longitudinal and transverse optical phonons in the center of Brillouin zone (one of the transverse phonons is soft), C is a constant, Q is Curie – Weiss temperature, ε_L is dielectric contribution, which depends only slightly on temperature [1].

The frequency of transverse and longitudinal optical phonons can be calculated from the equations:

$$\omega_T^2 = \frac{c}{m} - \frac{nq^2}{3\varepsilon_0 m} \cdot \frac{\varepsilon_\infty + 2}{3} \text{ and } \omega_L^2 = \frac{c}{m} + \frac{2nq^2}{3\varepsilon_0 m} \cdot \frac{\varepsilon_\infty + 2}{3} \tag{3}$$

where c is the elastic coupling parameter of phonons; m is reduced mass; q, n are ion charge and concentration; ε_0 is an electric constant [1].

The parameters c, n, ε_{opt} are temperature-dependent; they decrease with rising temperature due to thermal expansion (lattice anharmonicity). It is evident from system (3) that ω_L is a weak function of temperature because it is determined by the sum of two terms, whereas the dependence ω_T (T) may be strong since ω_T depends on the difference of two terms (see Eqs (3)). The variation of this difference as a function of temperature depends on which effect predominates: the variation of c/m (minuend) or the variation of the subtrahend, which depends on n, ε_{opt}. Depending on ω_T, the dielectric contribution $\Delta\ \varepsilon_{ir}$ also varies with temperature [1]:

$$\Delta\varepsilon_{IR} = \varepsilon_{MW} - \varepsilon_{Opt} = \frac{nq^2}{m\omega_T^2\varepsilon_0} \cdot \left(\frac{\varepsilon_{Opt} + 2}{3}\right). \tag{4}$$

The above analysis shows that the chemical composition can influence, in principle, the contribution of different polarization mechanisms and hence the value of permittivity and loss, as well as their variation in the MW range as a function of temperature.

The present paper considers the structure peculiarities, electrophysical properties and possible applications of inorganic microwave (MW) dielectrics based on oxide systems.

2. MW DIELECTRICS BASED ON (LA, CA)(TI, AL)O_3 SOLID SOLUTIONS

MW dielectrics are often synthesized on the basis of solid solutions, e.g. $Ba(Zn, Mg)_{1/3}(Nb, Ta)_{2/3}O_3$ [3, 4, 5], $(La, Ca)(Ti, Al)O_3$ [6, 7], etc. The substance of this approach is that solid solutions are formed by the interaction of phases belonging to the same crystal structure, which have in the MW range a different trend of the plot of permittivity against temperature and a low dielectric loss. Paraelectric is characterized by a low dielectric loss; for example, $CaTiO_3$, which crystallizes in perovskite structure, can be used in the $CaTiO_3$ – $LaAlO_3$ system as a phase with negative temperature coefficient of permittivity ($TC\varepsilon < 0$) [6]. At the same time, $LaAlO_3$ can be used as a phase with perovskite structure having $TC\varepsilon > 0$ [7]. By varying the ratio $CaTiO_3$ /$LaAlO_3$, one can control the value of TCε. Positive TCε in dielectrics in the MW range usually indicates the presence of a high-temperature phase transition, which is connected with the existence of spontaneously polarized state (ferroelectrics, antiferroelectrics). However, the materials in which spontaneous polarization exists have, as a rule, a considerable dielectric loss in the MW range, which is inadmissible for the creation of high–Q dielectrics. In $LaAlO_3$, there is no spontaneous polarization. It should be noted that there are very few materials having $TC\varepsilon > 0$ in the MW range and a low dielectric loss. Therefore, the development of high-Q MW dielectrics with high ε and positive temperature coefficient of permittivity ($TC\varepsilon > 0$) is of independent scientific and practical interest.

3. CONTROL OF THE TCE VALUE BY INFLUENCING THE PHONON SPECTRUM

As follows from the analysis of expressions (3), the trend of the plot of ε against temperature in the MW range can be controlled by influencing the phonon spectrum. One of the ways of influencing the phonon spectrum in some types of structures can be iso- and heterovalent substitutions in cation sublattices. As an example, we chose $La_{2/3-x}$ (Na, K)$_{3x}$ TiO_3 materials, which crystallize in defect-perovskite structure in a wide x range (Fig 1).

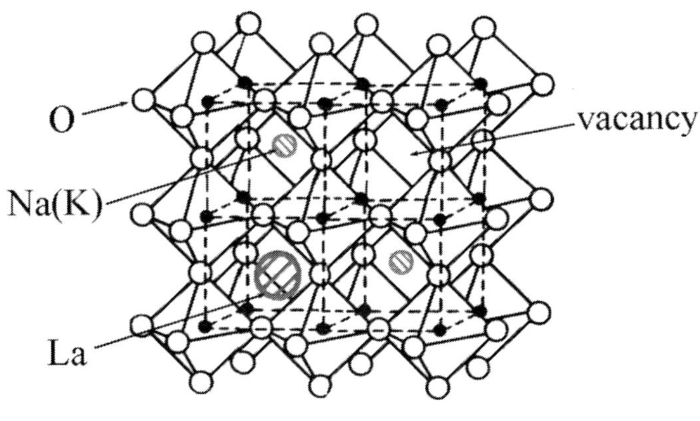

Figure 1. Crystal structure of $La_{2/3-x}$(Na, K)$_{3x}$$TiO_3$ perovskite

In this system, lanthanum ions in the oxidation state +3 are partially substituted by alkali (sodium or potassium) ions in the oxidation state +1. Substitution is performed so that the electroneutrality condition is satisfied. The value of x was varied from 1/24 to 1/6. In this case, lanthanum ions, alkali metal (sodium or potassium) ions and a structural vacancy (vacant crystal site) could be in one crystal sublattice at the same time. We hoped that in the case of such heterovalent substitution, the phonon spectrum, and hence the trend of the plot of ε against temperature in the MW range, had to change.

$La_{2/3-x}$ (Na, K)$_{3x}$ TiO_3 materials are not characterized by high temperature stability of dielectric parameters. The dielectric properties of these materials had been studied in a wide frequency range [8, 9].

It had been found that by decreasing the number of structural vacancies, using heterovalent substitution in sublattices and locating different ions in vacant crystal sites, one can influence greatly the dielectric loss level (Fig 2) [10].

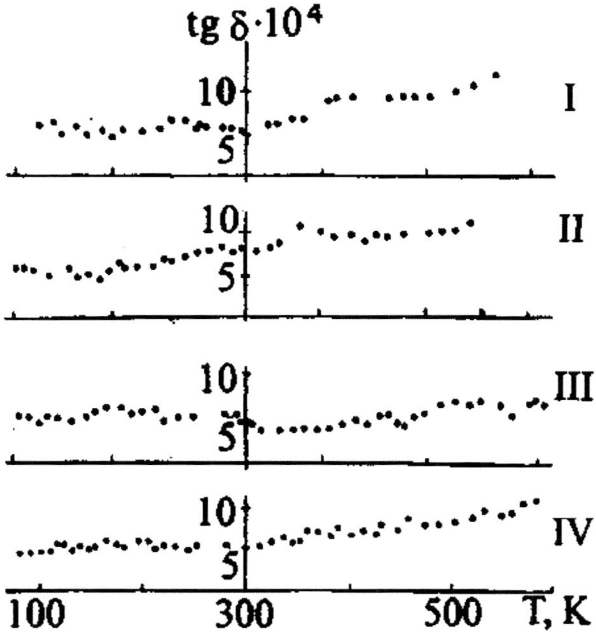

Figure 2. Plots of dielectric loss (tg δ) in the $La_{2/3-x}(Na, K)_{3x}TiO_3$ system at 1.2×10^{10} Hz against temperature: (I) $La_{1/2}Na_{1/2}TiO_3$, (II) $Nd_{1/2}Na_{1/2}TiO_3$, (III) $La_{1/2}Na_{1/4}K_{1/4}TiO_3$, (IV) $La_{7/12}Na_{1/4}\bullet_{1/6}TiO_3$, where • is the structural vacancy

Investigations showed that heterovalent substitutions in cation sublattices affect greatly the value of TCε too. To explain this effect, IR reflection spectra of $La_{2/3-x}$ M_{3x} TiO_3 materials have been analyzed [11]. The analysis of IR reflection spectra made it possible to calculate the parameters of dispersion oscillators (Table 1). It is known that in the materials that crystallize in perovskite structure, a low-frequency lattice vibration exists which is responsible for the high ε value in the MW range [12].

The partial heterovalent substitution of ions in crystal sublattice gives rise to a low-frequency vibration, which affects noticeably the value of permittivity. In this case, the temperature stability of dielectric parameters increases greatly. Thus, the proposed method of influencing the phonon spectrum can be employed in the development of novel MW dielectrics with high temperature stability of dielectric properties.

Table 1. Parameters of $(La_{2/3-x} M_{3x} \bullet_{1/3-2x})TiO_3$ dispersion oscillators*

$La_{1/2}Na_{1/4}K_{1/4}TiO_3$ (ε = 106)				$La_{7/12}Na_{1/4} \bullet_{1/6}TiO_3$ (ε = 87)			
ω_{TO}	ω_{LO}	ΔE	g	ω_{TO}	ω_{LO}	ΔE	g
cm^{-1}				cm^{-1}			
116	163	74.0	0.75	133	178	55.0	0.64
198	258	22.0	0.44	201	224	19.5	0.31
237	334	3.0	0.21	230	265	4.1	0.30
336	375	0.2	0.33	270	343	1.8	0.26
381	489	0.4	0.16	345	460	0.2	0.10
554	747	1.2	0.10	563	694	1.0	0.11
785	816	0.1	0.12	789	860	0.3	0.30

4. MW DIELECTRICS BASED ON $BA_{6-X}LN_{8+2X/3}TI_{18}O_{54}$ (LN = LA-GD)

The $Ba_{6-x}Ln_{8+2x/3}Ti_{18}O_{54}$ materials (Ln = La - Gd) (BLTss) have promise in the development on their basis of thermostable high-Q MW dielectrics with high permittivity (ε ≈ 80 - 100) [13, 14]. They crystallize in KW bronze structure (Fig 3), which includes elements of perovskite structure [15, 16, 17]. In this structure, the octahedra are linked, as in perovskite, by their apices into parallel rectilinear chains. Unlike perovskite structure, however, the oxygen octahedra are linked so that they form pentangular, quadrangular and triangular channels, in which A ions can be, having in this case the coordination numbers 15, 12 and 9 respectively. This structure allows one to perform iso- and heterovalent substitutions in cation sublattices in a wide range, to control the number of vacant crystal sites in the A sublattice, to influence the partial redistribution of A ions among the pentangular, quadrangular and triangular channels and hence to control the electrophysical properties in the MW range, including the temperature dependence of ε.

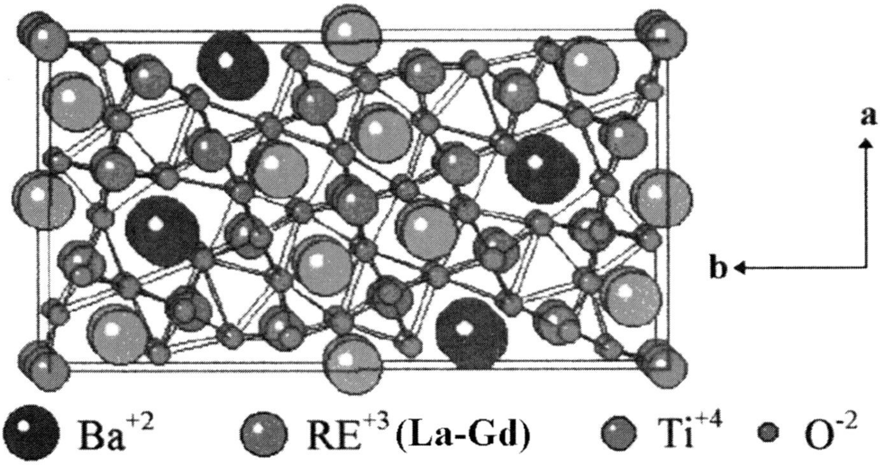

Figure 3. Unit cell of Ba_{6-x} $Ln_{8+2x/3}Ti_{18}O_{54}$ [17]

When investigating $Ba_{6-x}Ln_{8+2x/3}Ti_{18}O_{54}$ materials (Ln = La - Gd), which crystallize in KW bronze structure, a special attention was given to the study of formation reaction and the anomalous behavior of the temperature characteristics of dielectric parameters. It should be noted that the knowledge of the reactions proceeding during the synthesis of compounds can allow dielectric loss to be reduced. The formation of $Ba_{6-x}Ln_{8+2x/3}Ti_{18}O_{54}$ materials (Ln = Nd, Sm) when using the solid-state reaction method was studied on compositions with x = 0.75, 1.5, 2.0. $BaCO_3$, Sm_2O_3 and Tio_2 were used as starting reagents. It was shown that it is a multistage process, which is accompanied by the formation of intermediate phases, e.g. $Ln_2Ti_2O_8$, $BaTi_4O_9$, $BaTiO_3$ [18]. It has been found that independent of x value, the phase $Ba_{3.9}Ln_{9.4}Ti_{18}O_{54}$ is formed at first, which belongs to $Ba_{6-x}Ln_{8+2x/3}Ti_{18}O_{54}$ solid solutions and corresponds to the maximum x value. The formation of the other phases, which belong to the region of $Ba_{6-x}Ln_{8+2x/3}Ti_{18}O_{54}$ – type solid solutions, takes place as a result of interaction between intermediate $Ba_{3.9}Ln_{9.4}Ti_{18}O_{54}$ phases and barium metatitanate ($BaTiO_3$). The phase $Ba_{3.9}Ln_{9.4}Ti_{18}O_{54}$ crystallizes, as $Ba_{6-x}Ln_{8+2x/3}Ti_{18}O_{54}$ materials, in KW bronze structure, which makes the identification of the phase $Ba_{3.9}Ln_{9.4}Ti_{18}O_{54}$ only by the data of X-ray phase analysis impossible. Therefore, EDS and TEM analyses were used additionally [18]. The latter analysis showed that even when all $BaTiO_3$ had reacted, the homogeneity of materials was not reached yet (Fig 4(a)). If the ceramic sintering time was relatively short, a phase, e.g. $Ba_{3.9}Ln_{9.4}Ti_{18}O_{54}$, was present (Fig 4(a)), which had low and high x values within the limits of formation of $Ba_{6-x}Ln_{8+2x/3}Ti_{18}O_{54}$ solid solutions. The structural-defect concentration decreased, and the homogeneity of $Ba_{6-x}Ln_{8+2x/3}Ti_{18}O_{54}$ materials increased only in the case of long ceramic sintering time (t ≥ 3h) (Fig 4(b)).

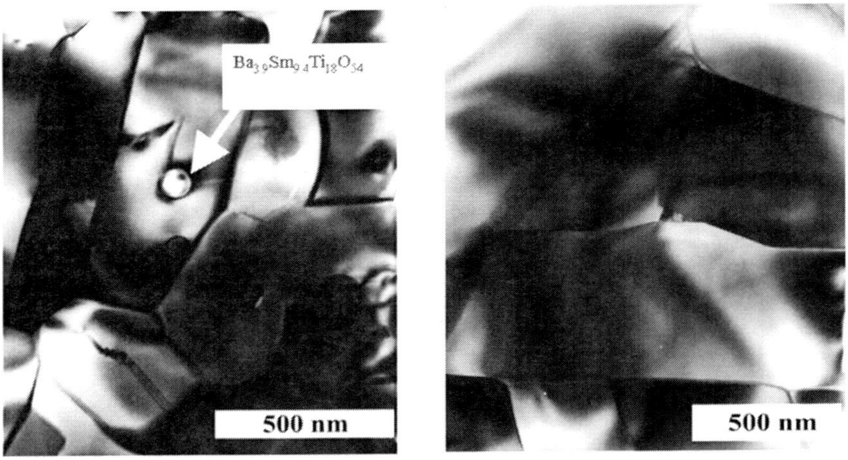

Figure 4. Results of a TEM analysis of $Ba_{6-x}Ln_{8+2x/3}Ti_{18}O_{54}$ ceramic; sintering time: 1 h (a), over 3 h (b)

The electrophysical characteristics of $Ba_{6-x}Ln_{8+2x/3}Ti_{18}O_{54}$ depend largely upon ions in the A sublattice (r_A) [14, 19, 20]. When the rare-earth ion in the A sublattice is changed from La to Gd, the permittivity (ε) value and loss-angle tangent in BLTss decrease. At the same time, the temperature coefficient of permittivity, TCε, increases and changes its sign in the series of rare-earth ions, which are in the A sublattice, in going from Nd to Sm [20, 21]. When investigating $Ba_{6-x}Ln_{8+2x/3}Ti_{18}O_{54}$ materials (where x = 1.5), we had found for the first time an anomaly on the plot of permittivity against temperature [21]. Later, anomalies on the plots of dielectric parameters (ε, tg δ) against temperature were detected in the other $Ba_{6-x}Ln_{8+2x/3}Ti_{18}O_{54}$ materials too [22]. The nature of these anomalies remained uncertain. Moreover, there was no information about the existence of temperature dependence anomalies of dielectric parameters in other barium-lanthanide analogs, including La-, Nd-, Gd-containing $Ba_{6-x}Ln_{8+2x/3}Ti_{18}O_{54}$. Therefore, we tried to find out the cause of the temperature dependence anomalies of permittivity and dielectric loss since this makes it possible to establish the nature of the thermostability of electrophysical properties in these systems.

It has been found that temperature dependence anomalies of dielectric parameters in La- and Nd-containing BLTss are observed at low temperatures in a wide frequency range, including the submillimeter-wave region (Fig 5) [23], the position of these anomalies on the temperature scale depending not on measurement frequency, but on chemical composition.

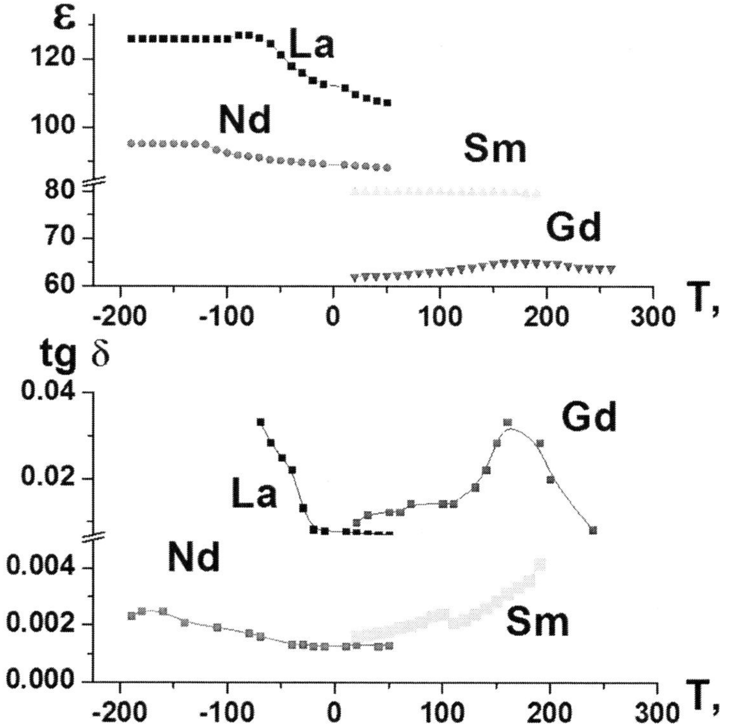

Figure 5. Plots of the dielectric parameters of $Ba_{6-x}Ln_{8+2x/3}Ti_{18}O_{54}$ solid solutions (Ln = La, Nd, Sm, Gd) at 10 GHz against temperature

In Sm- and Gd-containing systems, anomalies of dielectric parameters appear at temperatures above room temperatures (Fig 5). The plots of dielectric parameters against temperature have a similar trend when ferroelectric or antiferroelectric ordering occurs. In the case of BLTss, however, no hysteresis loops were observed, and the temperature dependence of ε did not obey the Curie-Weiss law, which indicated the absence of spontaneously polarized state in these materials. It may be supposed that the appearance of dielectric anomalies is due to the presence of unknown phase transitions. Therefore, $Ba_{6-x}Ln_{8+2x/3}Ti_{18}O_{54}$ systems (x = 1.0), in which dielectric anomalies of ε and tg δ were observed at 100-120 ^{0}C, have been studied by low-temperature differential scanning calorimetry (LT-DSC) and high-temperature X-ray structural analysis of samples. We did not find any phase transitions [22], which was confirmed by the authors of [24], who carried out synchrotron X-ray diffraction studies of $Ba_{4.5}Sm_9Ti_{18}O_{54}$ samples in the temperature range 10-295 K. These data indicate the absence of structural transitions in the temperature ranges where anomalies of dielectric parameters were observed. It can be concluded that the temperature dependence anomalies of dielectric parameters are not coupled with the peculiarities of sample preparation and the presence of structural transitions.

On the basis of an analysis it was assumed that the nature of the anomaly of dielectric parameters is coupled with harmonic and anharmonic BLTss lattice vibration, which is different in the character of influence on the temperature behavior of dielectric parameters [23].

Investigations showed that the plots of ε and tg δ against temperature depend largely upon harmonic and anharmonic lattice vibration modes. Therefore, using different hetero- and isovalent substitutions in cation sublattices, one can influence the lattice phonon spectrum and hence obtain materials with high temperature stability of dielectric parameters, which are used in modern decimeter and centimeter wave band communication systems [25, 26, 27, 28].

5. MW DIELECTRICS WITH "MOBILE SUBLATTICE"

The authors of [29] reported the development of novel MW dielectrics with high temperature stability of dielectric parameters based on solid solutions, where one of the phases (lithium-containing $La_{1/2}Li_{1/2}TiO_3$) had a positive temperature coefficient of permittivity (TCε > 0). However, there was no explanation of the nature of this fact in literature.

The phase $La_{1/2}Li_{1/2}TiO_3$, which crystallizes in perovskite structure, belongs to $Ln_{2.3-x}M_{3x}Tio_3$ solid solutions (Ln = rare-earth elements, M = alkali metal ion) (Fig 6). In this system, the M ions partially substitute for rare-earth ions. In this case, the electroneutrality condition is satisfied [30, 31]. If M = Na, K, materials crystallize in perovskite structure and are characterized by low dielectric loss (tg $\delta \leq 10^{-3}$) and high permittivity in the MW range (about 100) and TCε < 0 [32].

At the same time, when M = Li, a system of solid solutions is formed, in which rare-earth ions, lithium ions, structural vacancies, which are characterized by TCε > 0, are in one sublattice at the same time. Lithium ions can move along structural channels, ensuring a high lithium-ion conductivity [33, 34], positive temperature coefficient of permittivity (TCε > 0) and causing considerable dielectric loss in the MW range. The latter is inadmissible in the creation of high-Q dielectrics. Investigations showed, however, that the dielectric loss in the MW range can be greatly reduced by decreasing lithium ion conductivity. The latter is achieved by substituting rare-earth ions with smaller radius for lanthanum ions. This leads to a decrease in the size of structural channels, in which lithium ions are; this reduces dielectric loss in the MW range through a decrease in lithium ion mobility in the structure. In this case, TCε > 0 is retained. As a result, solid solutions have been obtained, in which rare-earth ions are simultaneously substituted by alkali metal ions with large (Na, K) and small (Li) radius and which have a high temperature stability of dielectric parameters in the MW range (Table 2) [35].

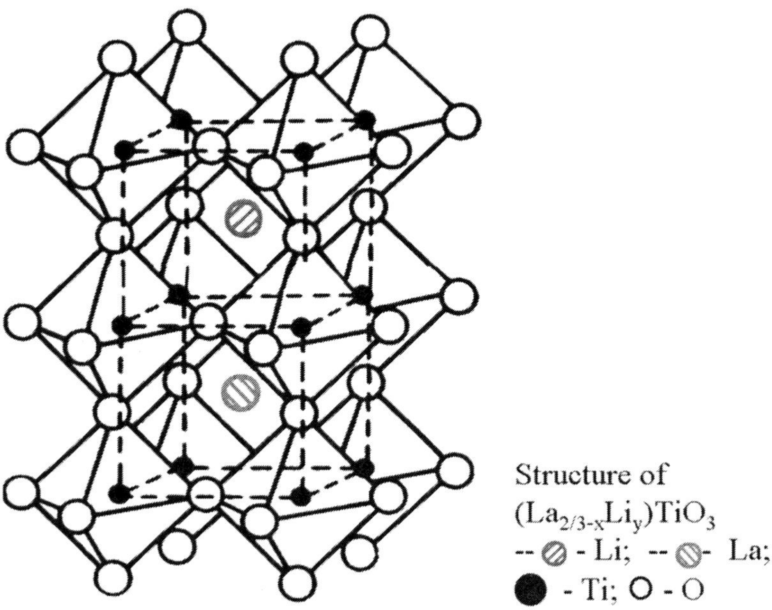

Figure 6. Perovskite structure of $La_{2/3-x}Li_{3x}TiO_3$

Table 2. Dielectric parameters of the $Sm_{1/2}Li_{1/2}TiO_3$ – (1-x) $Sm_{1/2}Na_{1/2}TiO_3$ system at 10 GHz

x	ε	TCε, ppm/°C (20–100°C)	Q (10 GHz)
0	80	–520	1300
0.30	95	–240	100
0.50	90	–140	150
0.55	85	–50	200
0.58	80	–5	200
0.60	75	+60	200
0.1	52	+580	100

6. MW DIELECTRICS BASED ON ANTIFERROELECTRICS-PARAELECTRICS

The permittivity value of the above-mentioned thermostable MW dielectrics is not over 80 – 100. To achieve higher permittivity values, use must be made of other polarization mechanisms connected with spontaneously polarized state. One of the possible ways of developing thermostable MW resonant elements is the creation of two-layer systems. Each of the layers must have a high Q ($Q = 1/ tg\ \delta$), a high permittivity and TCε of different sign. A large number of high-Q dielectrics which have TCε < 0 in the MW range is known (mainly paraelectrics), whereas there are few of them with TCε > 0 ($LiNbO_3$, $LiTaO_3$, $LiAlO_3$ single crystals).

Besides, they are characterized by relatively low ε values (ε < 50) in the MW range, and the use of them in this range makes it possible to obtain two-layer resonant dielectric elements with an effective permittivity (ε_{eff}) of not over 50. Therefore, we examined the possibility to use as high-Q materials with TCε > 0 antiferroelectrics based on tellurium-containing Pb_2BTeO_6-type perovskites (B = bivalent metal ions) [36, 37]. It has been found that the plot of ε against temperature for lead-cobalt tellurate (Pb_2CoTeO_6) in the MW range passes through a maximum (phase transition from the antiferroelectric to the paraelectric state) at 380 K (Fig 7). At room temperature, the materials of this group have high permittivity values (ε > 110). In the temperature range 220-350 K, the dependence ε (T) is close to linear one and has TCε ~ $700 \times 10^{-6}\ K^{-1}$. Two-layer resonant elements: Pb_2CoTeO_6 – TiO_2, Pb_2CoTeO_6 – $CaTiO_3$, Pb_2CoTeO_6 – $SrTiO_3$ have been prepared on the basis of paraelectrics (TiO_2, $CaTiO_3$, $SrTiO_3$) and the antiferroelectric Pb_2CoTeO_6. Known two-layer resonant dielectric $LiNBO_3$ – TiO_2 elements have been investigated for comparison. It can be seen from

Table 3 and Fig 8 that Pb_2CoTeO_6 - $SrTiO_3$ based two-layer resonant elements have in the MW range a high effective permittivity ($\varepsilon_{eff} \approx 135$), a high Q ($Q_{10GHz} \sim 900$) and a high temperature stability of dielectric parameters (TCε tends to zero) [37]. Of course, two-layer systems have a number of demerits since they require mechanical bonding of different layers.

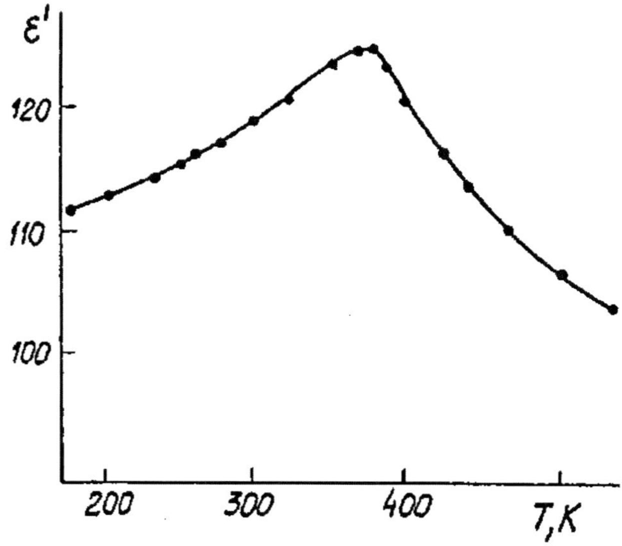

Figure 7. Plot of permittivity against temperature for Pb_2CoTe_6 at 10GHz

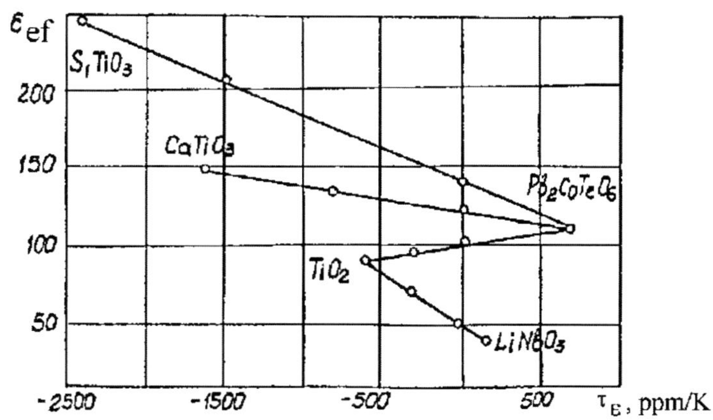

Figure 8. Dependence of the effective permittivity (ε_{ff}) of two layer resonant elements on the temperature coefficient of frequency (τ_ε) at 10 GHz

Table 3. Properties of two-layer resonant elements at 10GHz

$La_{1/2}Na_{1/4}K_{1/4}TiO_3(\varepsilon = 106)$				$La_{7/12}Na_{1/4} \bullet {}_{1/6}TiO_3(\varepsilon = 87)$			
ω_{TO}	$\omega\__{LO}$	ΔE	g	ω_{TO}	ω_{LO}	ΔE	g
cm^{-1}				cm^{-1}			
116	163	74.0	0.75	133	178	55.0	0.64
198	258	22.0	0.44	201	224	19.5	0.31
237	334	3.0	0.21	230	265	4.1	0.30
336	375	0.2	0.33	270	343	1.8	0.26
381	489	0.4	0.16	345	460	0.2	0.10
554	747	1.2	0.10	563	694	1.0	0.11
785	816	0.1	0.12	789	860	0.3	0.30

*$\varepsilon_\infty = 5.1$.

Thermostable MW dielectrics with high permittivity ($\varepsilon \sim 430$) and Q ($Q_{1GHz} \sim 700$) have been obtained by the authors of [38] on the basis of the Ag (Nb, Ta) O_3 system, which is characterized by the spontaneously polarized state. However, in the materials in which the spontaneously polarized state is present, increase in permittivity is always accompanied by an increase in dielectric loss, which impairs the technical characteristics of MW elements made on their basis.

7. MW DIELECTRICS BASED ON SPONTANEOUSLY POLARIZED PHASES

The materials in which there is a phase transition from the spontaneously polarized to the unpolarized state at high temperatures are characterized by positive TCε. In the phase transition region, the tg δ values in the MW range are, as a rule, large, which is due to the presence of mobile domain walls (ferroelectrics). A salient feature of antiferroelectrics is the immobility of domain walls. This results in the fact that antiferroelectric (e.g. Pb_2CoTeO_6) is characterized by a relatively low tg δ value at room temperatures in the MW range [37]. In some cases, ferroelectrics, too, have low tg δ values, e.g. single-domain $LiNbO_3$ single crystal, in which phase transition is observed at high temperature (> 1200 ^{0}C).

It was investing to find out whether it is possible to create thermostable MW dielectrics on the basis of solid solutions firmed by ferroelectrics and/or antiferroelectrics, which are characterized by high phase transition temperature, and materials having a defect crystal structure (with vacancies). To this end, we investigated $Ln_{2/3-x}Na_{3x}Nb_2O_6$ materials (Ln =

La, Nd), which were formed by interaction between the $La_{2/3} \bullet _{4/3}Nb_2O_6$ phase with defect-perovskite structure (Fig 9) and the $NaNbO_3$ phase with perovskite structure (Fig 10), in which transition from the spontaneously polarized to the unpolarized state is observed at high temperature (> 520 ^{0}C).

An analysis of X-ray data for polycrystalline $La_{2/3-x}Na_{3x} \bullet _{4/3-2x}Nb_2O_6$ samples (Ln = La, Nd) showed that depending on x, solid solutions having three different space groups are formed. The space group changes in the order Pmmm → Pmmn Pbcn with increasing x. In the interval $0 \leq x \leq 0.24$, the solid solutions have, independent of Ln, a defect-perovskite structure ($La_{2/3} \bullet _{4/3}Nb_2O_6$), where is a vacancy in the cation sublattice with the space group Pmmm [39].

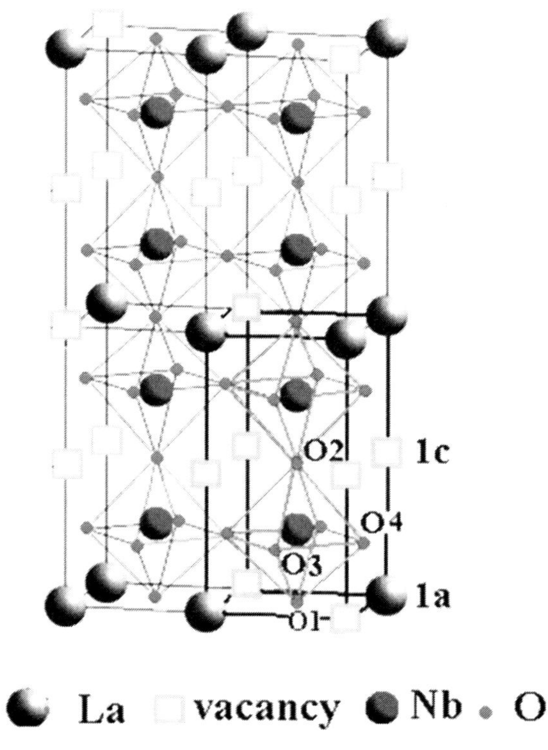

Figure 9. Defect-perovskite structure $Ln_{2/3} \bullet _{4/3}Nb_2O_6$ (space group Pmmm). Atomic positions: La (1a) 000; Nb (2t) ½ ½ z; 0 (1) (1f) ½ ½ 0; 0 (2)

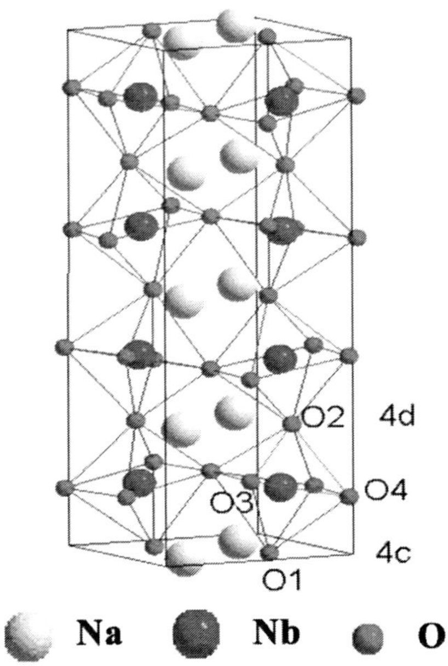

Figure 10. Perovskite structure of $NaNbO_3$ at room temperature (space group Pbcm). Atomic positions: Na (1) (4c) x ¼ 0; Na (2) (4d) x y ¼; Nb (8c) x y z; O (1) (4c) x ¼ 0; O (2) (4d) x y ¼ ; O (3) (8c) x y z; 0 (4) (8c) x y z

When x is increased, the space group changes, independent of the kind of Ln, from Pmmm to Pmmn. In neodymium-containing solid solutions, the change of space group doespractically lead to change in normalized unit cell volume V/Z; in lanthanum-containing solid solutions, monotonic variation of V/Z as a function of x is accompanied by a slight inflection in the sodium concentration range corresponding to the change from Pmmm to Pmmn. The observed differences in the trend of plots of V/Z (x) are probably due to the fact that the ionic radius of Na^+ is smaller than that of the substituted La^{3+} ion, whereas the difference in the ionic radii of Na^+ and Nd^{3+} is small. The closeness of the ionic radius values of Na^+ and Nd^{3+} leads to the existence of a wider concentration range ($0.24 \leq x \leq 0.54$), which corresponds to the space group Pmmm, for neodymium-containing solid solutions as compared with lanthanum-containing ones ($0.24 \leq x \leq 0.45$).

In the intervals $0.54 \leq x \leq 0.66$, (for neodymium) and $0.45 \leq x \leq 0.66$ (for lanthanum), the crystal structure of $La_{2/3-x}Na_{3x} \bullet {}_{4/3-2x}Nb_2O_6$ solid solutions has the space group Pbcn, which is typical of $NaNbO_3$ at room temperature [40]. In fact, when the sodium content of the system is increased, a decrease in the symmetry of solid solutions from Pmmm to Pmmn and to Pbcn is observed.

$La_{2/3}$ Nb_2O_6 and $Nd_{2/3}$ Nb_2O_6 materials (x = 0) are characterized by a high permittivity value (130 and 160 respectively) and a relatively low dielectric loss (in both cases, tg δ is of the order of 2 – 5 × 10^{-3} at a frequency of 10 GHz). There is no permittivity dispersion. The plot of ε (T) for $La_{2/3}$ Nb_2O_6 and $Nd_{2/3}$ Nb_2O materials exhibits deflections in the MW range. In the low-temperature range, the ε value varies only slightly with rising temperature. As the x values in $La_{2/3-x}Na_{3x}$ • $_{4/3-2x}Nb_2O_6$ materials (Ln = La, Nd) increases, TCε changes its sign from negative to positive. Plots of dielectric parameters against concentration in the MW range are shown in Fig 11. In the interval 0 ≤ x ≤ 0.24 (space group) Pmmm, increasing the sodium concentration leads to a slight increase in permittivity independent of the kind of rare-earth element (La or Nd), which is accounted for by increase in cation vacancy concentration.

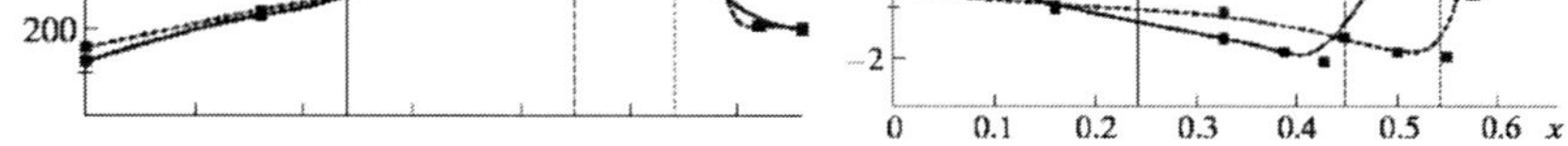

Figure 11. Plots of permittivity (a) and the temperature coefficient of permittivity (b) against the sodium content of the solid solutions $La_{2/3-x}Na_{3x}$ • $_{4/3-2x}Nb_2O_6$ (1) and $Nd_{2/3-x}Na_{3x}$ • $_{4/3-2x}Nb_2O_6$ (2)

Within the limits of the space group Pmmn, the permittivity value decreases greatly with increasing x, passing through a maximum, when the space group changes from Pmmn to Pbcn. Investigations showed that it is possible to create thermostable dielectrics based on the system in $Ln_{2/3}Na_{3x}$ •$_{4/3-2x}Nb_2O_6$ (Ln = La, Nd), which have a high permittivity (ε ~ 300-600) and a relatively low dielectric loss (tg δ ~ 2 – 7 × 10^{-3}) in the MW range [41, 42, 43, 44].

The systems considered above have a relatively high thermostability of electrophysical properties (TCε ~ 10^{-5} - $10^{-6}K^{-1}$), Q × f ≤ 12000 and relatively high permittivity values (ε ≥ 80 600) in the MW range. This makes it possible to develop on their basis elements for decimeter wave band communication systems, where the problems of microminiaturization, for the solution of which high ε values are required, are especially important.

In the centimeter wave band and especially in the millimeter wave band, however, materials with relatively low permittivity (10 -30) are required, which must possess very high Q valies (Q × f ≥ 80000 - 100000). Let us consider some systems, which have promise in gaining these purposes.

8. $M_{1+X}NB_2O_6$ – BASED MW DIELECTRICS (M = MG, CO, ZN) WITH COLUMBITE STRUCTURE

Among the MW dielectrics known to date, M^{2+} Nb_2O_6 niobates (M = Mg, Co, Zn) are of considerable interest. The crystal structure of A^{2+} Nb_2O_6 columbite is infinite zigzag chains of oxygen linked by shared edges (Fig 12) [45]. For this structure, redistribution of crystal sites, which are in the oxygen octahedral, among A^{2+} ion size and the cation ratio Nb^{5+} : A^{2+} in the unit cell will affect the crystallographic distortions of the columbite structure and hence the phase composition and electrophysical properties of synthesized materials.

The $ZnNb_2O_6$ and $MgNb_2O_6$ materials have a high Q (3000 and 9400 respectively) and permittivity (23 and 20 respectively) [46, 47, 48]. In contrast to magnesium- and zinc-containing niobates, the literature data on the dielectric properties of cobalt niobate ($CoNb_2O_6$) are very contradictory. For instance, Ref [46] reported low Q values for $CoNb_2O_6$ (Q × f = 40000). Earlier it was shown [49, 50, 51, 52, 53] that the making of single-phase M^{2+} Nb_2O_6 materials depends largely upon their synthesis conditions. For instance, the difficulty of making single-phase magnesium niobate with columbite structure is accounted for, in particular, by the simultaneous formation of two phases: $MgNb_2O_6$ columbite and $Mg_4Nb_2O_9$ corundum [51]. It should be noted that in the Mg-Nb-O system, a number of compounds: $MgNb_2O_6$, $Mg_4Nb_2O_9$, $Mg_5Nb_4O_{15}$, $Mg_{1/3}Nb_{11\,(1/3)}$ are formed [52, 53]; however, only the phases $MgNb_2O_6$ (columbite structure) and $Mg_4Nb_2O_9$ (corundum structure) are stable at room temperature [54]. Therefore, even after long heat treatment at highj temperature (T > 1100 ^0C) [50], there were intermediate phases in the end product (generally a corundum phase). In such cases, the phase composition and electrophysical properties can be greatly affected even by a small deviation from stoichiometry. In view of this, we have studied the effect of small deviations from stoichiometry in $A^{2+} Nb_2O_6$ materials (A = Co, Mg, Zn) with columbite structure on phase composition, microstructure and MW properties [55, 56, 57].

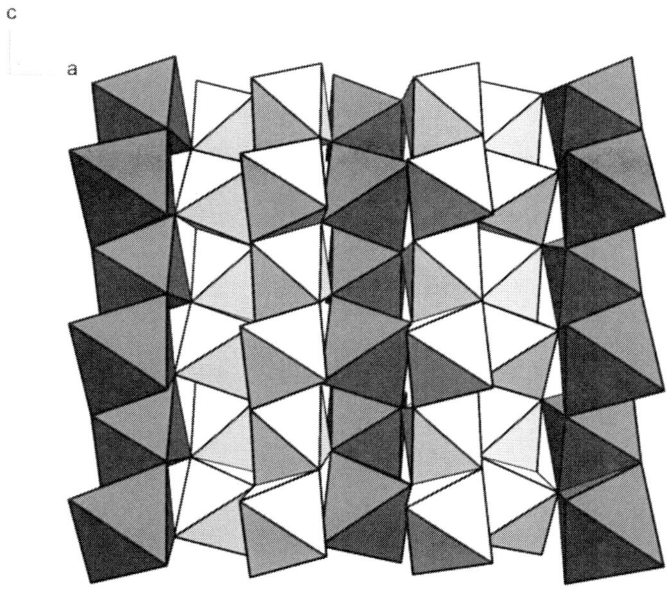

Figure 12. Columbite structure of $A^{2+} Nb_2O_6$

We have shown that when the solid-state reaction method is used, the formation of cobalt- and magnesium-containing niobates with columbite structure is a multistage process. In this case, two concurrent processes of formation of niobates with columbite structure ($A^{2+} Nb_2O_6$) and corundum structure ($A_4^{2+}Nb_2O_9$) ($A^{2+} = Co^{2+}, Mg^{2+}$) take place:

$$2A_3^{2+}O_4 + 6Nb_2O_5 \rightarrow 6A^{2+}Nb_2O_6 + O_2;$$

$$4A_3^{2+}O_4 + 3Nb_2O_5 \rightarrow 3A_4^{2+}Nb_2O_9 + 2O_2.$$

At higher temperatures (> 1000 ^{0}C), the formation of columbite structure took place by interaction between the $A_4^{2+}Nb_2O_9$ phase and unreacted Nb_2O_5:

$$A_4^{2+}Nb_2O_9 + 3Nb_2O_5 \rightarrow 4A^{2+}Nb_2O_6 \ (A^{2+} - Co^{2+}, Mg^{2+}).$$

At the same time, the synthesis of Zn Nb_2O_6 with columbite structure takes place in the temperature range 500-800 ^{0}C without formation of intermediate products.

In the case of deviation from stoichiometry in the A_{1+x}^{2+} Nb_2O_6 system (A^{2+} = Mg^{2+}, Co^{2+}, Zn^{2+}), when x < 0, samples contained two phases: the main phase A^{2+} Nb_2O_6 with columbite structure and the Nb_2O_5 phase, whose concentration increased with x (Fig 13). At x > 0, a narrow concentration range, in which samples are single-phase ones, exists in all three systems. On further deviation from stoichiometry in the direction of increasing excess of cobalt, magnesium or zinc, extra phases are formed.

The results of investigating electrophysical properties in the MW range turned out unlooked-for. At x < 0, when there were traces of the minor phase Nb_2O_5, the samples had a low Q. At the same time, extremely high Q values (Q × f) were observed at x > 0 (Fig 14). For example, in $Mg_{1+x}Nb_2O_6$, Q × f reached a value of 128000 at x ≥ 0.03 – 0.05 in multiphase samples, in which the phase $Mg_4Nb_2O_9$ (corundum structure) was present together with the main phase $MgNb_2O_6$ (columbite structure).

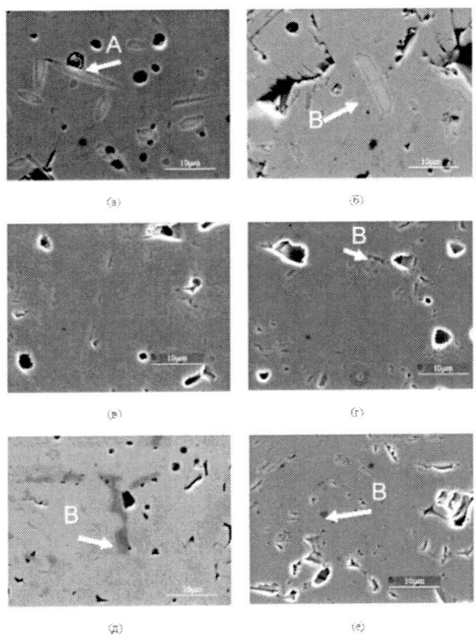

Figure 13. Micrographs of microsections of polycrystalline $Co_{1-x}Nb_2O_{6-x}$ samples with x = 0.05 (a), x = 0.03 (b), x = 0 (c), x = - 0.03 (d), x = - 0.05 (e, f), sintered at 1500 ^{0}C for 1h (a - f), 6 h (e): A = Nb_2O_5, B = $Co_4Nb_2O_9$

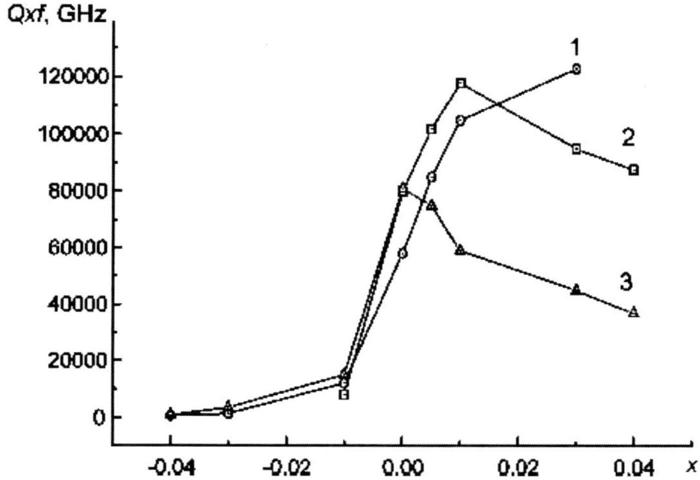

Figure 14. Plots of the product Q × ƒ of $A_{1-x}Nb_2O_6$ samples (where A = Mg (1), Zn (2), Co(3)) against concentration. The samples were sintered in air for 8 h at 1400 ^{0}C (1 and 3) and 1300 ^{0}C (2).

This is accounted for by the high Q values (Q × f ~ 230000) of the extra phase $Mg_2Nb_4O_9$. However, further deviation from stoichiometry leads to a considerable decrease in permittivity, which is due to the fact that ε of the corundum phase was 11.

This instance is interesting in that each of the phases $MgNb_2O_6$ and $Mg_2Nb_4O_9$ characterized by certain merits and demerits. Only multiphase materials based on them have both a relatively high permittivity (ε ~ 20) and a highQ(Q × f ~ 128000) in the MW range.

9. ZRO_2-TIO_2-SNO_2 – BASED MW DIELECTRICS

The first information about $ZrTiO_4$ as a promising high-Q dielectric was presented in Ref [58]. Later, in the 1950s, investigations of solid solutions in the ZrO_2-TiO_2-SnO_2 system were carried out [59]. It was shown that the composition $Zr_{0.8}Sn_{0.2}TiO_4$ has the highest Q values [60, 61].

$ZrTiO_4$ crystallizes in orthorhombic structure (space group Pbcn) [62, 63, 64] with the space lattice parameters: a = 4.806Å, b = 5.447Å, c = 5.032Å. The unit cell contains two formula units, theoretical density 5.15 g/cm^3. It should be noted that in $ZrTiO_4$ there is an order-disorder phase transition in the temperature range 1100-1200°C [65, 66, 67, 68, 69]. When the temperature is decreased, this transition is from an α-PbO_2 – type high-temperature phase of which disordered arrangement of Zr and Ti ions is typical, to a low-temperature phase with ordered arrangement of Zr and Ti ions [70, 71]. Addition of Sn to $ZrTiO_4$ results in the stabilization of the disordered distribution of cations. The variation of the lattice parameters in the $Zr_{1-x}Sn_xTiO_4$ system with increasing x is shown in Fig15. As is seen from Fig 15, there are no noticeable changes in the behavior of the parameters a and b in the phase transition region (1100-1200°C) with increasing Sn content. At the same time, as Sn ions are added, a noticeable change in the dependence of the parameter c in the phase transition region is observed [69].

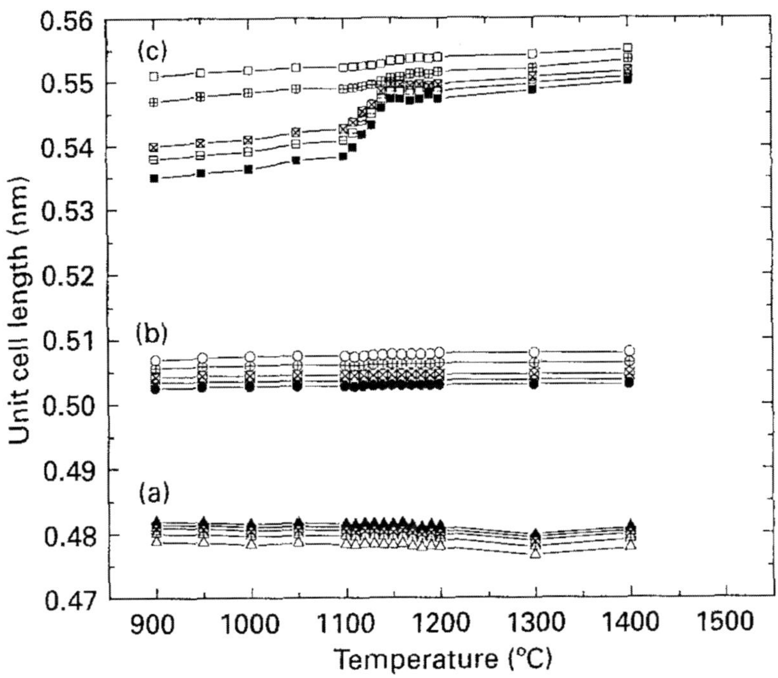

Figure 15. Variation of the lattice parameters of $Zr_{1-x}Sn_xTiO_4$ materials as a function of temperature:(■) - $ZrTiO_4$; (⊟) $Zr_{0,95}Sn_{0,05}TiO_4$; (⊠)$Zr_{0,9}Sn_{0,1}TiO_4$; (⊞) $Zr_{0,8}Sn_{0,2}TiO_4$; (□) $Zr_{0,7}Sn_{0,3}TiO_4$ [69].

The phase transition is greatly affect by the cooling rate of $Zr_{1-x}Sn_xTiO_4$ samples (Fig. 16.). At high cooling rates, the high-temperature disordered phase is frozen in the sample. As the cooling rate is decreased, a noticeable change in the behavior of the parameter *c* is observed, which is connected with increase in the degree of cation ordering [69].

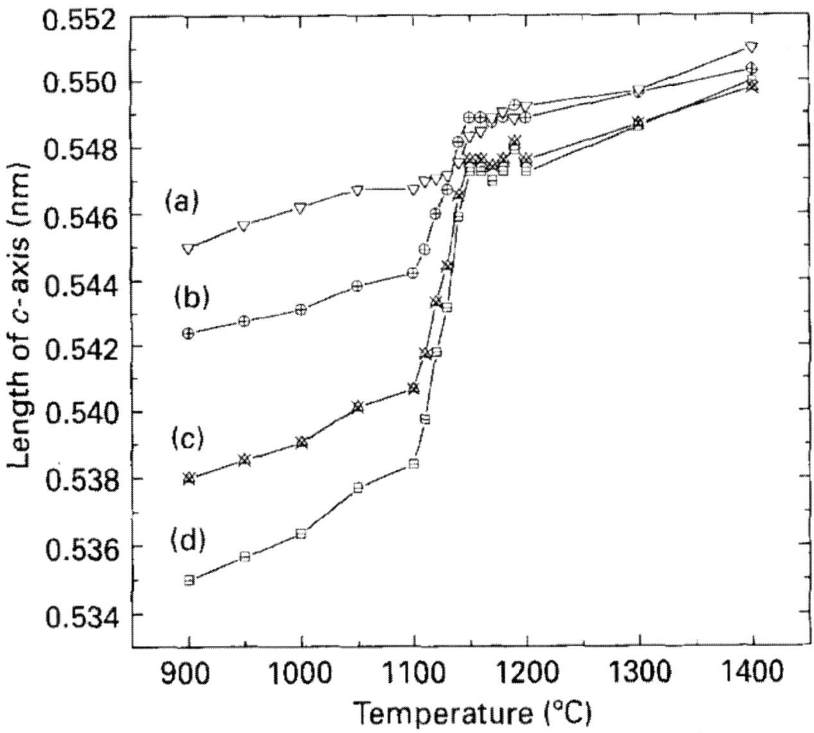

Figure 16. Variation of the parameter c of $Zr_{1-x}Sn_xTiO_4$ ceramic as a function of temperature at different cooling rates: (a) 100°C/h, (b) 15°C/h, (c) 5°C/h, (d) 1°C/h [69].

In the system $Zr_xSn_yTi_zO_4$, where x+y+z = 2, single-phase materials are formed in a limited region [62]. The partial substitution of Sn ions for Zr ions stabilized the high-temperature phase with disordered distribution of cations [72] and extends the temperature range of phase transition in $ZrTiO_4$ [73, 74].

Many authors studied the dielectric properties of $ZrTiO_4$ in the MW range and showed it to the have the following parameters: $\varepsilon = 42$, $Q \times f = 28\ 000$ GHz, $\tau_f = 58$ ppm/°C [62, 63, 75, 76]. The partial substitution of Sn ions for Zr ions affect greatly the dielectric properties of $Zr_{1-x}Sn_xTiO_4$ materials.

The substitution of Sn ions for Zr ions results in the formation of $Zr_{0.8}Sn_{0.2}TiO_4$, which has good dielectric properties in the MW region: $\varepsilon = 38$, $Q \times f = 49\ 000$ GHz, $\tau_f = 0$ ppm/°C [77]. This made is possible to use widely this composition in engineering. The use of anatase as a starting reagent instead of rutile made it possible to increase the value of $Q \times f$ [78]. It should be noted that cation ordering in the $ZrTiO_4$ structure leads to an increase $Q \times f$. It was shown that addition of tin leads to a decrease in cation ordering [62, 72, 79, 80, 81], but in spite of this, the value of $Q \times f$ increases greatly. Cation ordering is also observed in tin-containing samples, but the ordering domain size decrease with increasing tin content [82]. These data indicate that the increase in $Q \times f$ on the partial substitution of tin ions for zirconium ions cannot be attributed to cation ordering. The authors of [80] suggested that segregation of tin ions takes place at domain boundaries, which decrease their contribution to dielectric loss in ceramic.

$Zr_{0.8}Sn_{0.2}TiO_4$ – based dielectric materials are widely used in the manufacture of various MW equipment elements [63, 83, 84, 85]. How ever, the manufacture of high-density $Zr_{0.8}Sn_{0.2}TiO_4$ ceramic is a big technological problem even at high temperatures (above 1600°C). Therefore, many authors studied the effect of small additions on the sintering temperature and density of $Zr_{0.8}Sn_{0.2}TiO_4$ [85,86]. The dopants ZnO, CuO, Y_2O_3 are generally used. Addition of small amounts, e.g. of ZnO, results in the formation of a liquid phase at grain boundaries, which increases greatly the ceramic density thanks to fast mass transport through the liquid phase and considerable decrease in sintering temperature.

10. BA($M^{2+}_{1/3}$ $M^{5+}_{2/3}$)O_3 – BASED MW DIELECTRICS (M^{2+} = MG, ZN, CO, NI; M^{5+} = TA, NB) WITH EXTREMELY HIGH Q

Ba($B^{2+}_{1/3}$ $B^{5+}_{2/3}$) O_3 compounds where B^{2+} = Mg, Zn, Co, Ni; B^{5+} = Ta, Nb (perovskite crystal structure) had been synthesized for the first time by the authors of [89,90, 91, 92]. In these compounds, a 2:1- type ion ordering in the B sublattice is observed, in which two layers filled with B^{5+} ions alternate with a layer filled with B^{2+} ions. The authors [93, 94] showed that tantalum-containing materials possess a high Q value in the MW range. It should be noted that the synthesis of these materials involves many problems. Ceramics sinter at a high temperature, which may result in considerable evaporation of constituents (cobalt, zinc) and hence in the impairment of electrical properties. In the case of synthesis by the solid-state reaction method, extra phases $Ba_5Ta_4O_{15}$, $Ba_4Ta_2O_9$ are often present in ceramics [95], which affect adversely the Q value. To prevent this, the authors of [96] carried out synthesis from solutions, where solutions containing Mg^{2+}, Ta^{5+} were used as starting substances, to which a solution of ammonia with oxyquinoline was added.

It had been found that in this case, the single-phase product $Ba(Mg_{1/3}Ta_{2/3})O_3$ is formed above 1300 ^0C without intermediate phases. Single-phase $Ba(Mg_{1/3}Ta_{2/3})O_3$ had been obtained by the solid-state reaction method too, using highly active reagents as starting substances [95-97].

When synthesizing tantalum-containing materials, the preparation of high-density ceramics was a difficult problem. Therefore, Nomura with coauthors [98] proposed to prepare dense ceramics ($\rho \approx 7520$ kg/m^3) by using additionally manganese impurities. Matsumoto and Hinga [99] used fast heating (330 ^0C/min) for the same purposes, which made it possible to achieve 96% of the theoretical density. To increase the rate of sintering and to order ions in the B sublattice, the authors of [100, 101] proposed preliminarily synthesized MTa_2O_6(M = Mg, Zn) as starting reagents. Renoult with coworkers [102] had synthesized fine-grained $Ba(MgTa)O_3$ by the sol-gel method. In that case, dense ceramics could be obtained without additives.

The Q value in $Ba(B^{2+}{}_{1/3}\ B^{5+}{}_{2/3})O_3$ perovskites is greatly affected by the type and degree of ion ordering in the B sublattice [103]. It had been found that by the partial substitution of Zr^{4+}, Ti^{4+}, W^{6+} ions for ions in the B sublattice, one can increase the degree of 1:2 - type ion ordering and hence increase the Q value [104, 105].

Ion ordering in the B sublattice is also affected by slight substitutions of ions in the A sublattice. For instance, Ref [104] showed that partial substitution of La^{2+} ions for Ba^{2+} ions in magnesium-barium tantalate results in the change of ion ordering in the B sublattice from the 2:1 type (space group Pm3I) to the 1:1 type (space group Fm3m). Similar changes of the type of ion ordering in the B sublattice were found in the case of partial substitution of lanthanum ions for barium ions in zinc-barium niobate with perovskite structure [105]. This result shows that the type of ion ordering in the B sublattice is very sensitive to chemical composition and preparation technique. The 1:1-type ion ordering in the B sublattice in $Ba(B^{2+}{}_{1/3}\ B^{5+}{}_{2/3})O_3$ compounds was explained in terms of a "space-charge" model [106, 107], according to which only B^{2+} and B^{5+} ions can occupy the sites in the ordered $Ba(\beta'_{1/2}\ \beta''_{1/2})O_3$ structure (1:1 type). Since in this case the electroneutrality condition is not satisfied in the crystalline phase, it may be assumed that the domains of the ordered crystalline phase (ordering type 1:1), which has an uncompensated charge, are in a disordered matrix rich in B^{5+} ions, as a result of which the electroneutrality condition is satisfied throughout the sample volume.

However, when investigating the system of solid solutions $(1\text{-}x)Ba(Zn_{1/3}Nb_{2/3})O_3\text{-}xLa((Zn_{1/3}Nb_{2/3})O_3$ $(0 \le x \le 0.6)$, it was shown that 1:2 –type ordering persists in the interval $0 \le x \le 0.5$ [105], whereas in the interval $0.05 \le x \le 0.6$, 1:1 ion ordering in the B sublattice is observed. In this case, there is no aggregation. The investigation of the microstructure did not reveal the existence of disordered perovskite phase region, which was assumed in the "space-charge" model. Therefore, to describe the 1:1 ion ordering in the B sublattice, a "random-site" model was proposed [105, 108, 109, 110]. According to this model, 1:1 ordering in the above systems is described as follows. There are two alternating crystal planes, in which the B sublattice ions reside. One of them is occupied with B^{5+} cations and the other with B^{2+} cations and the remaining B^{5+} cations, which are disordered in this crystal plane.

Thus, in $Ba(M^{2+}{}_{1/3}\ M^{5+}{}_{2/3})O_3$ compounds with perovskite structure, the B sublattice ions may be fully disordered as well as have 1:2 or 1:1 ordering. A calculation of the lattice energy of ordered and disordered structures showed [111] that ordered structure is characterized by lower Madelung energy, indicating this structure to be stable. Investigations of the electrophysical properties of $Ba(B^{2+}{}_{1/3}\ B^{5+}{}_{2/3})O_3$ compounds showed that ion ordering in the B sublattice affects greatly the Q value [112].

It should be noted that $Ba(B^{2+}{}_{1/3}\ B^{5+}{}_{2/3})O_3$ compounds with perovskite structure and 1:1 cation ordering in the B sublattice have, as a rule, a relatively low Q [113]. The largest Q value is observed in the $Ba(Mg_{1/3}Ta_{2/3})O_3$ and $Ba(Zn_{1/3}Ta_{2/3})O_3$ compounds. In these compounds, the B cations are stoichiometrically ordered in the hexagonal unit cell Pm3I (2:1ordering), in which the layers of Ta^{5+} and $Zn(Mg)^{2+}$ cations are sequentially arranged along the (111) crystal plane (Fig 17). The layers of cations are separated by oxygen layers, which are displaced in the direction of small pentavalent tantalum cations [112]. The Q value is very sensitive to ordering in the B sublattice [93, 114]. It can be greatly increased by using additional annealing. For instance, the $Q \times f$ value increased from 60000 to 168000 after additional annealing at 1350 ^{0}C for 120 h [93].

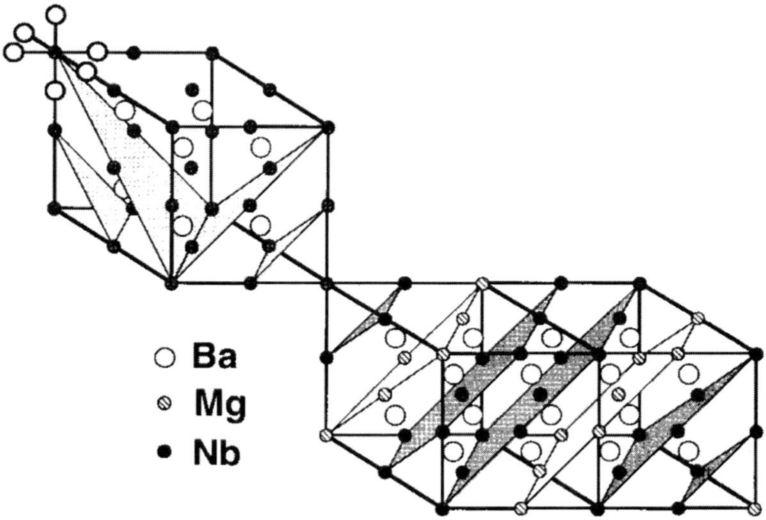

Figure 17. Schematic representation of 1:2 cation ordering in $Ba(Zn_{1/3}Ta_{1/3})O_3$. On the top left are shown two possible (111) directions for Zn and Ta orientation in the perovskite structure; on the right below is shown one of the possible variants of 1: 2 ordering. The oxygen ions were omitted for clarity [108].

Tamura et al [115] showed that Q in $Ba(Zn_{1/3}Ta_{2/3})O_3$ can be improved by adding $BaZrO_3$(BZ) of low concentration (< 4 mol %). In this case, the ceramic sintering time is greatly reduced, which is required for the attainment of high Q values ($Q \times f = 105000$ had been attained by adding 4 mol % BZ).

This was accounted for by the formation of defects in the B sublattice, the presence of which increased the rate of cation ordering in this sublattice. As the BZ concentration was increased, the type of ordering changed (when substitution reached 4 mol %), the system came to have 1:1 ordering and a double cell. Similar regularities were observed when small amounts of $BaWO_4$ [116] and $BaSnO_3$ [117] were added to $Ba(Mg_{1/3}Ta_{2/3})O_3$. The individual compounds $Ba(Mg_{1/3}Ta_{2/3})O_3$ and $Ba(Zn_{1/3}Ta_{2/3})O_3$ allow one to achieve high Q values, but their electrophysical properties have a low thermostability. Therefore, to increase the thermostability of electrical properties, materials are synthesized on the basis of solid solutions, where the end members have temperature dependences of permittivity of different sign. The materials based on solid solutions possess a higher thermostability, but the Q value is lower as compared with individual compounds, which may be attributed, in particular, to decrease in cation ordering in the B sublattice. On the basis of a solid solution, e.g. $Ba(Zn_{1/3}Ta_{2/3})O_3$, materials have been obtained which have a high level of electrophysical properties: $\varepsilon = 30\text{-}40$, $TC\varepsilon = (0\text{-}28) \times 10^{-6}\,{}^0C^{-1}$, $Q_{8GHz} = 15000$.

The materials based on tantalum-containing perovskites possess today the highest Q values in the MW range among dielectrics with increased ε value. However, the difficulty of their synthesis: high sintering temperatures ($T_{sint} > 1600\ {}^0C$), the necessity of long additional annealing ($T_{ann} \approx 1500\ {}^0C$, 20-120 h), low reproducibility of properties, as well as the high price of tantalum (in 2000, the price of reagents containing tantalum increased by 500%) calls for search for new promising systems. Therefore, in recent years, attention has been given just to niobium-containing $Ba(Nb_{2/3}B^{2+}{}_{1/3})O_3$ compounds (where B^{2+} = Mg, Zn, Co), which crystallize in perovskite structure [118, 119, 120].

These materials sinter at lower temperature, and cheaper reagents are needed for them. However, their main disadvantage is a lower *Q* value as compared with tantalum-containing perovskites [121, 122]. We have shown for the first time that $Ba(Mg_{1/3}Nb_{2/3})O_3$ – based materials can have in some cases an extremely high Q ($Q \times f \approx 150000$) [123]. New papers appeared [5, 124, 125], which show the creation of high-Q MW dielectrics based on niobium-containing perovskites to be worth-while.

$Ba(Co_{1/3}Nb_{2/3})O_3$ (BCN) offers a particularly attractive combination of properties. Polycrystalline $Ba(Co_{1/3}Nb_2/_3)O_3$ has high dielectric permittivity (ε =32) and Qf = 40000–60000 GHz [126, 127]. At the same time, centimeter and millimeter wave applications require higher *Q* values. The properties of BCN are very weak functions of temperature. In particular, the temperature coefficient of its resonant frequency (τf) lies in the range −10 to −7 ppm/K, suggesting that it can be used in the production of high-Q/low-τf microwave materials [128, 129, 130]. The electrical properties of BCN, especially its *Q* (or its *Qf* product, where *f* is frequency), strongly depend on preparation conditions, in particular on the sintering temperature, heat-treatment time, and heating/cooling rate.

It is reasonable to assume that the observed variations in the properties of BCN are related to the ceramic microstructure evolution during the fabrication process, cation ordering, and lattice distortions. Such distortions can be produced in BCN via slight changes in its cation composition.

Therefore, we have studied the effect of partial nonstoichiometry in cation sublattices on the phase composition, microstructure and electrophysical properties of BCN.

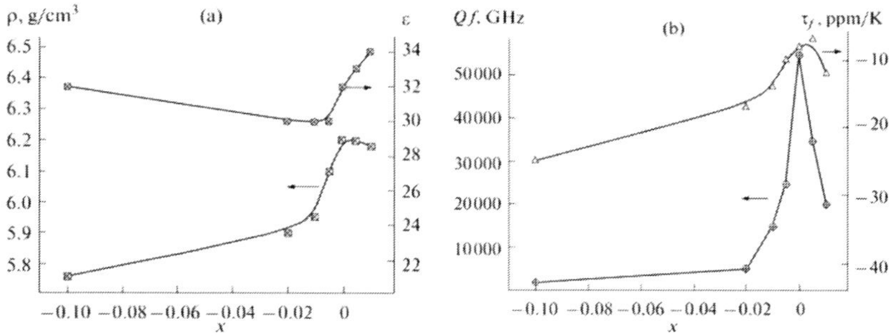

Figure 18. Composition dependences of the (a) apparent density (ρ), permittivity (ε), (b) Qf product, and temperature coefficient of resonant frequency (τ_f) for $Ba_{3+3x}CoNb_2O_{9+3x}$ materials; measurements at10 GHz.

When studying nonstoichiometry in barium sublattice ($Ba_{1+x}Co_{1/3}Nb_{2/3}O_{3+x}$ at -0.03 <x <0.03), an extra phase $Ba_9CoNb_{14}O_{45}$ appears, which makes a noticeable decreases greatly (Fig. 18(a)). Therefore, the highest Q values are observed for stoichiometric composition ($BaCo_{1/3}Nb_{2/3}O_3$) (Fig. 18(b)).

Interesting properties are observed in the case of nonstoichiometry in the cobalt sublattice ($BaCo_{1/3+y}Nb_{2/3}O_{3+y}$, where -0.05 <y <0.01).

Increasing the cobalt content of this system with reference to stoichiometry leads to a monotonic increase in relative density and corresponding slight increase in permittivity ε in the range 32 to 34 (Fig. 19a). At higher cobalt deficiencies (y < −0.02), the samples contained an additional phase $Ba_8CoNb_6O_{24}$. In the range −0.03 ≤ y ≤ 0.01, the quality factor varies nonlinearly, with a maximum at −0.03 ≤ y ≤ −0.02. The Qf product exceeds that of stoichiometric BCN by 30–50%, reaching 80000–85000 GHz (Fig. 19b).

Electron diffraction data for the $BaCo_{1/3+y}Nb_{2/3}O_{3+y}$ samples indicates that cobalt deficiencies in the range −0.03 ≤ y ≤ −0.02 are favourable for 1 : 2 B-site cation ordering in the cobalt-containing perovskite (Fig. 20). As mentioned above, cation ordering is accompanied by an increase in quality factor, as observed in this system (Fig. 18b). The reduction in quality factor at large deviations from stoichiometry (y < −0.1) is due to the presence of a significant amount of $Ba_8CoNb_6O_{24}$.

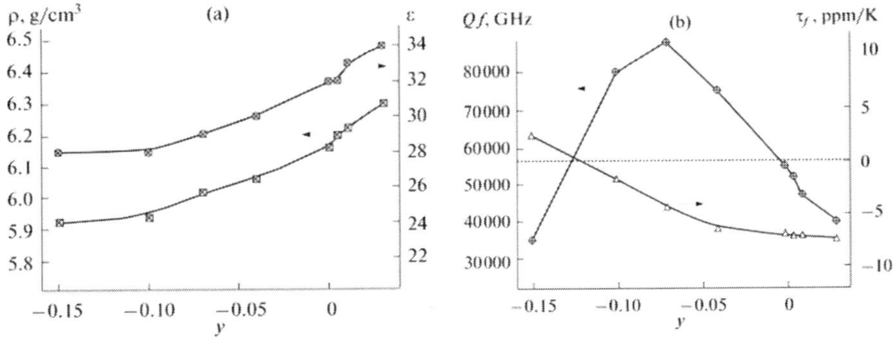

Figure 19. Composition dependences of the (a) apparent density (ρ), permittivity (ε), (b) Qf product, and temperature coefficient of resonant frequency (τ_f) for $Ba_3Co_{1+y}Nb_2O_{9+y}$ materials; measurements at 10 GHz.

The increase in $Ba_8CoNb_6O_{24}$ content with increasing cobalt deficiency ($y < 0$) leads to a result of practical interest: the temperature coefficient of resonant frequency switches sign (Fig. 18 b). The reason for this is that $BaCo_{1/3}Nb_{2/3}O_3$ and $Ba_8CoNb_6O_{24}$ differ in the sign of τ_f: −7 and +16 ppm/K, respectively [129, 131]. Because of this, the $BaCo_{1/3+y}Nb_{2/3}O_{3+y}$ materials exhibit a temperature compensation effect, whose magnitude can be tuned by varying the cobalt content. The present results, therefore, suggest that cobalt-deficient $Ba(Co_{1/3}Nb_{2/3})O_3$ is an attractive host for engineering advanced temperature-stable microwave dielectric materials with Qf on the order of 80000–90000 GHz and τ_f = −2 to +3 ppm/K.

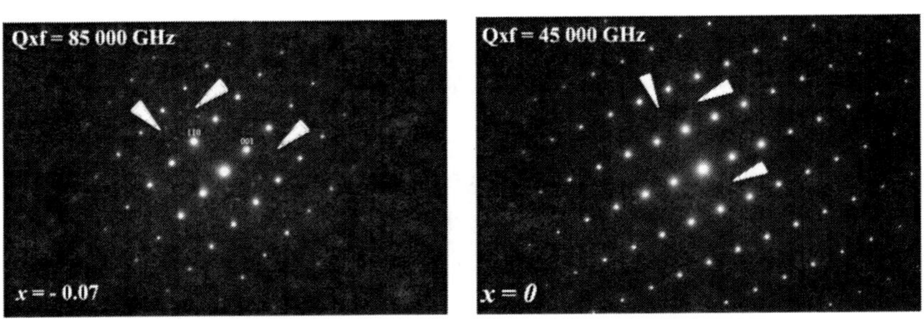

Figure 20. [110] electron diffraction patterns of the $Ba_3Co_{1+y}Nb_2O_{9+y}$ samples with y = (a) −0.07 and (b) 0. The arrows mark superlattice reflections.

11. MULTIPHASE MW DIELECTRICS

High Q value is usually observed in single-phase systems. In the case of complex cation sublattices, it is necessary that the ions should be ordered by a definite type [105, 106]. In multiphase systems, which are chemically inhomogeneous, considerable dielectric loss (relatively low Q) is generally observed. When investigating barium polytitanates, however, we showed that multiphase systems having a high Q and thermostability of electrophysical properties can be formed. When zinc oxide is added to barium polytitanates, an extra $BaZn_2Ti_4O_{11}$ phase is formed [132] which does not interact chemically with the main phase. A multiphase system is formed, in which the main and extra phases have the dependence ε(T) of different sign, which ensures the realization of the volume temperature compensation effect and hence a high thermostability of electrophysical properties (TCε = $\pm 2 \times 10^{-6} K^{-1}$) in the MW range. The multiphase dielectrics obtained have a high Q ($Q_{10GHz} \sim$ 6500-7000).

One more example of multiphase MW dielectrics is TiO_2 materials, viz the compounds $MgTiO_3$ and Mg_2TiO_4, which have a high Q ($Q_{10GHz} \sim$ 5000-10000) and permittivity (14 and 16 respectively) [133]. A demerit of these materials is the temperature instability of electrophysical parameters (TCε = $(40\text{-}50) \times 10^{-6} K^{-1}$). To increase the Q value, cobalt ions were partially substituted for magnesium ions, and to increase the temperature stability of electrophysical properties, small amounts of the paraelectric phase $CaTiO_3$, which has a high negative value of TCε, were added. Investigations showed that in this case multiphase systems of chemically noninteracting phases are formed (Fig 21) [134, 135, 136].

This made it possible to obtain MW dielectrics with a permittivity of 18-20, high Q values ($Q \times f \geq$ 5000-10000) and thermostable electrical properties. It may be supposed that high Q values are due to the fact that the size of chemical inhomogeneity is much smaller than the electromagnetic wavelength in dielectric and does not cause, therefore, noticeable electromagnetic scattering.

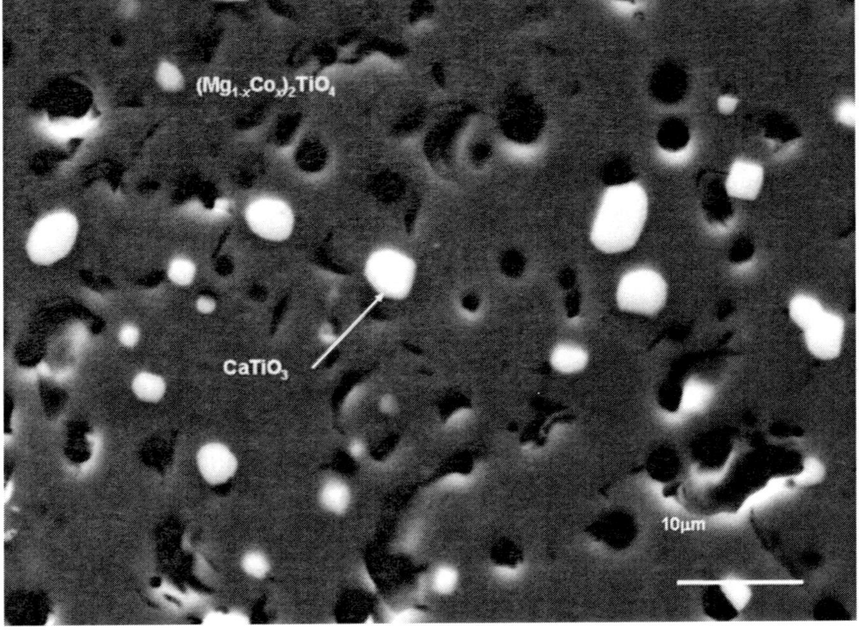

Figure 21. Micrograph of the microsection of 0.93 [$0.98Mg_2TiO_4$ – 0.02 Co_2 TiO_4] -0.07$CaTiO_3$ ceramic

12. ANALYSIS OF THE PHYSICAL PROPERTIES OF MW DIELECTRICS

The main attention in the analysis of the physical properties of MW dielectrics is given to the temperature coefficient of permittivity (TCε) and to dielectric loss (tg δ). The expression for TCε can be derived directly from the Clausius-Mossotti equation [137]:

$$TC\varepsilon = \frac{(\varepsilon-1)(\varepsilon+2)}{\varepsilon}\left[\frac{\nu}{\alpha}\left(\frac{\partial\alpha}{\partial\nu}\right)_T a_l + \frac{1}{3\alpha}\left(\frac{\partial\alpha}{\partial T}\right)_\nu - a_l\right] \tag{4}$$

where a_l is linear thermal expansion coefficient.

It should be noted that polarization a in Eq (4) is equal to the sum of polarizations of all atoms of cell, whose volume is ν, only if all atoms of the structure have a cubic environment. This is the case, e.g. for alkali halide crystals. In more complex structures, effective polarization a_{eff} is used [138]. For example, for perovskite structure, effective polarization a_{eff} is obtained by introducing an ionic component of polarization, Δa^i, in addition to the electronic and ionic polarization of all atoms in the unit cell. In this case, the last term a_l in expression (4) must be written as $a_l \Delta a^i$ [137]. An analysis of the TCε value of different materials as a function of chemical composition showed [139] that the large positive value of TCε in alkali halide crystals probably arises from high ionic polarizability a^i and great thermal expansion a_l. There are compounds, e.g. $LaAlO_3$, $SrZrO_3$, the behavior of whose ε(T) differs from that of paraelectrics. This is attributed to the presence of nonferroelectric phase transitions [137], which are coupled with the rotation of oxygen octahedra. Similar nonferroelectric phase transitions were found, e.g. in the system $(Ba_xSr_{1-x})(Zn_{1/3} - Nb_{2/3})O_3$ [140]. It was shown that structural transitions are accompanied by the reversal of the TCε sign, though the ε quantity does not undergo noticeable changes, which are typical of spontaneously polarized state. Similar dependences were also found in the system $(Ba_xSr_{1-x})(Mg_{1/3} - Nb_{2/3})O_3$ [141]. When x is changed, a correlation between the value of tolerance factor (t) and the inclination of oxygen octahedra is observed, which results in the reversal of the TCε sign.

When MW dielectric loss is studied, three sources of loss are considered: 1) loss in perfect crystal, which is coupled with anharmonicity, which is due to interaction between crystal phonons, resulting in optical-phonon attenuation; such loss is usually called intrinsic loss; 2) loss in real homogeneous material, which is caused by deviation from lattice or defect periodicity (point defects, dopant atoms, vacancies or twin defects, which give rise to quasi-bound states); such defects give rise to phonon scattering; 3) loss in real inhomogeneous crystalline materials, which is caused by the existence of dislocations, grain boundaries, including minor phases; this loss is usually called extrinsic loss.

ε Dispersion for infrared polarization is usually described by the Drude-Lorentz equation:

$$\varepsilon(\omega) = \varepsilon_\infty + \frac{\varepsilon_0 - \varepsilon_\infty}{1 - \left(\frac{\omega}{\omega_T}\right)^2 + i\Gamma\frac{\omega}{\omega_T}} \tag{5}$$

where ε_∞ is permittivity at optical frequencies, ω_T is transverse optical mode frequency, $(\varepsilon_0 - \varepsilon_\infty)$ is dielectric oscillator strength, Γ is relative attenuation.

Using Eq (5), we can estimate dielectric loss (tg δ) pertaining to phonon attenuation in the MW range (intrinsic loss) ($\omega \gg \omega_T$):

$$tg\delta \approx \Gamma \frac{\omega}{\omega_T^4} \frac{\varepsilon_0 - \varepsilon_\infty}{\varepsilon_0} \tag{6}$$

Even in perfect crystal, loss may arise from the anharmonicity of vibration. In this case, three-phonon and four-phonon interactions may predominate, in which $\Gamma \sim T$ and $\Gamma \sim T^2$ respectively [142]. Equation (6) shows that if intrinsic loss predominates, the product of Q (Q = 1/tg δ) and frequency f($\omega = 2\pi f$) is a constant, which is employed for the analysis of MW dielectrics.

To determine intrinsic loss, IR spectroscopy is usually used [143, 144] since the value of intrinsic loss at IR frequencies is much larger than that of extrinsic loss. Having determined intrinsic loss at IR frequencies, one can approximate the quantity $Q \times f$ = const to MW frequencies and calculate thereby extrinsic loss, which is due to ceramic imperfection; this loss can be reduced by improving the technology.

One of the important questions concerning the determination of dielectric loss in MW ceramics is the possibility of determining intrinsic loss solely from IR spectroscopic data. To this end, the authors of [145, 146] investigated several $Ba(B'_{1/2}B''_{1/2})O_3$ compounds in a wide frequency ($10^2 - 10^{14}$ Hz) and temperature (20-600 K) range. On the basis of the data obtained, they came to the following conclusions:

- MW loss calculated from IR spectra and Eq (6) can be interpreted as a lower limit of intrinsic loss.
- MW loss extrapolated from IR data is systematically lower than that determined in accordance with the microscopic theory [147].
- The value of intrinsic loss correlates with tolerance factor (t); increase in intrinsic loss with decreasing t indicates that the main contribution to it is made by low-frequency vibration mode.
- The value of intrinsic loss varies with the ε value as tg $\delta \sim \varepsilon^a$, where a = 4.

It is not less important to elucidate the effect of cation ordering on MW loss. It was shown earlier [58] that cation ordering in $Ba(Zn_{1/3}Ta_{2/3})O_3$ and $Ba(Mg_{1/3}Ta_{2/3})O_3$ allows MW loss to be reduced. When investigating $(Zr_{1-x}Sn_x)$-TiO_4 materials, however, it was found that the substitution of Sn^{4+} ions for Zr^{4+} ions results in the suppression of cation ordering [148], and that the Q value increases in this case [140]. It is likely that in the system $(Zr_{1-x}Sn_x)TiO_4$, the increase in Q on the substitution of Sn^{4+} ions with smaller radius for Zr^{4+} ions with larger radius is due to a decrease in intrinsic loss.

13. APPLICATIONS OF MW DIELECTRICS

MW dielectrics are used in modern communication systems, for the manufacture of dielectric resonators (DR) of various types, substrates for MW hybrid integrated circuits. On the basis of DRs, radio-frequency filters are developed; they are also used in the manufacture of solid state oscillators. In the frequency range 150 MHz – 3 GHz, coaxial resonators are often used, whose surface is metallized (Fig 22). The height of quarter-wave coaxial resonator is determined from the formula:

$$l = \frac{\lambda_0}{4}\frac{1}{\sqrt{\varepsilon}} \tag{7}$$

where λ_0 is free-space electromagnetic wavelength, ε is permittivity in the operating frequency range.

At higher frequencies, open resonators are generally used (Fig 23), whose diameter is determined from the formula:

$$D = \frac{\lambda_0}{\sqrt{\varepsilon}} \tag{8}$$

The characteristic modes for coaxial and open dielectric resonators are usually the modes TE_{01s} and H_{01s}.

The Q value of coaxial resonator is determined both by dielectric loss and by loss in the resonator metal coating, which may be high; therefore, the Q value of coaxial resonators is, as a rule, under 1000, which is their demerit. At the same time, Q of open resonators is determined only by dielectric loss (intrinsic and extrinsic). Therefore, Q of open resonators is over 1000. Since open resonators have a high Q value, they can be used in the decimeter wave band (about 1 GHz), though in this case their size becomes large.

Figure 22. Coaxial dielectric resonators

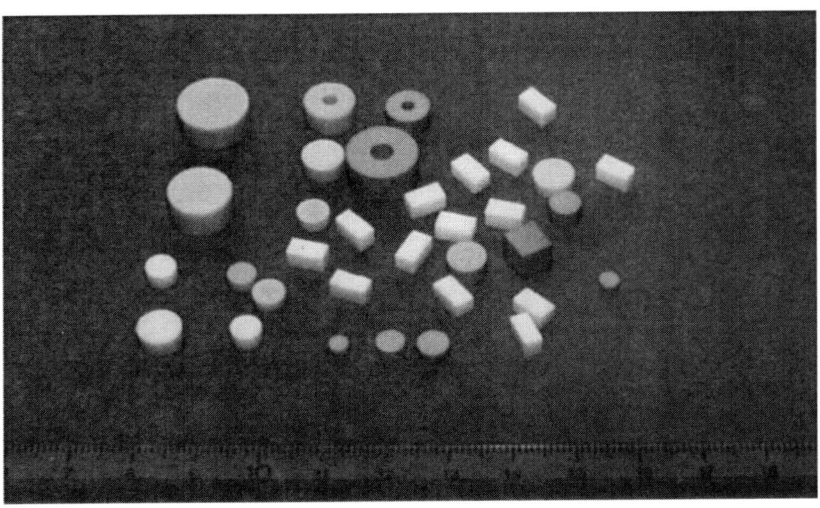

Figure 23. Open dielectric resonators

The size of open dielectric resonators operating on the characteristic modes TE_{01s} and H_{01s} becomes very small at frequencies above 30 GHz, which makes their use in this band impossible. Therefore, at frequencies above 30 GHz, it is expedient to use extraordinary vibration modes such as whispering gallery modes.

The modern communication systems in which dielectric resonators are used operate in a temperature range of – 40-80 ^{0}C. Therefore, high thermostability of dielectric resonator resonance frequency is required. It is necessary that the temperature coefficient of frequency should tend to zero; it is defined as:

$$TK\varepsilon = \frac{1}{f_p}\frac{\Delta f_p}{\Delta T} \tag{9}$$

where f_r is the resonance frequency of dielectric resonator, Δf_r is change in the resonance frequency of dielectric resonator in the temperature range ΔT.

Dielectric resonators are used for the frequency stabilization of oscillators, which are used, in turn, in radars, various communication systems. At frequencies below 200 MHz, quartz resonators are often used; in the frequency range 1000-3000 MHz, coaxial resonators and above 3000 MHz open resonators are used.

On the basis of dielectric resonators, miniature bandpass filters, frequency separators are developed (Fig 24) [149].

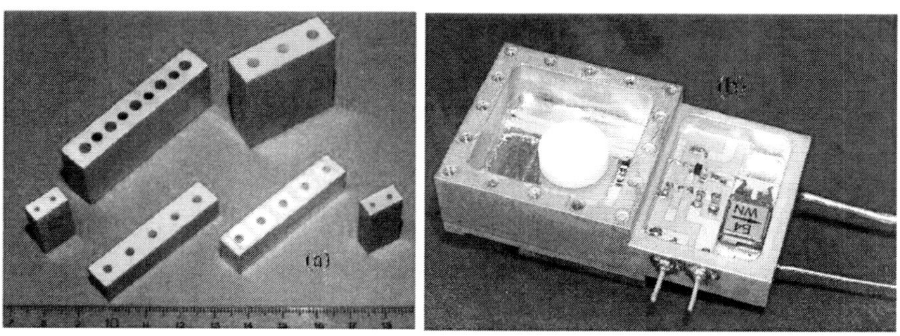

Figure 24. Monolithic ceramic blocks for the utilization in radiofilters operating in the decimetre wavelength band (a), and low-noise microwave oscillator for the frequency of around 9 GHz (b)..

Dielectric resonators are also used in the creation of antennas of the new generation. The advantages of such antennas are: small size, simplicity, relatively broad emission band, simple scheme of coupling with all commonly used transmission lines; possibility to obtain different radiative characteristics using different resonator modes.

As was mentioned above, the use of dielectric resonators operating on the TE_{01s} and H_{01s} modes is limited in the millimeter wave band since the size of resonators becomes too small. Therefore, it is expedient to use in the millimeter wave band dielectric resonators operating on whispering gallery modes [150]. Besides, it is relatively easy to suppress spurious modes in such resonators. It should be noted that the Q value in the resonators using whispering gallery modes is limited only by intrinsic loss in the material in contrast to coaxial and open resonators.

14. CONCLUSION

High-Q MW dielectrics with high thermostability of electrophysical properties can be developed on the basis of single-phase and multiphase systems. Single-phase MW dielectrics are produced on the basis of solid solutions [6, 15, 98] using heterovalent substitutions in one of the crystal sublattices and influencing thereby the phonon spectrum [10] and by making one of the sublattices "mobile" [151]. At the same time, high-Q thermostable MW dielectrics based on multiphase systems are developed using the volume temperature compensation effect [55-57, 124, 132, 135].

During the last decade, MW dielectrics with increased permittivity ($\varepsilon \geq 10$) contribute, to a larger measure than other factors, to considerable miniaturization and reduction in the price of modern communication systems. It should be noted that there is still a great potential for further microminiaturization and reduction of prices of modern communication systems thanks to the use of components made on the basis of MW dielectrics.

Depending on the frequency range of modern communication systems, MW dielectrics with different properties are needed. In the decimeter wave band, high permittivity values ($\varepsilon \geq 100$) are required along with the high thermostability of electrophysical properties and high Q, which enables effective solution of microminiaturization problems. At the present time, solid solutions based on barium-lanthanide titanates ($Ba_{6-x}Ln_{8+2x/3}Ti_{18}O_{54}$ (Ln = La-Gd)), which have a potassium-tungsten bronze structure and $\varepsilon \approx$ 80-100, meet best these requirements. However, the nature of the thermostability of the electrophysical properties of these solid solutions has not been elucidated definitively; there are only qualitative explanations, which greatly restrains the search for new promising MW dielectrics with high permittivity ($\varepsilon \geq 100$). The presence of spontaneous polarization in dielectrics causes, along with increase in ε, a considerable increase in dielectric loss (the Q value decreases), which impairs greatly the technical characteristics of communication system elements based on them. Therefore, the acquirement of fundamental knowledge, which is required for obtaining thermostable high-Q materials with $\varepsilon \geq 150$-200, is the most important problem in developing modern decimeter wave band communication systems.

In the centimeter and millimeter wave bands, where the electromagnetic wavelength is much smaller as compared with the decimeter wave band, thermostable MW dielectrics with extremely high Q values are required. In this case, permittivity values may be relatively low (ε = 15-30). To date, tantalum-containing perovskites possess the highest Q values. However, the difficulty of their preparation and the high price call for search for new promising compounds, and it is going on in several directions. In particular, research is now under way to develop niobium-containing perovskites and to create multiphase systems, in which volume temperature compensation effect is realized. It is these directions that will probably be major directions in the next few years in developing high-Q centimeter and millimeter wave band MW dielectrics, though the search for new promising compounds will always be vital. Quite a number of problems pertaining to solid-state physics and chemistry will have to be solved. For instance, it is necessary to investigate the nature of extrinsic loss, which is coupled with various structural defects, as well as with the presence of grain boundaries, ordering of crystal sublattices and domain nanostructure. Research aimed at developing thermostable dielectrics, which will be used as millimeter wave band dielectric resonators using whispering gallery modes, will be of special scientific and practical interest. This requires considerable increase of the chemical and structural homogeneity of dielectrics.

An important problem is the creation of retunable resonant elements. To this end, multilayer bulk and film materials, which will contain a thermostable dielectric phase and a nonlinear magnetic or electrical phase at the same time, will probably have to be developed.

Thus, the synthesis of novel high-Q MW dielectrics and the investigation of their structure and properties are an important scientific-technical trend in solid-state chemistry.

REFERENCE

1. Poplavko Yu.M., (1980) Physics of Dielectrics (in Russian), Vyshch. Shk., Kyiv 325.
2. Poplavko Yu.M., Belous A.G., (1984) Physical background of the temperature stability of microwave dielectrics. Dielektriki I Poluprovodniki. 25: 3-15.
3. Chool-Woo Ahu, Hyun-Jung Jang, Sahn Nahm et al. (2003) Effects of microstructure on the microwave dielectric properties of Ba(Co1/3Nb2/3)O3 and (1−x)Ba(Co1/3Nb2/3)O3–xBa(Zn1/3Nb2/3)O3 ceramics J. Eur. Ceram. Soc. 23: 2473-2474.
4. J-Nan Lin, Chih-Ta Chia, Hsiang-Lin Liu et al. (2002) Dielectric Properties of xBa(Mg1/3Ta2/3)O3-(1-x)Ba(Mg1/3Nb2/3)O3 Complex Perovskite Ceramics. Jpn. J. Appl. Phys. 41: 6952-6956.
5. Cheol-Woo Ahn, Sahn Hahm, Seok-Jin Yoou et al. (2003) Microstructure and Microwave Dielectric Properties of (1−x)Ba(Co1/3Nb2/3)O3–xBa(Zn1/3Nb2/3)O3 Ceramics Jpn. J. Appl. Phys. 42: 6964-6968.
6. Nenesheva E. A., Mudroliuba L. P., and Kartenko N. F. (2003) Microwave dielectric properties of ceramics based on CaTiO3–LnMO3 system (Ln=La, Nd; M=Al, Ga). J. Eur. Ceram. Soc. 23: 2443–2448.
7. Moon J. H., Jung H. M., Park H. S., Shin J. Y., and Kim H. S. (1999) Sintering behaviour and microwave dielectric properties of (Ca,La)(Ti,Al)O3 ceramics. Jpn. J. Appl. Phys. 38: 6821–6827.
8. Belous A.G., Butko V.I., Novitskaya G.N. et al. (1985) Dielectric spectra of the perovskites La2/3-XM3XTiO3 Fizika Tv. Tela. 27: 2013-2016.
9. Belous A.G., Butko V.I., Polyanetskaya S.V. (1984) Electrical parameters of the solid solutions of rare-earth titanates Ukr. Khim. Zhurn. 50: 1139-1142.
10. Belous A.G (1998) Physicochemical aspect of the development of new functional materials based on heterosubstituted titanates of rare-earth elements with the perovskite structure. Teoret. I Eksperim. Khimiya. 34: 301-318.
11. Butko V.I., Belous A.G., Yevtushenko N.P. (1986) Vibration spectra of the perovskites La2/3-xM3xTiO3 Fizika Tv. Tela. 28: 1181-1183.

12. Knyazev A.S., Poplasvko Ye.M., Zakharov V.P., Alekseev V.V. (1973) Soft mode in the vibration spectrum of CaTiO3 Fiz. Tverd.Tela. 15: 3006-3010.

13. Ohsato H., Nishigaki S., Okuda T. Superlattice and Dielectric Properties of BaO-R2O3-TiO2 (R=La, Nd and Sm) Microwave Dielectric Compounds //Jpn. J. Appl. Phys. -1992. -31. –P. 3136-3140.

14. Negas N., Davies P.K. (1995) Influence of chemistry and processing on the electrical properties of Ba6-3xLn8+2xTi18O54 solid solutions Materials and processing for wireless communications. Ceram. Trans. 53: 170-196.

15. Matveeva R.G., Varfolomeev M.B., Ilyushenko L.S. (1984) Refinement of the composition, and the crystal structure of Ba3.75Pr9.5Ti18O54. Zhurn. Neorgan. Khimii. 29: 31-34.

16. Rawn C.J., Birnie D.P., Bruck M.A. et al. (1998) Structural investigation of Ba6–3xLn8+2xTi18O54 (x = 0.27, Ln = Sm) by single crystal x-ray diffraction in space group Pnma. J. Mater. Res. 13: 187-196.

17. Ubic R., Reaney I.M., Lee William E. (1999) Space Group Determination of Ba6-3xNd8+2xTi18O54 J. Amer. Ceram. Soc. 82: 1336-1338.

18. Belous A., Ovchar O., Valant M., Suvorov D. (2001) Solid-state reaction mechanism for the formation of Ba6–xLn8+2x/3Ti18O54 (Ln = Nd, Sm) solid solutions J. Mater. Res. 16: 2350-2356.

19. Ohsato H., T. Ohhashi, S. Nishigaki, T. Okuda, K. Sumiya, and S. Suzuki. (1993) Formation of solid solutions of new tungsten bronze type microwave dielectric compounds Ba6–3xR8.2xTi18O54 (R=Nd,Sm 0_x_1). Jpn. J. Appl. Phys. 32: 4323–4326.

20. Valant M., Suvorov D., Rawn C.J. (1999) Intrinsic Reasons for Variations in Dielectric Properties of Ba6-3xR8+2xTi18O54 (R= La–Gd) solid solutions. Jpn. J. Appl. Phys. 38: 2820-2826.

21. Butko V.I., Belous A.G., Nenasheva Ye. A. et al. (1984) Microwave dielectric properties of barium lanthanide tetratitanates Fizika Tv. Tela. 26: 2951-2955.

22. Belous A., Ovchar O., Valant M., Suvorov D. (2000) Anomalies in the temperature dependence of the microwave dielectric properties of Ba6–xSm8+2x/3Ti18O54 Appl. Phys. Lett. 77: 1707-1709.

23. Belous A., Ovchar O., Valant M., Suvorov D. (2002) Abnormal behavior of the dielectric parameters of Ba6–xLn8+2x/3Ti18O54 (Ln=La–Gd) solid solutions J. Appl. Phys. 92: 3917-3922.

24. Tang C.C., Roberts M.A., Azough F. et al. (2002) Synchrotron X-ray Diffraction Study of Ba4.5Nd9Ti18O54 Microwave Dielectric Ceramics at 10–295 K. J. Mater. Res. 17: 675-682.

25. Declaration patent no 58005 A, H 01B 3/12. Published 15.07.2003.

26. Declaration patent no 58007 A, H 01B 3/12. Published 15.07.2003.

27. Declaration patent no 58009 A, H 01B 3/12. Published 15.07.2003.

28. Declaration patent no 58008 A, H 01B 3/12. Published 15.07.2003.

29. Takahashi H., Baba Y., Ezaki K. et al. (1992) Dielectric Characteristics of (A1/21+A1/23+)TiO3 ceramics at microwave frequencies. Jpn. J. Appl. Phys. 30: 2339-2342.

30. Belous A.G., Novitskaya G.N., Polyanetskaya S.V. (1987) Study of the oxides Ln2/3-xM3xTiO3 (Ln - Gd-Lu, M-Li,Na,K). Izv. AN SSSR. Ser. Neorgan. Materialy. 23: 1330-1332.

31. Belous A.G., Novitskaya G.N., Polyanetskaya S.V., Gornikov Yu.I. (1987) Investigation ox complex oxide of the composition Ln2/3-xLi3xTiO3. Izv. AN SSSR. Ser. Neorgan. Materialy. 23: 470-472.

32. Belous A.G., Novitskaya G.N., Polyanetskaya S.V., Gornikov Yu.I. (1987) Crystal-chemical and electro-physical properties of complex oxides Ln2/3-xM3xTiO3. Zurn. Neorgan. Khimii. 32: 283-286.

33. Belous A.G., Butko V.I., Novitskaya G.N. et al. (1986) Electrical conductivity of the perovskites La2/3-xM3xTiO3. Ukr. Fiz. Zurn. 31: 576-581.

34. Belous A.G., Gavrilova L.G., Polyanetskaya S.V. et al. (1984)Stabilization of the perovskite structure of lanthanum titanate Ukr. Khim. Zurn. 50: 460-461.

35. Belous A., Ovchar O. (2003) Temperature compensated microwave dielectrics based on lithium containing titanates. J. Eur. Ceram. Soc. 23: 2525-2528.

36. Belous A.G., Poplavko Yu.M. (1976) Dielectric properties of the tellurium containing perovskites in the microwave region Fizika Tv. Tela. 18: 2248-2451.

37. Belous A.G., Politova Ye.D., Venevtsev Yu.N. et al. (1981) Lead-cobalt telluride – a material for microwave dielectric resonators Elektronnaya Tekhnika. Ser. Elektronika, SVCh. 331: 45-46.

38. Valant M., Suvorov D., Hoffman C., Sommariva H. (2001) Ag(Nb,Ta)O3-based ceramics with suppressed temperature dependence of permittivity. J. Eur. Ceram. Soc. 21: 2647-2654.

39. Trunov V.K., Frolov A.M., Averina I.M. (1981) Refinement of the structure La0.33NbO3 Kristallografiya. 26: 189-191.

40. Sakowski-Cowley A.C.,Lurasszewich K., Megaw H.D. (1969) The structure of sodium niobate at room temperature, and the problem of reliability in pseudosymmetric structures Acta Crystallogr. Sect. B. 25: 851-856.

41. Mishchuk D.O., Vyunov O.I., Ovchar O.V., Belous A.G. (2004) Structural and dielectric properties of solid solutions of sodium niobate in lanthanum and neodymium niobates Inorganic Materials. 40: 1324-1330.

42. Belous A.G. (2006) Microwave dielectrics with enhanced permittivity J. Eur. Ceram. Soc. 26: 1821-1826.

43. Declaration patent no 54166 A, H 01P 7/10. Published 17.02.2003.

44. Declaration patent no 54167 A, H 01P 7/10. Published 17.02.2003.

45. dos Santos C.A., Zawislak L.I., Antonietti V. et al. (1999) Iron oxidation and order-disorder in the (Fe2+,Mn)(Ta,Nb)2O6 to (Fe2+,Mn)Fe3+(Ta,Nb)2O8 transition. J. Phys. Condens. Matter. 11: 7021-7033.

46. Lee H.J., Hong K.S., Kim I.T. (1997) Dielectric properties of MNb2O6 compounds (where M = Ca, Mn, Co, Ni, or Zn) Mater. Res. Bull. 32: 847-855.

47. Zhang Y.C., Yue Z.X., GuiZ., Li L.T. (2003) Microwave dielectric properties of (Zn1−xMgx)Nb2O6 ceramics. Mater. Lett. 57: 4531-4534.

48. Pullar R.C., Breeze J.D., Alford N. (2005) Characterization and microwave dielectric properties of M2+Nb2O6 ceramics // Journal of the American Ceramic Society. 88: 2466-2471.

49. Joy P.A., and Sreedhar K. (1997) Formation of Lead Magnesium Niobate Perovskite from Niobate Precursors Having Varying Magnesium Content J. Am. Ceram. Soc. 80: 770–772.

50. Ananta S. (2003) Effect of calcination condition on phase formation characteristic of magnesium niobate powders synthesized by the solid-state reaction. CMU. Journal. 2: 79–88.

51. Ananta A., Brydson R. N., Thomas W. (1999) Synthesis, formation and Characterisation of MgNb2O6 Powder in a Columbite-like Phase J. Eur. Ceram. Soc. 19: 355–362.

52. Norin P., Arbin C.G., Nalander B. (1972) Note on the Phase Composition of the MgO-Nb2O5 System. Acta Chim. Scand. 26: 3389–3390.

53. Paqola S., Carbonio R.E., Alonso J.A. Femandez-Diaz M.T. (1997) Crystal structure refinement of MgNb2O6 columbite from neutron powder diffraction data and study of the ternary system MgO–Nb2O5–NbO, with evidence of formation of new reduced pseudobrookite Mg5−xNb4+xO15−δ (1.14≤x≤1.60) phases. Journal of Solid State Chemistry. 134: 76–84.

54. You Y.C., Park H.L., Song Y.G., Moon H.S., Kim G.C. (1994) Stable phases in the MgO-Nb2O5 system at 1250°C. J. Mater. Sci. Lett. 13: 1487–1489.

55. A.G. Belous, O.V, Ovchar, A.V, Kramarenko et al, (2007) Synthesis and microwave dielectric properties of Zn1+xNb2O6+x Inorganic Materials. 43: 326-330.

56. Belous A.G., Ovchar O.V., Mishchuk D.O. et al. (2007) Synthesis and properties of columbite-structure Mg1−xNb2O6−x Inorganic Materials. 43: 477-483.

57. Belous A.G., Ovchar O.V., Kramarenko A.V. et al. (2006) Effect of nonstoichiometry on the structure and microwave dielectric properties of cobalt metaniobate. Inorganic Materials. 42: 1369-1373.

58. Rath W. (1941) Keramische sindermassen fur die elektronik fortschritter auf dem greblet der keramischen isollerstoff fur die electroteknik. Keram. Radsch. 49: 137–139.

59. Brit Patent No. 692468 (1952).

60. Wakino K., Nishikawa T., Tamura S., and Ishikawa Y. (1975) Microwave band pass filters containing dielectric resonators with improved temperature stability and spurious response. Proc.IEEE MTT Symposium (New York) 63–65.

61. Wakino K., Nishikawa T., Tamura S., and Ishikawa Y. (1978) Miniaturised band pass filters using half wave dielectric resonators with improved spurious response. Proc IEEE MTT Symposium 230–232.

62. Wolfram W. and Gobel H.E. (1981) Existence range, structural and dielectric properties of ZrxTiySnzO4 ceramics (x.y.z=4). Mater. Res. Bull. 16: 1455–1463.

63. Tamura H. (1994) Microwave loss quality of (Zr0.8Sn0.2)TiO4. Am. Ceram. Soc. Bull. 73: 92–95.

64. Blasse G. (1966) Compounds with α-PbO2 structure. J. Anorg. Allg. Chem. 345: 222–224.

65. Newnham R.E. (1967) Crystal structure of ZrTiO4. J. Am. Ceram. Soc. 50: 216.

66. Mc Hale A.E. and Roth R.S. (1983) Investigation of the phase transition in ZrTiO4 and ZrTiO4–SnO2 solid solutions. J. Am. Ceram. Soc. 66: 18–20.

67. Mc Hale A.E. and Roth R.S. (1986) Low-temperature phase relationships in the system ZrO2–TiO2. J. Am. Ceram. Soc. 69: 827-832.

68. Ikawa H., Iwai A., Hiruta K., Shimojima H., Urabe K., and Udagawa S. (1988) Phase transformation and thermal expansion of zirconium and hafnium titanate and their solid solutions. J. Am.Ceram. Soc. 71: 120–127.

69. Park Y. (1995) Influence of order disorder transition on microwave characteristics of tin-modified zirconium titanate. J. Mater. Sci. Lett. 14: 873–875.

70. Ikawa H., Shimojima H., Ukrabe K., Yamada T., and Udagawa S. (1988) Polymorphism in ZrTiO4. Science of Ceramics. D. Taylor (Ed.). Institute of Ceramics, Shelton, Uk. 509–514.

71. Christerfferson R. and Davies P. K. (1992) Structure of commensurate and incommensurate ordered phase in the system ZrTiO4-Zr5Ti7O24. J. Am. Ceram. Soc. 75: 563–569.

72. Han K. R., Jang J.-W., Cho S.-Y., Jeong D.-Y., and Hong K.-S. (1998) Preparation and dielectric properties of low temperature-sinterable (Zr0.8Sn0.2)TiO4 powder. J. Am. Ceram. Soc. 81: 1209–1214.

73. Park Y. and Kim Y. (1996) Order-disorder transition of tin-modified zirconium titanate. Mater.Res. Bull. 31: 7–15.

74. Park Y., Kim Y., and Kim H. G. (1996) Structural-phase transition and electrical conductivity in tin-modified zirconium titanate. Solid State Ionics. 90: 245–249.

75. Wakino K., Nishikawa T., Tamura S., and Ishikawa Y. (1978) Miniaturised band pass filters using half wave dielectric resonators with improved spurious response. Proc IEEE MTT Symposium. 230–132.

76. Azough F., Freer R., Wang C.L., and Lorimer G.W. (1996) The relationship between the microstructure and microwave dielectric properties of zirconium titanate ceramics. J. Mater. Sci. 31: 2539–2549.

77. Wakino K., Minai K., and Tamura H. (1984) Microwave characteristics of (Zr,Sn)TiO4 and BaO–PbO–Nd2O3–TiO2 dielectric resonator. J. Am. Ceram. Soc. 67: 278–281.

78. Heiao Y.C., Wu L., and Wei C.C. (1988) Microwave dielectric properties of (ZrSn)TiO4 ceramic. Mater. Res. Bull. 23: 1687–1692.

79. Khairulla F. and Phule P. (1992) Chemical synthesis and structural evolution of zirconium titanate. Mater. Sci. Eng. B. 12: 327-336.

80. Christofferson R., Davies P.K., Wei X., and Negas T. (1994) Effect of Sn substitution on cation ordering in (Zr1–xSnx)TiO4 microwave dielectric ceramics. J. Am. Ceram. Soc. 77: 1441–1450.

81. Iddles D. M., Bell A. J., and Moulson A. J. (1992) Relationship between dopants, microstructure and the microwave dielectric properties of ZrO2–TiO2–SnO2 ceramics. J. Mater. Sci. 27: 6303–6310.

82. Davies P.K. (1994) Influence of internal interfaces on the dielectric properties of ceramic dielectric resonators. Res. Soc. Symp. Proc. 357: 351–361.

83. Wersing W. (1991) High frequency ceramic dielectrics and their applications for microwave components. In: Electroceramics, B. C. H. Steele (Ed.), Elsevier Applied Sciences, London and New York. 67–119.

84. Wakino K. (1989) Recent developments of dielectric resonator materials and filters in Japan. Ferroelectrics. 91: 69–86.

85. Azough F. and Freer R. (1989) The microstructure and low frequency dielectric properties of some zirconium titanate stannate (ZTS) ceramics. Proc. Br. Ceram. Soc. 42: 225–233.

86. Huang C.-L., Weng M.-H., and Chen H.-L. (2001) Effects of additives on microstructures and microwave dielectric properties of (Zr,Sn)TiO4 ceramics. Mater. Chem. Phys. 71: 17–22.

87. Takada T., Wang S. F., Yoshikawa S., Jang S.-J., and Newnham R. E. (1994) Effects of glass on (Zr,Sn)TiO4 for microwave applications. J. Am. Ceram. Soc. 77: 2485–2488.

88. Ioachin A., Banau M. G., Toacsan M. I., Nedelcu L., Ghetu D., Alexander H. V., Toica G., Annino G., Cassettari M., and Martnelli M. (2005) Nickel doped (Zr0.8Sn0.2)TiO4 for microwave and millimeter wave applications. Mater. Sci. Eng. B. 118: 205–209.

89. Galasso F., Pule J. (1963) Ordering in compounds of the A(B'0.33Ta0.67)O3 type. Inorg Chem. 2: P. 482–484.

90. Galasso F., Katz L. (1959) Substitution in the octahedrally coordinated cation positions in compounds of the perovskite type J. Amer. Ceram. Soc. 81: 820-823

91. Galasso F., Pule J. (1962) Preparation and study of ordering in A(B'0.33Nb0.67)O3 perovskite-type compounds J. Phys. Chem. 67: 1561–1562.

92. Roy R. (1954) Multiple Ion Substitution in the Perovskite Lattice J. Amer. Ceram. Soc. 37: 581-588.

93. Kawashima S., Nichida M., Ueda I., Oici H. (1983) Ba(Zn1/3Ta2/3)O3 Ceramics with Low Dielectric Loss at Microwave Frequencies J. Amer. Ceram. Soc. 66: 421–423.

94. Nomura S., Kaneta K. (1984) Ba(Mn1/3Ta2/3)O3 Ceramic with Ultra-Low Loss at Microwave Frequency Jpn. J. Appl. Phys. 23: 507–508.

95. Chen X.M., Suzuki Y., Sato N. (1994) Sinterability improvement of Ba(Mg1/3Ta2/3)O3 dielectric ceramics. Journal of Materials Science. Materials in Electronics. 5: 244–247.

96. Kakegawa K., Wakabayashi T., Sasaki Y. (1986) Preparation of Ba(Mg1/3Ta2/3)O3 Using Oxine J. Amer. Ceram. Soc. 69: 82–89.

97. Chen X.M., Wu Y. J. (1995) A low-temperature approach to synthesize pure complex perovskite Ba(Mg13Ta23)O3 powders Materials Letters. 26: 237–239.

98. Nomura S., Toyoma K. and Kaneta K. (1982) Ba(Mg1/3Ta2/3)O3 ceramics with temperature-stable high dielectric constant and low microwave loss. Jpn. J. Appl. Phys. 21: 624–626.

99. Matsumoto K., Hiuga T., Takada K. et al. (1986) BA(M1/3TA2/3)O3 Ceramics with ultra-low loss at microwave-frequencies. IEEE Transactions on ultrasonics ferroelectrics and frequency control. 33: 802-802.

100. Novitskaya G.N., Yanchevskii O.Z., Polyanetskaya S.V., Belous A.G. (1991)Formation of the phases (phase formation) in the systems BaCO3-(Nb,Ta)2O5-ZnO. Ukr. Khim. Zhurn. 57: 801-802.

101. USSR inventor's certificate 1837599, IPC C 04 B 35, H 01 B 3/12. Published 13.10.1992.

102. Renoult O., Bollot J.-P., Chaput F., papierik R., Hubert-Pfalzgraf L.G., Leycune M. (1992) Sol–Gel Processing and Microwave Characteristics of Ba(Mg⅓Ta⅔)O3 Dielectrics J. Amer. Ceram. Soc. 75: 3337-3340.

103. Kolodiazhnyi T., Petric A., Johari G. Belous A. (2002) Effect of preparation conditions on cation ordering and dielectric properties of Ba(Mg1/3Ta2/3)O3 ceramics J. Eur. Ceram. Soc. 22: 2013–2021.

104. Youn H.J., Hong K.S., Kim H. (1997) Coexistence of 1 2 and 1:1 long-range ordering types in La-modified Ba(Mg0.33Ta0.67)O3 ceramics. J. Mater. Res. 12: 589–592.

105. Akbas M. A., Davies P. K. (1998) Cation Ordering Transformations in the Ba(Zn1/3Nb2/3)O3-La(Zn2/3Nb1/3)O3 System J. Amer. Ceram. Soc. 81: 1061–1064.

106. Chen J., Chan H.M., Harmer M.P. (1989) Ordering structure and dielectric properties of undoped and La/Na-Doped Pb(Mg1/3Nb2/3)O3 J. Amer. Ceram. Soc. 72: 593–598.

107. Hilton A.D., Barber D.J., Randall A., Shrout T.R. (1990) On short range ordering in the perovskite lead magnesium niobate. J. Mater. Sci. 25: 3461–3466.

108. Akbas M. A., Davies P. K. (1998) Ordering-induced microstructures and microwave dielectric properties of the Ba(Mg1/3Nb2/3)O3–BaZrO3 system. J. Amer. Ceram. Soc. 81: 670–676.

109. Chai L., Akbas M. A., Davies P. K., Parise J.B. (1997) Cation ordering transformations in Ba(Mg13Ta23)O3-BaZrO3 perovskite solid solutions Mater. Res. Bull. 32: 1261–1269.

110. Chai L., Davies P. K. (1997) Formation and Structural Characterization of 1:1 Ordered Perovskites in the Ba(Zn1/3Ta2/3)O3–BaZrO3 System J. Amer. Ceram. Soc. 80: 3193–3198.

111. Djuniadi A., Sagala N. (1992) Lattice Energy Calculations for Ordered and Disordered Ba(Zn1/3Ta2/3)O3. Journal of the Physical Society of Japan. 61: 1791–1797.

112. Mehmet A., Akbas M.A., Davies P.K. (1998) Ordering-Induced Microstructures and Microwave Dielectric Properties of the Ba(Mg1/3Nb2/3)O3–BaZrO3 System J. Amer. Ceram. Soc. 81: 670–676.

113. Mitsuhiro Takata, Keisuke Kageyama (1989) Microwave Characteristics of A(B3+1/2B5+1/2)O3 Ceramics (A = Ba, Ca, Sr; B3+= La, Nd, Sm, Yb; B5+= Nb, Ta) J. Amer. Ceram. Soc. 72: 1955–1959.

114. Matsumoto K., Hiuga T., Takada K., and Ichimura H.. (1986) Ba(Mg1/3Ta2/3)O3 ceramics with ultralow loss at microwave frequencies. Proc. 6th IEEE Intl. Symp. On Applications of Ferroelectrics IEEE (NY). 118.

115. Tamura H., Konoike T., Sakabe Y., and. Wakino K. (1984) Improved High-Q Dielectric Resonator with Complex Perovskite Structure J. Am. Ceram. Soc. 67: C.59–61.

116. Yon Ki H., Dong P.K., Kim E.S. (1994) Annealing effect on microwave dielectric properties of Ba(Mg1/3Ta2/3)O3 with BaWO4 Ferroelectric. 154: 337–342.

117. Matsumoto H., Tamura H., Wakino K. (1991) Ba(Mg,Ta)O3–BaSnO3 High-Q dielectric resonator. Jpn. J. Appl. Phys. 30: 2347-2349.

118. Hughes H., Iddles D., Reaney I.M. (2001) Niobate-based microwave dielectrics suitable for third generation mobile phone base stations Appl. Phys. Letters. 79: 2952–2954.

119. Cheng-Ling Huang, Ruei-jsung Lin (2002) Liquid Phase Sintering and Microwave Dielectric Properties of Ba(Mg1/3Ta2/3)O3 Ceramics. Jpn. J. Appl. Phys. 41: 712–716.

120. Azough F., Leach C., Freeer R. (2006) Effect of nonstoichiometry on the structure and microwave dielectric properties of Ba(Co1/3Nb2/3)O3 ceramics. J. Eur. Ceram. Soc. 26: 2877–2884.

121. Liu H.X., Tian Z.Q., Wang H., Yu H.T., Ouyang S.X. (2004) New microwave dielectric ceramics with near-zero τf in the Ba(Mg1/3Nb2/3)O3-Ba(Ni1/3Nb2/3)O3 system. J. Mater. Sci. 39: 4319–4320.

122. Seo-Yong Cho, Hyuk-Joon Youn, Kug-Sun Hong (1997) A new microwave dielectric ceramics based on the solid solution system between Ba(Ni1/3Nb2/3)O3 and Ba(Zn1/3Nb2/3)O3 J. Mater. Res. 12: 1558–1562.

123. Kolodiazhnyi T., Petric A., Belous A., V'yunov O., Yanchevskij O. (2002) Synthesis and dielectric properties of barium tantalates and niobates with complex perovskite structure. J. Mater. Res. 17: 3182–3189.

124. Scott R.I., Thomas M., Hampson C. (2003) Development of low cost, high performance Ba(Zn1/3Nb2/3O3) based materials for microwave resonator applications. J. Eur. Ceram. Soc. 23: 2467–2471.

125. Davies P.K., Borisevich A., Thirunal M. (2003) Communicating with wireless perovskites: cation order and zinc volatilization. J. Eur.Ceram.Soc. 23: 2461-2466.

126. Davis P.K., Tong, J., and Negas, T., (1997) Effect of Ordering-Induced Domain Boundaries on Low-Loss Ba(Zn1/3Ta2/3)O3-BaZrO3 Perovskite Microwave Dielectrics. J. Am. Ceram. Soc. 80: 1724–1740.

127. Molodetsky, I. and Davies, P.K., (2001) Effect of Ba(Y1/2Nb1/2)O3 and BaZrO3 on the Cation Order and Properties of Ba(Co1/3Nb2/3)O3 Microwave Ceramics. J. Eur. Ceram. Soc. 21: 2587–2591.

128. Endo, K., Fujimoto, K., and Murakawa, K. (1987) Dielectric Properties of Ceramics in Ba(Co1/3Nb2/3)O3–Ba(Zn1/3Nb2/3)O3 Solid Solutions. J. Am. Ceram. Soc. 70: 215–218.

129. Ahn C.W., Jang H.J., Nahm S., et al. (2003) Effect of Microstructure on the Microwave Dielectric Properties of Ba(Co1/3Nb2/3)O3 and (1–x)Ba(Co1/3Nb2/3)O3–xBa(Zn1/3Nb2/3)O3 Solid Solutions. J. Eur. Ceram. Soc. 23: 2473–2478.

130. Davies P.K., Borisevich A., and Thirumal M. (2003) Communicating with Wireless Perovskites: Cation Order and Zinc Volatilization. J. Eur. Ceram. Soc. 23: 2461–2466.

131. Mallinson, P.M., Allix, M.M., Claridge, J.B., et al. (2005) Ba8CoNb6O24: A d0 Dielectric oxide host containing ordered d7 cation layers 1.88 nm apart. Angew. Chem. Int. Ed. 44: 7733–7736.

132. Belous A., Ovchar O., Macek-Krzmanc M., Valant M. (2006) The homogeneity range and the microwave dielectric properties of the BaZn2Ti4O11 ceramics. J. Eur. Ceram. Soc. 26: 3733–3739.

133. C.-L. Huang, Chung-Long Pan 2002 Low-temperature sintering and microwave dielectric properties of (1 - x)MgTiO3-xCaTiO3 ceramics using bismuth addition. Jpn. J. Appl. Phys. 41: 707–711.

134. Belous A., Ovchar O., Durilin D., Macek-Krzmanc M., Valant M., Suvorov D. (2006) High-Q Microwave Dielectric Materials Based on the Spinel Mg2TiO4 J. Am. Ceram. Soc. 89: 3441–3445.

135. O.V. Ovchar, O.I. Vyunov, D.A. Durilin et al. (2004) Synthesis and microwave dielectric properties of MgO–TiO2–SiO2 ceramics. Inorganic Materials. 40: 1116-1121.

136. Belous A.G., O.V. Ovchar, D.A. Durilin et al. (2006) Microwave composite dielectrics based on the system MgO–CaO–SiO2–TiO2. Abstr. Book "Microwave Materials and Their Applications" (12-15 June). –Oulu (Finland) 97.

137. Wersing W. (1991) Electronic Ceramics. Ed. By Steele BCH. – London and New York: Elsevier Appl. Science. 67-119.

138. Heywang W. (1951) Zur Wirksamen Feldstarke im kubischen Gitter. Zeitschrift fur Naturforschung section a-a journal of physical sciences. 6: 219, 220.

139. Wersing W. (1996) Microwave ceramics for resonators and filters. Current Ohinion in Solid State and Materials Science. 1: 715-731.

140. Colla I.L., Reaney I.M., Setter N. (1993) Effect of structural changes in complex perovskites on the temperature coefficient of the relative permittivity. J. Appl. Phys. 74: 3414-3425

141. Sugiyama M., Nagai T. (1993)Anomaly of dielectric-constant of (Ba1-xSrx)(Mg1/3Ta2/3)O3 solid-solution and its relation to structural-change. Japanese Journal Of Applied Physics Part 1-Regular Papers Short Notes & Review Papers. 32: 4360-4363.

142. Gurevich V.L., Tagantsev A.K. (1986) Intrinsic dielectric losses in crystals - low-temperatures. Sov. Phys. JETP. 64: 142-151.

143. Kudesiak K., Mc Itale A.E., Condrate R.A., Sr. Snyder R.L. (1993) Microwave characteristics and far-infrared reflection spectra of zirconium tin titanate dielectrics. J. Mater. Sci. 28: 5569-5575.

144. Fukuda K., Kitoh R. (1994) Far-Infrared Reflection Spectra of Dielectric Ceramics for Microwave Applications J. Amer. Ceram. Soc. 77: 149-154.

145. Zurmuhlen R., Colla E., Dube D.C. et al. (1994) Structure of Ba(Y+31/2Ta+51/2)O3 and its dielectric properties in the range 102–1014 Hz, 20–600 K J. Appl. Phys. 76: 5864-5863.

146. Zurmuhlen R., Petzelt J., Komba S. et al. (1995) Dielectric-spectroscopy of Ba(B1/2'B1/2'')O3 complex perovskite ceramics - Correlations between ionic parameters and microwave dielectric-properties .I. Infrared reflectivity study (1012-1014 HZ). J. Appl. Phys. 77: 5341-5350.

147. Gurevich V.L., Tagantsev A.K. (1991) Intrinsic dielectric loss in crystals. Adv. Phys. 40: 719-767.

148. Christoffersen R., Davies P.K., Wie X. (1994) Effect of Sn Substitution on Cation Ordering in (Zr1–xSnx)TiO4 Microwave Dielectric Ceramics J. Amer. Ceram. Soc. 77: 1441-1450.

149. Tsykalov V.G., Belous A.G., Ovchar O.V., Stupin Y.D. (1997) Monolithic filters and frequency-separation devices based on the ceramic resonators. 27th Europ. Microwave Conf. Proceed. 544-600.

150. Han Q., Kogami Y., Tomabechi Y. (1994) Resonance characteristics of circularly propagating mode in a coaxial dielectric resonator. IEICE Trans Electron. 77: 1747-1751.

151. Belous A.G., Ovchar O.V. (1995) The origin of the temperature stabilization of dielectric permittivity in the system

5

CHAPTER

EFFECT OF HOT CARRIER ON AMPLITUDE MODULATION AND DEMODULATION OF GAUSSIAN HIGH POWER HELICON WAVE IN HOMOGENEOUS LONGITUDINALLY MAGNETIZED STRAIN DEPENDENT DIELECTRIC MATERIAL

Shivani Saxena[1], Sanjay Dixit[1], Sanjay Srivastava[2]

[1]*Department of Physics, Govt. M.V. M College, Barkatullah University, Bhopal, India*

[2]*Department of Material Science & Metallurgical Engineering, Maulana Azad National Institute of Technology, Bhopal, India*

ABSTRACT

In the present communication, the hydrodynamic model is used to investigate the amplitude modulation as well as demodulation of an electromagnetic wave of high power helicon pump wave into another helicon wave in strain dependent dielectric material incorporating carrier heating (CH) effects. The consideration of CH in modulation and demodulation is prime importance for the adding of new dimension in analysis of amplification of acoustic helicon wave. By using the dispersion relation, threshold pump electric filed and growth rate of unstable mode from the modulation and demodulation of the high power helicon wave well above from the threshold value will be discussed in the present analysis. The numerical analysis is applied to a strain dependent dielectric material, BaTiO3 at room temperature and irradiated with high power helicon wave of frequency 1.78 × 1014 Hz. This material is very sensitive to the pump intensities, therefore during studies, Gaussian shape of the helicon pump wave is considered during the propagation in stain dependent dielectric material and opto-acoustic wave in the form of Gaussian profile (ω_0, k_0) is induced longitudinally along the crystallographic plane of BaTiO3. Its variation is caused by the available magnetic field (ω_c), interaction length (z) and pulsed duration of interaction (τ). From the analysis of numerical results, the incorporation of CH effect can effectively modify the magnitude of modulation or demodulation of the amplitude of high power helicon laser wave through diffusion process. Not only the amplitude modulation and demodulation of the wave, the diffusion of the CH effectively modifies the growth rate of unstable mode of frequency in BaTiO3. The propagation of the threshold electric field shows the sinusoidal or complete Gaussian profile, whereas this profile is found to be completely lost in growth of unstable mode. It has also been seen that the growth rate is observed to be of the order of 108 - 1010 s−1 but from diffusion of carrier heating, and that its order is enhanced from 1010 - 1012 s−1 with the variation of the magnetized frequency from 1 to 2.5 × 1014 Hz.

Keywords: Amplitude/Frequency Modulation, High Power Laser Wave, Hot Carrier Effect, Plasma Effect in Strain Dependent Dielectric Material

1. INTRODUCTION

A number of extensive researches have been carried out from the different researchers in worldwide to investigate the effect of modulation and demodulation of high power pump wave from nano-pulsed laser action. From this interaction of electromagnetic wave in solid-state plasmas, the numbers of application have been raised to diagnose the metals, semimetals and semiconductor [1] . Out of these materials, the number of collective modes arises in magnetized semiconductor plasma from nonlinear interaction at relatively large amplitude of the propagating pump wave. When an un-modulated electromagnetic wave starts to propagate through a semiconductor plasma or ionized gaseous plasma with periodically varying parameters, it gets modulated or demodulated amplitude or frequency. The modulation and demodulation of the pump wave can be considered with respect to either amplitude or in terms of frequency. The periodic variation in "z" and "t" dependent exponential parameter $\exp\left[i\left(k_0 z - \omega_0 t + m\theta\right)\right]$ may be induced by the propagation of the acoustic wave in semiconductor plasma, where m defines the modulation parameter. The periodic modulation caused by the time varying changes in carrier density and electron-electron collision frequency. This parameter arises from the periodic modulation of the magnetic field, power of the rf discharge or propagation of the acoustic wave inside the semiconducting material [2] . The frequency, intensity, and direction of optical beam are controlled from the scattering of light from sound or low frequency electromagnetic wave. In nonlinear acoustics an important field of study is amplification and frequency mixing of waves in semiconductors [3] . This method has been used for the design of acousto-optic modulators, based on the interaction of an acoustic wave or a low frequency wave with the incident laser beam. Various piezoelectric materials (like InSb, and high strain dielectric material such as BaTiO3) have been used to investigate the effect of modulation and demodulation of the plane wave in nonlinear dispersive medium by using the different approximation [4] . Lashmore-Davies [5] reported the mechanistic approaches for arising spontaneous beak up of a shear Alfven wave to study the modulation instability of finite amplitude.

Much more attention has been paid by the researcher to find out the basic problem of the frequency and amplitude modulation in the gaseous plasma. The charge carriers due to ionization of the gas, from the interaction of the high power laser wave, are responsible for arising the plasma in gaseous state. In most of study of modulation and demodulation, interaction with nonlocal effects such as diffusion of the charge exciting the change in responsible parameters such as nonlinear refractive index has been ignored during calculation. Nonlinear polarizations due to acousto-optic interaction in dielectric and semiconducting materials are playing an increasing role in optical modulation, demodulation and beam splitting [6] [7] . But in presence of electro-strictive effect, the nonlinearity is due to slow movement of free charge carrier from the diffusion effect which shows short distance travel before recombining. Hence, the charge carrier can be easily moved in nonlinear dispersive medium, particularly high mobility III-V type of semiconductor.

The charge carrier is easily excited in nonlinear dispersive media when an intense laser beam passes through it, and considerable heating raises the steady state temperature which is higher than that of lattice temperature. The carrier heating provides the momentum transfer collision frequency to an electron, thus modifying the mobility and diffusion of charge carriers, as well as conductivity of the dispersive medium by adsorption of ions from the gaseous state and hence it shows the refinement effect on modulation of amplitude or frequency. It is found that increasing the diffusion through charge carrier heating makes it more difficult to reflect or transmit the light from the local equilibrium which represents the unstable and stable TE and TM linear surface wave [8] . Ghosh and Rishi [9] at first time studied the diffusion induced modulation and demodulation of the acousto-optic frequency in magnetized semiconductor. The frequency modulated beam in less dispersive acoustic media is easily amplified due to excess charge carrier from enhanced diffusion. They reported the excitation of an acousto-helicon wave modulation of plasma wave in longitudinally (k, k0, k1 || || Bs) magnetized semiconducting plasma [10] . The scattering of the wave occurs from ordinary and the other extraordinary wave, and lies on the surface of a circular cone i.e., $k_p^o + k_p^e = k + k_0$ with a phase angle of $\varphi' = \varphi + \pi$. This imparts the effect of geometry of the dispersive media which supports the propagation of the helicon wave with the interaction from acoustic and transverse induced non-helicon wave through polarization. The considerable gain in frequency is obtained from the nonlinear interaction of pump-helicon wave (ω_0, k_0) with a transversely acoustic wave (ω, k) and another helicon wave (ω_1, k_1) in cubic semiconductor plasma [11] .

In this present article, our analysis can be employed to see the effect of hot carriers on the amplitude modulation instability of an intense helicon pump wave due to acoustic-optic interaction in diffusive strain dependent dielectric constant. The dispersion relation can be solved by considering the complex relation for diffusion dependent dispersion $\omega(= \omega_r + i\omega_i)$ taking with the real and positive value of (k_1, k_0) throughout the present analysis. The intense pump source interacts with semiconductor plasma generating the acoustic wave with the interaction from the diffusion of free carrier (through electron plasma wave) and acoustic phonons (through material vibration). The numerical analysis was made by selecting the different parameter of z and t and also assumed that the propagation of high power helicon wave inside the material followed the Gaussian profile. The propagation mode of high power helicon wave will be completely unstable;

only when ω_i is less than ω_r; $|\Omega_1|$ represents the growth-rate of unstable mode arising from the modulation and demodulation of the pump- wave. The interaction of free carrier and acoustic phonon induces a strong threshold electric field that modulates or demodulates the pump wave from unstable growth rate. Thus, the applied electric field generates the sinusoidal threshold electric field and acoustic wave in acousto-optic modulator which can enhance the hyperbolic growth rate for the amplitude modulation and demodulation of the acoustic wave frequency in diffusive strain dependent dielectric constant in presence of strong magnetic field and hot carrier.

2. DISPERSION RELATION

The hydro-dynamical model is considered for the numerical calculation for the propagation of high power helicon pump wave in one-component homogeneous n-type piezoelectric semiconductor plasma of infinite extent with electrons as major charge carriers, along the direction z of externally applied static magnetic field, $B_S \parallel k_0 \parallel \hat{z}$.

$$E = E_0 \exp\left[i\left(k_0 z - \omega_0 t + m\theta\right)\right] \tag{1}$$

where, $E_{0x} = -iE_{0y}$. The time harmonic Maxwell's equation by assuming there is no external current, can be expressed as $\nabla \times E = i\omega\mu_0 H$, $\nabla \times H = -i\omega\varepsilon_0 \bar{K} \cdot E$, where $\nabla \times H = -i\omega\varepsilon_0 \bar{K} \cdot E$, where $\bar{K}$ is the equivalent dielectric. The magnetic field strength and plasma density gives the dependency of dispersion relation of the helicon wave. The well-known cold plasma dispersion tensor can be expressed by the following equation

$$\bar{K} = \begin{bmatrix} S - n_z^2 & -iD & n_z^2 n_\perp^2 \\ iD & S - n^2 & 0 \\ n_z^2 n_\perp^2 & 0 & P - n_\perp^2 \end{bmatrix} \tag{2}$$

where $S = 1 - \frac{\omega_p^2}{\omega^2 - \omega_c^2}$, $D = -\frac{\omega_p^2 \omega_c}{\omega\left(\omega^2 - \omega_c^2\right)}$, $P = 1 - \frac{\omega_p^2}{\omega^2}$, $n_z = \frac{ck_z}{\omega}$, $n_\perp = \frac{ck_\perp}{\omega}$,

and $\omega_p = \sqrt{\frac{n_e e^2}{m_e \varepsilon_0 \varepsilon_L}}$ is the

electron plasma frequency. By using analysis of Fourier analysis, the modified Maxwell's equations for unbounded

plasma is given by $\eta \times (\eta \times E) + \frac{\omega^2}{c^2} \bar{K} \cdot E = 0$, where $\eta = kc/\omega$, is the index of refraction vector, whose direc-

tion is the direction of the wave vector k and whose magnitude is the index of refraction. The basic equation involve in the analysis are the zeroth and first-order momentum transfer, Maxwell equation, continuity equation and the equation of motion of the lattice of a piezoelectric semiconductor. The classical equation of motion for carriers of charges e and effective mass m is

$$\frac{\partial v_0}{\partial t} + \left(v_0 \cdot \nabla\right) v_0 + \nu v_0 = \frac{e}{m}\left(E_t + v_t \times B_t\right) \tag{3}$$

$$\frac{\partial n}{\partial t} = N_0 \left(\nabla V\right) \tag{4}$$

If we assume a constant drift velocity υ_0 in the z-direction and let $E_t = E_0 + E\mathrm{e}^{i\omega t}$ and $B_t = B_0 + B\mathrm{e}^{i\omega t}$.

Equation (3) yields

$$\left(i\omega\upsilon + \upsilon_0 \cdot \frac{\mathrm{d}\upsilon}{\mathrm{d}z}\right)\mathrm{e}^{i\omega t} = \frac{e}{m}\left(E_0 + \upsilon_0 \times B_0\right) + \frac{e}{m}\left(E + \upsilon_0 \times B + \upsilon \times B_0\right)\mathrm{e}^{i\omega t} + \frac{e}{m}\left(\upsilon \times B\right)\mathrm{e}^{2i\omega t} \tag{5}$$

Since the applied magnetic field B parallel to the υ_0, and since the carriers are drifting, the time independent term are vanished. Helicon waves are basically low-frequency bounded whistler waves. In an ionosphere, whistler wave are right-hand-circularly polarized electromagnetic wave which are propagating along parallel to the magnetic field. The two normal are opposite in nature and travel with different phase velocity.

$$\upsilon_0^{\pm} = \frac{\upsilon_0}{1 \pm \omega_c/\omega} \tag{6}$$

In general, when high intensity wave interacts with semiconductor which contained high mobility charge, they gain momentum and energy as a result from electron collision. This collision develops the heat from momentum transfer (MTCF) through the relation

$$\upsilon_0^{\pm} = \upsilon_0 \left(\frac{T_e}{T_0} \right)^{1/2} \tag{7}$$

The power absorbed per electron from the pump electric field becomes

$$\frac{e}{2} \mathrm{Re}\left(\vartheta \cdot E_0^* \right) = \frac{e^2 \upsilon_0}{2m} \frac{\left(\omega_c^2 + \omega_0^2 \right)}{\left[\left(\omega_c^2 - \omega_0^2 \right)^2 + 4\upsilon_0^2 \omega_0^2 \right]} E_0 E_0^* \tag{8}$$

where * denotes the complex conjugate of the quantity and re denotes the real part. This power is dissipated in collision of electron from the acoustic phonon in the Brillouin active medium. Following Conwell [12] , the power dissipation per electron in collision with the polar optical phonon (POP) may be expressed as

$$\langle P \rangle_{POP} = \left(\frac{2k_B \theta_D}{m\pi} \right)^{1/2} eE_{POP} x_e^{1/2} K_0 \left(\frac{x_e}{2} \right) \exp\left(\frac{x_e}{2} \right) \frac{\exp\left(x_0 - x_e \right) - 1}{\exp\left(x_0 \right) - 1} \tag{9}$$

where $x_{0,e} = \hbar\omega_l / k_b T_{0,e}$, in which $\hbar\omega_l$ is the energy of POP given by $\hbar\omega_l = k_B \theta_D$, and θ_D is the Debye temperature of the medium. $E_{PO} = \frac{me\hbar\omega_0}{\hbar^2}\left[\frac{1}{\varepsilon_\infty} - \frac{1}{\varepsilon_L} \right]$ Is the field of POP scattering potential in which ε_L and

ε_∞ are the static and high-frequency dielectric permittivity of the medium, respectively. $K_0(x_0/2)$ Is the zeroth-order Bessel function of first kind. The Einstein equation is the basic relation between the ionic mobility μ

to the ionic self-diffusion coefficient D (in cm2·s−1) $\mu = \frac{eD}{k_b T}$. Therefore, for moderate heating of carrier, using

(7) and (8), we obtain the expression for electron temperature and modified diffusion coefficient as

$$\frac{T_e}{T_0} = 1 + \frac{e^2 \upsilon_0}{2m} \frac{\tau\left(\omega_c^2 + \omega_0^2\right)}{\left[\left(\omega_c^2 + \omega_0^2\right) + 4\upsilon_0^2 \omega_0^2\right]} E_0 E_0^* \tag{10}$$

$$D = \frac{k_B T_0}{e}\left[1 + \frac{e^2 \upsilon_0}{2m} \frac{\tau\left(\omega_c^2 + \omega_0^2\right)}{\left[\left(\omega_c^2 + \omega_0^2\right) + 4\upsilon_0^2 \omega_0^2\right]} E_0 E_0^*\right]\mu_e \tag{11}$$

The equation of motion for an element of volume dxdydz and density ρ is

$$\frac{\partial v_1}{\partial t} + \left(v_0 \nabla\right) v_1 = \frac{e}{m}\left(E_1 + v_1 \times B_s + v_0 \times B_1\right) - \frac{K_B T}{m n_0} \nabla n_1 - \nu \upsilon_1 \tag{12}$$

The propagation of the acoustic wave (AW) in a crystal with SDDC is just possible by only longitudinal EKWs due to the piezoelectric effect, which is slightly induced by the transverse electric field of the helicon. By using the Poisson equation $(\nabla D) = -en$, the piezoelectric field strength can be expressed by

$$\frac{\partial E_1}{\partial z} = \frac{e n_1}{\varepsilon_0 \varepsilon} + \varepsilon_L \varepsilon_0 g \frac{\partial}{\partial z}\left(\frac{\partial u}{\partial z} E\right) \tag{13}$$

The piezo-electrically excited longitudinal plasma oscillation can be obtained from Equation (13) by considering (3) and (4)

$$\frac{\partial^2 n_1}{\partial t^2} + \upsilon \frac{\partial n_1}{\partial t} - D \frac{\partial^2 n_1}{\partial x^2} - \varepsilon_0 \varepsilon_L g N_0 \frac{e}{m} \frac{\partial}{\partial z}\left(\frac{\partial u}{\partial z} E\right) + \omega_p^2 n = 0 \tag{14}$$

where $D=\frac{k_b T}{m}\mu_e$, N0 being the equilibrium electron concentration and n1 the electron concentration perturba-

tion. Assuming that $n\sim\exp\left[i\left(kx-\omega t\right)\right]$, we get

$$n=\frac{e\varepsilon_0\varepsilon_L g N_0}{mQ\left(\Omega,k\right)}\frac{\partial}{\partial z}\left(\frac{\partial u}{\partial z}E\right) \tag{15}$$

where $Q\left(\omega,k\right)=-\omega^2-i\omega\upsilon+\omega_p^2\left(1+\lambda_D^2 k^2\right)$, λ_D^2 is the Debye screening length. Then the AW wave propagation may be written as

$$\rho\frac{\partial^2 u_x}{\partial t^2}=\frac{\partial}{\partial x}\left\{\left[C_s+\Delta\hat{C}_s\left(E^2\right)-\frac{g^2\Omega_p^2\varepsilon\varepsilon_0}{Q\left(\omega,k\right)}\left(E\cdot E\right)\right]\frac{\partial u}{\partial z}\right\} \tag{16}$$

If the AW frequency $\omega\leq\omega_p^2$ and $\lambda_D^2 k^2\leq 1$, then $\frac{\omega_p^2}{Q}\approx 1+\frac{i\omega}{\omega_M}$, where $\omega_M\approx\frac{\omega_p^2}{\upsilon}$ -the Maxwell relaxation frequency. The following procedure adopted by Ghosh and Agarwal [13] and using Equations (2) to (9) in collision dominated regime $\upsilon_0\gg\Omega$, $k\upsilon_0$, one obtain.

$$\frac{\partial^2 n_1}{\partial t^2}+v\frac{\partial n_1}{\partial t}+k^2 v_{th}^2 n_1+\left[\mp\frac{\omega_p^2\omega_c\bar{\omega}}{\omega\left(\bar{\omega}-iv\mp\omega_c\right)}-\frac{\omega_p^4\bar{\omega}^2}{\omega\left(k^2\omega_L^2-\omega^2\right)\left(\bar{\omega}-iv\mp\omega_c\right)}\right]n_1+\left[\frac{n_0 e\varepsilon_0 g D E_0 k^2\omega\bar{\omega}}{m\varepsilon_L\left(k^2\vartheta_L^2-\omega^2\right)}\right]U\mp$$
$$=i\left(k+k_0\right)\overline{E_0}n_1 \tag{17}$$

in which $\bar{E}_0=\pm\frac{\varepsilon_L\rho\bar{\omega}E_0}{m\varepsilon_0 g^2\left(\bar{\omega}-iv\mp\omega_c\right)^2}$, $\bar{\omega}=\omega+k\cdot\upsilon_{th}$ and $\omega_p=\left(\frac{N_0 e^2}{m\varepsilon_0\varepsilon_L}\right)^{1/2}$ is the electron plasma frequency, $\omega_c=eB_0/m$ the electron cyclotron frequency where $B_0=k_0\times E_0/\Omega_0$. $\vartheta_L=\left(\mu_0\varepsilon_0\varepsilon_L\right)^{-1/2}$ Is the electromagnetic wave velocity in the crystal with lattice dielectric constant ε_L, $v_{th} = (k_B T/m)^{1/2}$ is electron thermal velocity.

As the density perturbation in the plasma has been assumed to vary as $\exp[i(\omega t - kz)]\exp[(i\Omega t - kz)]$. Hence the initial helicon pump beam can now beat with these perturbations to produce forced wave disturbance at upper $(\omega_0 + \omega)$ and lower $(\omega_0 - \omega)$ side-band frequencies. These side-bands are forced waves and one may write down the expression for them from Equation (10) as follows:

$$n_1(\omega_\pm, k_\pm) = \left[-\frac{\omega_p^2 \bar{\omega}(\omega^2 - k^2V_t^2 - K^2k^2V_t^2)}{(k^2c^2 - \omega^2)(\bar{\omega} - iv_\mp \omega_C)} \cdot \frac{\rho\varepsilon}{e\varepsilon_0 g E_0} + \frac{n_0 e\varepsilon_0 D g E_0}{m\varepsilon_L} \frac{k^2 \omega\bar{\omega}}{(k^2c^2 - \omega^2)} \right]$$
$$\cdot \left[(\omega_0 \pm \omega)^2 - i(\omega_0 \pm \omega)\nu - \omega_R^2 + i(k_0 \pm k)\bar{E}_0 \right]^{-1} u_\mp \tag{18}$$

in which $V_t = (C_s/\rho)^{1/2}$ traverse acoustic wave velocity in crystal, $\omega_R^2 = \left(\frac{\omega_p^2 \bar{\omega}}{\omega} + k^2 v_{th}^2 \right)$ the frequency for dispersive electron plasma wave and $K^2 = \varepsilon_0 g E_0 / \varepsilon_0 C$ the dimensionless electromechanical coupling coefficient. Equations (9) and (4) one may write $n_1 = n_1(\omega_+, k_+) + n_1(\omega_-, k_-)$, where $\omega_\pm = \omega_0 \pm \omega$ and $k_\pm = k_0 \pm k$ and the side band waves vary as $\exp[i(\omega_\pm t - k_\pm z)]$. The only nonlinear term present in Equation (13) is the one on the RHS and it is this term which couples the density variations at frequencies $\omega_0 \pm \omega$ back to the acoustic wave of frequency ω.

$$\frac{(\omega^2 - k^2V_t^2 - K^2k^2V_t^2)}{(\omega_p^2 K^2 k^2 V_t^2 \omega^2)} \left[k^2V_t^2 - \omega^2 + \frac{\omega_p^2 \bar{\omega}}{(\bar{\omega} - iv \mp \omega_c)} \right] = -\left[\frac{2\left(\delta + \frac{ik_+ \overline{E_0}}{2(\omega_0 - \omega_c)} \right)}{\omega_R(\omega - \delta - iv)\left\{ \omega + \delta - iv + \frac{ik_+ \overline{E_0}}{(\omega_0 - \omega_c)} \right\}} \right] \tag{19}$$

where $\delta = \omega_o - \omega_R$.

3. GROWTH RATE AND THRESHOLD ELECTRIC FIELD

In the slow wave limit, it is the quasi-static approx $\left(k^2\vartheta_L^2 \gg \Omega^2\right)$ and obtain the dispersion relation in simplified

$$\frac{\left(\omega^2 - k^2V_t^2 - K^2k^2V_t^2\right)}{\left(\omega_p^2K^2Dk^2V_t^2\omega^2\right)}\left[k^2V_t^2 + \frac{\omega_p^2\bar{\omega}}{\left(\bar{\omega} - iv \mp \omega_c\right)}\right] = -\left[\frac{2\left(\delta + \frac{ig\varepsilon_0k_+\overline{E_0}}{2\left(\omega_0 - \omega_c\right)}\right)}{\omega_R\left(\omega - \delta - iv\right)\left\{\omega + \delta - iv + \frac{ik_+\overline{E_0}}{\left(\omega_0 - \omega_c\right)}\right\}}\right] \tag{20}$$

From Equation (15) in absence of piezoelectric coupling coefficient $\varepsilon_0gE_0 = 0\left(K^2 = 0\right)$ one obtains

$$\omega^2 = k^2V_t^2 \tag{21}$$

We assume $\omega_R = kV_t$ and $\omega_i \ll \omega_R$ which is agreement with Steele and Vural [14] . Thus separating the real and imaginary part from Equation (5), we can obtain the expression at high magnetic fields from [1]

$$\omega_r^2 - \omega_i^2 = \left(k^2V_t^2 + K^2k^2V_t^2\right)\left[1 + \frac{2\delta\frac{\omega_p^2\omega^2}{\omega_RH}\left(G + \frac{k_+^2\overline{E_0^2}}{2\left(\omega_0^2 - \omega_c^2\right)}\right)}{\left(G^2 + \delta^2\frac{k_+^2\overline{E_0^2}}{2\left(\omega_0^2 - \omega_c^2\right)}\right)}\right] \tag{22}$$

and

$$\omega_r\omega_i = -\frac{K^2g^2k^2V_t^2D\omega^2}{H}\frac{\omega_P^2}{\omega_R^2}\frac{k_+\overline{E}_0}{\left(\omega_0 - \omega_c\right)^4}\left(G^2 + \delta^2\frac{k_+^2\overline{E_0^2}}{2\left(\omega_0^2 - \omega_c^2\right)}\right) \tag{23}$$

where $G = \nu^2 + \delta^2 - \nu \frac{kE_0}{(\omega_0 - \omega_c)}$, $\bar{G} = \nu^2 - \delta^2 - \nu \frac{k\overline{E_0}}{(\omega_0 - \omega_c)}$,

$H = k^2 C_L^2 - (\Omega_p^2 / \Omega^2) k^2 V_t^2 \gamma$, $\gamma = 1 - (V_t / V_{th})$

Since v_0 the oscillatory electron fluid velocity is much larger than V_t, it always get $\gamma < 0$ and consequently, H > 0. Hence above equation shows that the excited mode will be unstable only when $E_0 \neq 0$ and $\bar{G} < 0$. Now $\bar{G} < 0$ can be obtained

$$\nu^2 < \delta^2 - \frac{\nu \omega_p^3 k_+ \overline{E_0}}{(\omega_0 - \omega_c)^2} \tag{24}$$

In the presence of laser field one can obtain the threshold value of the electric field necessary for the onset instability by putting $\Omega_i = 0$ one can get

$$\bar{G} = \nu^2 - \delta^2 - \frac{\nu k_+ \overline{E_0}}{(\omega_0 - \omega_c)} = 0 \tag{25}$$

So the threshold value of the electric field is obtain by [1]

$$|\Omega_i| = \left\{ \frac{\omega_p \omega_0 \varepsilon_L K^2 k^4 V_t^3}{\left[k_0^2 \vartheta_L^2 + (\omega_p^2 / \omega_c^2) k^2 V_t^2 \gamma \right]} \right\} \times \left\{ \frac{4e}{m} \frac{E_{0x}}{(\omega_0 - \omega_c)^3} \right\} \times \left\{ \frac{\bar{G}^2}{G^2 + \delta^2 \frac{k^2 + \bar{E}_0^2}{\omega_0^2}} \right\} \times \left[1 + \frac{D \upsilon k_+^2}{\omega_p^2} \right]^{1/2} \tag{26}$$

Above equation shows that for instability the condition $\delta^2 > \nu^2$ must be satisfied. By adjusting the carrier concentration, we can easily find the suitable condition which satisfied for BaTiO3 high dielectric constant material. For a value of E_{0x} greater than $(E_{0x})_{th}$ we obtain the instability of the mode with a finite growth-rate given by utilizing above equations

$$|\Omega_i| = \left\{ \frac{\omega_p \omega_0 \varepsilon_L K^2 k^4 V_t^3}{\left[k_0^2 \vartheta_L^2 + \left(\omega_p^2 / \omega_c^2 \right) k^2 V_t^2 \gamma \right]} \right\} \times \left\{ \frac{4e}{m} \frac{E_{0x}}{\left(\omega_0 - \omega_c \right)^3} \right\} \times \left\{ \frac{\bar{G}^2}{G^2 + \delta^2 \frac{k^2 + \bar{E}_0^2}{\omega_0^2}} \right\} \times \left[1 + \frac{D\upsilon k_+^2}{\omega_P^2} \right]^{1/2} \quad (27)$$

4. RESULTS AND DISCUSSION

In this section, the numerical results of the possibility of modulation instability and the amplification of the acousto-helicon wave which arising from interaction of the pump helicon wave with acousto-helicon wave have been analyzed through Equations (28) and (29) for strain dependent dielectric material. The carrier heating on acousto-helicon interaction modifies the dependent parameter such as electron momentum transfer collision frequency (MTCF) (Equation (7)) and diffusion of charge carrier at the different temperature (Equation (10)) and hence consequently modifies the threshold electric field and modulation of a high power helicon wave effectively. The modulation instability and the amplification of acousto-helicon waves from Equations (26) and (27) was solved numerically as results from the transfer of modulation helicon wave to acousto-helicon wave for the different values of the semiconductor plasma parameters such as Ω_p, Ω_c, and $\upsilon_0^{\pm}$ at the different temperature. As a typical case, the numerical calculation was performed for the BaTiO3 cubic crystals at 300K, with following typical constant are taken: $n_0 = 10^{23}\ \mathrm{m}^{-3}$, $m = 0.045 m_0$, $\varepsilon = 2000$, $\rho = 4.0 \times 10^3\ \mathrm{kg/m^3}$, $V_t = (C_{44}/\rho)^{1/2} = 2.5 \times 10^3\ \mathrm{m/s}$, $\Omega_p = \left(\frac{N_0 e^2}{m \varepsilon_0 \varepsilon_L} \right)^{1/2}$, $K^2 = \frac{\varepsilon_0^2 g^2 E^2}{\varepsilon_L C_{44}}$, $\vartheta_L = \left(\mu_0 \varepsilon_0 \varepsilon_L \right)^{-1/2}$, $\upsilon = 5 \times 10^{14}\ \mathrm{s}^{-1}$, $\gamma = 1 - (V_t / V_{th})$. The crystal is subjected to irradiate with 10.6 μm nanosecond laser. The beauty of BaTiO3 is that its crystals structure is easily modified from the addition of impurities or creating the defect inside the lattice structure. Its easy tendency is due to formation of the solid solution with foreign atoms of the same size or fitted in the octahedral hole of the lattice arrangement. The electric field amplitude which is considered in the present investigation can be directly calculated from the pump intensity I0, expressed by the following expression: $I_0 = \frac{c \varepsilon_0 \varepsilon_L |E_{ox}|^2}{2\eta}$.

Now let us consider the important case where the mechanistic approaches of spatial lattice formation is due to the diffusion process of photo-excited electrons or holes from the surface of semiconducting material by irradiation of the pump wave. The intensity distribution from the diffusion process between the threshold electric field can be expressed

$$E_{ox}^{th} = \pm i\left(E_{ox}^{th}\right)^{D} \frac{I_0^D}{I_0} \tag{28}$$

where $\left(E_{ox}^{th}\right)^{D} = \frac{\Omega_{1,2} T}{e}$ is known as the diffusion field, T is the temperature, and e is the electronic charge. The

fastest imaginable diffusion process from the charge carriers may be due to free flight of the particles between the different lattice site of stain dependent material with the upper limit D, which can be expressed by the diffusion coefficient of an ideal gas $D_{gas} = \frac{\lambda \bar{\upsilon}}{3}$, where λ and $\bar{\upsilon}$ are the mean free path and mean speed respective-

ly. The thermo-dynamical equilibrium actually controlled the diffusion of the charge carriers by the jumping from one lattice position to other place by the interaction of acousto-helicon wave, then the residence time τ of the particle on its site is given by $\Gamma = \frac{1}{z\tau}$, where z define the coordination number indicating the number of

neighboring present sites. The hot carriers easily diffuse in entire lattice of BaTiO3. The random walk theory in a 3d can be considered for the self diffusion process and it is calculated from the following expression

$$D = \frac{z}{6} \cdot f \cdot (1-c) \cdot a_0^2 \cdot \vartheta_0 \cdot \exp\left\{\frac{\Delta G_m}{kT}\right\} \tag{29}$$

The entire diffusion process changes the modulation and amplification of the acousto-helicon wave which results from transfer of momentum and energy through pump wave. The numerical results are plotted in Figure 1 and Figure 2. The variation of threshold electric field dependent on interaction length and cyclotron frequency is depicted in Figure 1 and Figure 2 with including the carrier effect. The interaction length varies from 0 to 20 μm and the range of the cyclotron frequency is taken to be less than Ω_0. The produced threshold electric field varies sinusoidal with interaction length and cyclotron frequency. The threshold electric filed shows the maxima and minima with an increase the interaction length. It may be inferred from Figure 1 and Figure 2 that the amplitude modulation of the pump helicon in Figure 2 is found to be 100 times more than that from Figure 1. Due to diffusion of the charge carrier (CH), the surface nature of the threshold electric filed is not changed, but changed the magnitude only. The nature of variations of $|\Omega_i|$ dependent on $(E_{ox})_{th}$ with k+ are plotted in Figure 2 where Figure 2(a) shows the variation of $|\Omega_i|$ without CH and Figure 2(b) with diffusion of CH. Figure 2(a) indicates that the growth rate changes sharply with the increase in the cyclotron frequency ω_c. The surface profile of the growth rate is not similar to surface profile of the threshold electric field. The surface of the electric field directly varies with $\omega_0 - \omega_c$, whereas $|\Omega_i|$ have an inverse relation of $(\omega_0 - \omega_c)^3$ by assuming that there is no possibility of diffusion. Due to this relationship, they lose the sinusoidal nature of the propagation. Figure 2(b) presents the variation of the growth rate with diffusion of the carrier heating. There problem is solved by using Equation (28) after assuming that growth rate is directly proportional to $\left[1+\frac{D\upsilon k_+^2}{\omega_p^2}\right]^{3/2}$.

Figure 3 and Figure 4 shows the variation of the threshold electric filed and growth rate with the interaction length and cyclotron frequency which varies from 1 to 2.5 × 1014 Hz. These modulations tend to be minimum at $\omega_c \approx \omega_0$. On further increasing or decreasing the cyclotron frequency from the pump frequency $\omega_0 < \omega_c > \omega_0$, they abruptly change the nature of propagation. The propagation of the threshold electric field below and greater than ω_0 is the inverse of each other. The profile shape is preserved in the propagation, although the width distribution changes with the magnetic field strength. Below and above cyclotron frequency from the pump frequency, only sign of the width is changed but nature of propagation is almost same.

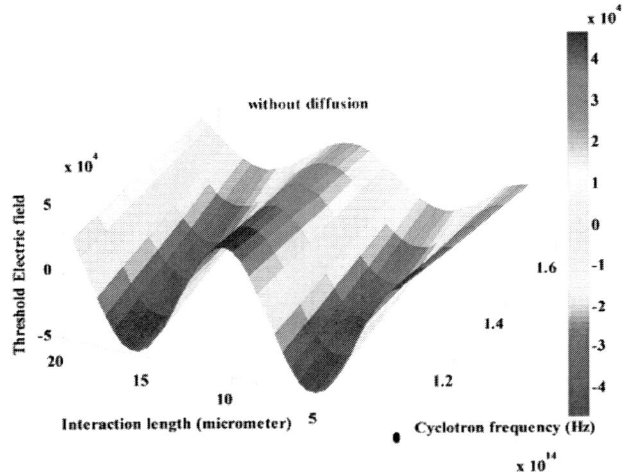

(a)

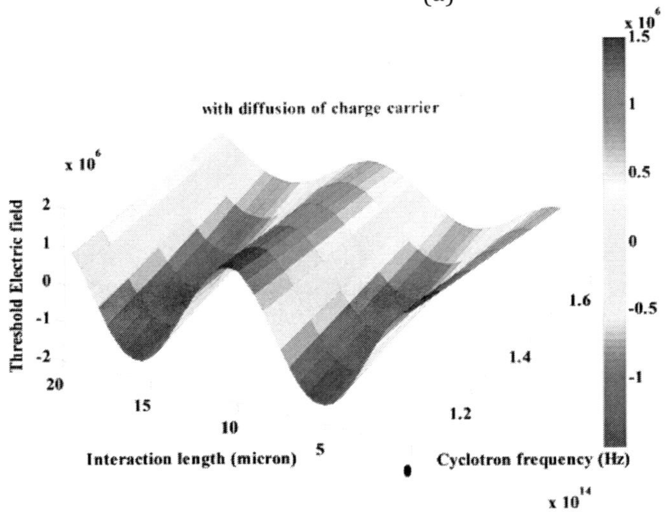

(b)

Figure 1. Variations of the amplitude modulation from the pump wave with interaction length and cyclotron frequency. Curve (a) without CH effect; (b) with CH effect.

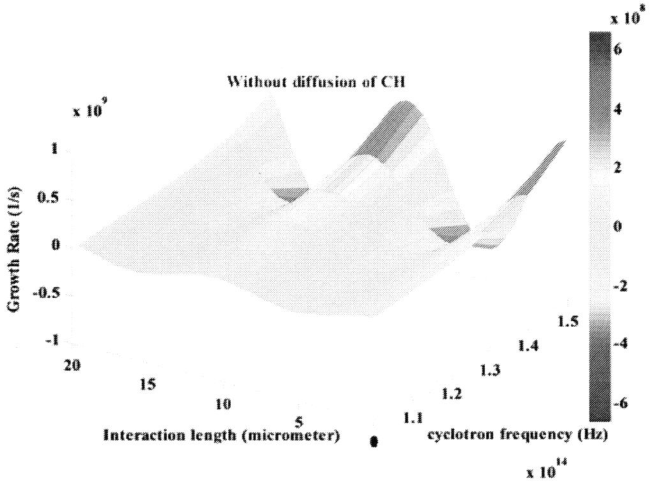

(a)

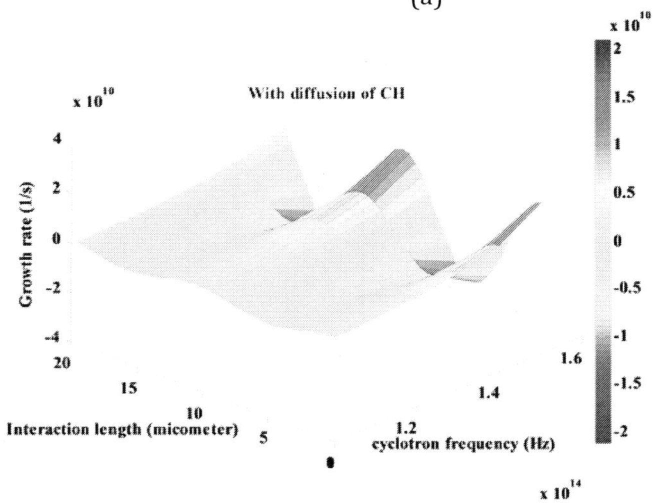

(b)

Figure 2. Dependence of growth rate from the pump wave with interaction length and cyclotron frequency. Curve (a) without CH effect; (b) with CH effect.

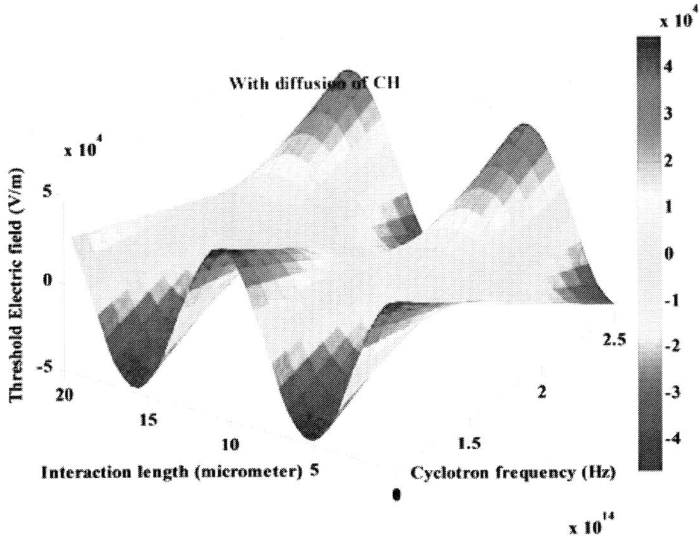

(a)

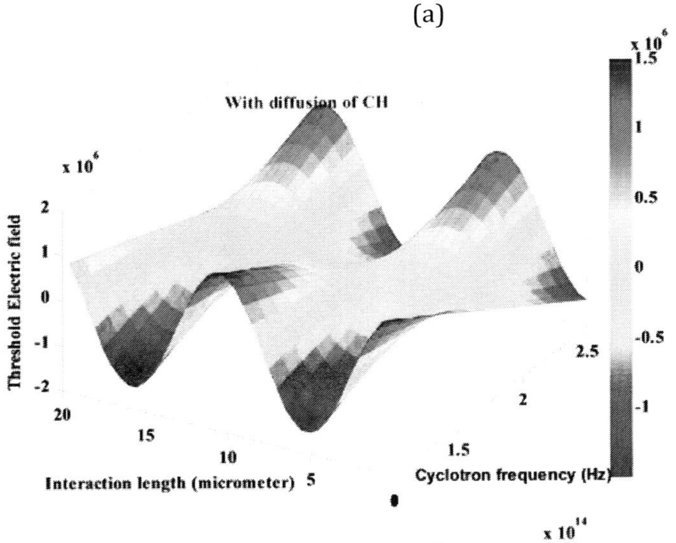

(b)

Figure 3. Variations of the amplitude modulation from the pump wave with interaction length and cyclotron frequency $\omega_0 < \omega_c > \omega_0$. Curve (a) without CH effect; (b) with CH effect.

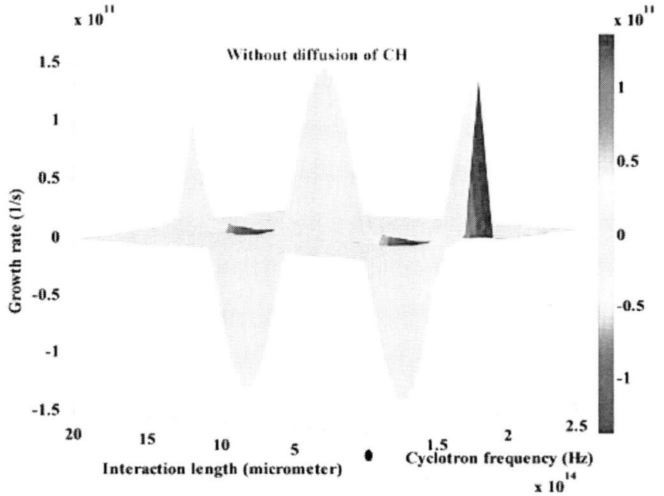

(a)

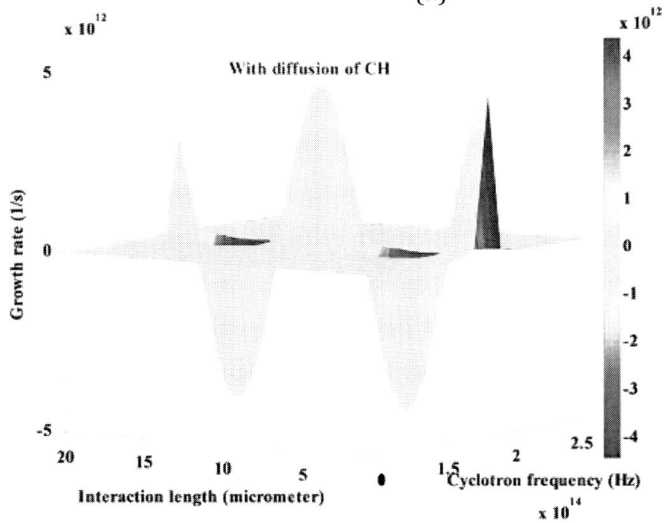

(b)

Figure 4. Dependence of growth rate from the pump wave with interaction length and cyclotron frequency varies from $\omega_0 < \omega_c > \omega_0$. Curve (a) without CH effect; (b) with CH effect.

The growth rate having the same profile like threshold electric field but their variation is not uniform above and below the propagating plane. They show the slight variation in magnetite below and above the plane at $\omega_c \approx \omega_0$. The negative amplitude shows from the electron-electron collision and positive corresponding to the holes migration. This is due to shifting of one-quarter of the period of the lattice corresponding from the pump intensity distribution. The carrier heating effect shows the same profile with enhanced magnetite. These important studies of nonlinear interactions revels that the inclusion of heating effects modifies the mobility of the semiconducting materials by diffusion process. This can be explained by Equation (29); we can easily see that the partial modulation of the amplitude of the fundamental lattice vibration, corresponding to the parametric pair k and k+, is expressed by sum of the different phase based on the diffusion, as shown in Figure 4(a) and Figure 4(b).

Pulse duration is important parameter to interact with the pump wave with semiconducting material. Not only change the interaction parameter, but also change the diffusion process of carrier heating by transferring the maximum amount of energy and momentum. This interaction changes the surface profile of modulation and amplification of the pump wave. The pulse duration modifies the carrier-lattice interaction by means of a collision time approximation. This effect is replaced Ω to $(\bar{\omega} - i\nu \mp \omega_c)$, which not altered the resultant behavior of the surface profile of modulation and amplification of the pump wave. The carrier heating and its controlled is by the external electric field which give rise to the carrier drift velocity, directly related with the mobility of the diffusing species. Thus for analyzing the behavior of threshold electric field, Equation (29) are used to solve for the numerical results. Their effect is shown in Figure 5(a) and Figure 5(b); in this case, the cyclotron frequency is less than the pump frequency. In order to investigate the results, the pulse duration varies up to 25 ns. Their results are plotted in terms of ω_c and pulse duration. Figure 5(a) shows the variation of the threshold electric field with pulse duration and cyclotron frequency, whereas the diffusion of the carrier heating enhances the magnitude of the amplitude modulation, as shown in Figure 5(b). The threshold electric field shows the sinusoidal nature of the propagation with small width and pulse height.

Similarly, Equation (29) is again solved for calculating the surface profile of threshold electric field for a BaTiO3 structure; in this case the profile is plotted from 1014 Hz to 2.5 × 1014 Hz. Figure 6(a) and Figure 6(b) show the propagation of threshold electric field with and without the presence of CH. The surface behavior is almost same but gives the entire profile of propagation. The carrier heating modifies the propagation of modulation

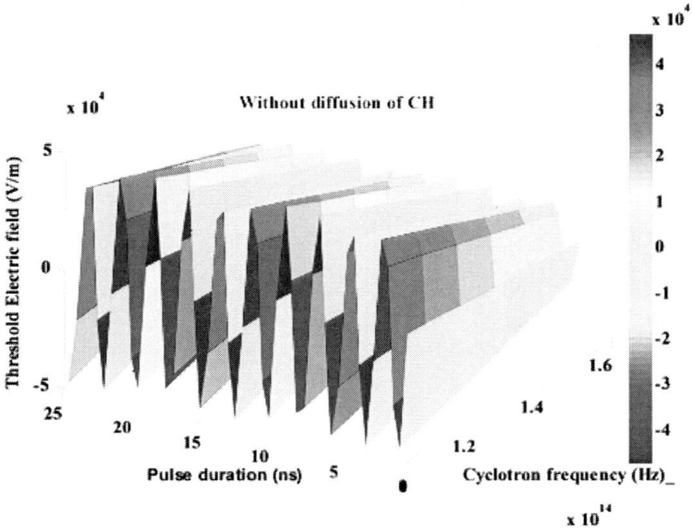

(a)

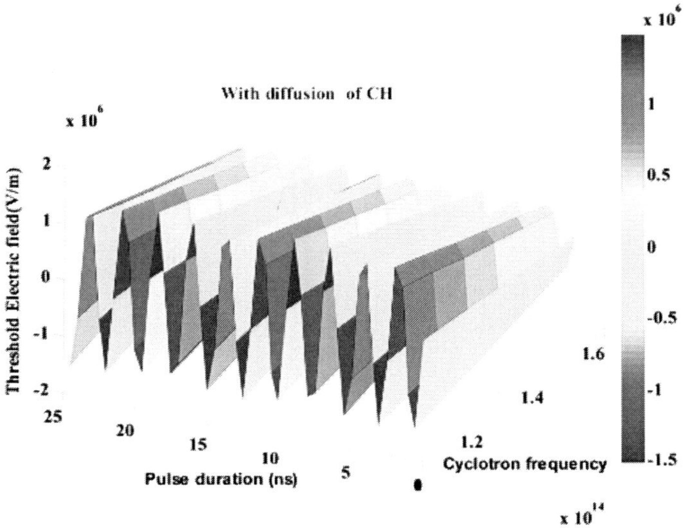

(b)

Figure 5. Dependence of growth rate from the pump wave with pulse duration and cyclotron frequency. Curve (a) without CH effect; (b) with CH effect.

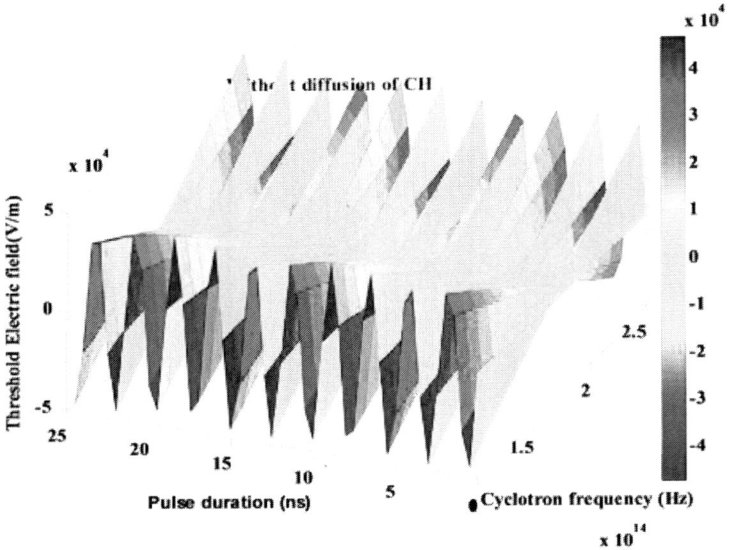

(a)

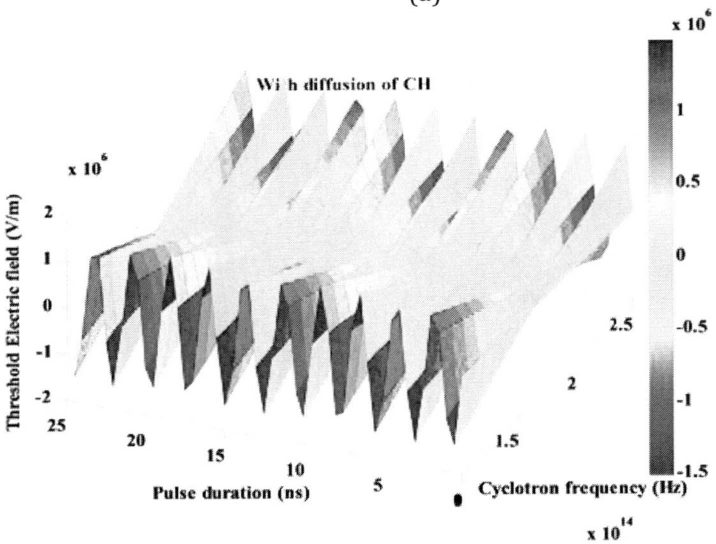

(b)

Figure 6. Variations of the amplitude modulation from the pump wave with interaction length and cyclotron frequency varies from $\omega_0 < \omega_c > \omega_0$. Curve (a) without CH effect; (b) with CH effect.

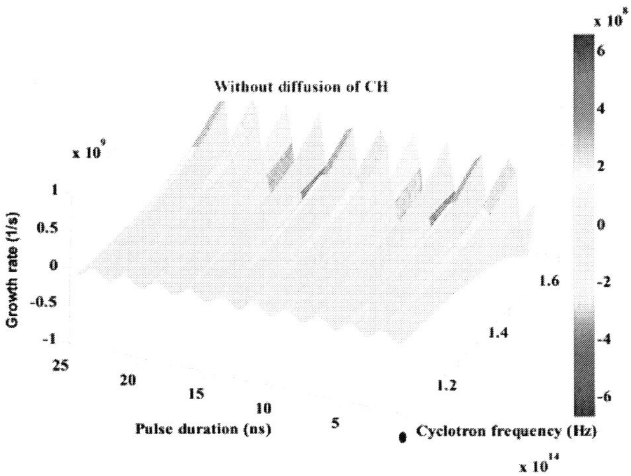

(a)

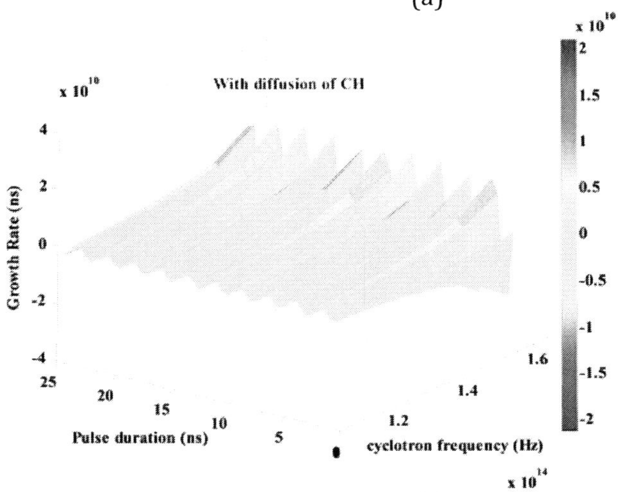

(b)

Figure 7. Dependence of growth rate from the pump wave with pulse duration and cyclotron frequency. Curve (a) without CH effect; (b) with CH effect.

by introducing the diffusion related parameter $1+\frac{D\upsilon k_{+}}{\omega_{P}^{2}}$ and it is related to the number of collision events per cycle from electron-electron interaction, which changed the nature beyond the pump frequency $\omega_0 = 1.78 \times 10^{14}\ s^{-1}$. In order to analyze growth rate in BaTiO3 with and without carrier heating

(CH), Equation (29) are numerically used against the cyclotron frequency ω_c and the pulse duration. During the investigation, the results of growth rate are again plotted below the pump frequency. Figure 7(a) and Figure 7(b) show the propagation of growth rate with pulse duration and cyclotron frequency up to 1.5 × 1014 Hz.

Similarly, Equation (29) is again presented graphically in Figure 8(a) and Figure 8(b) for a strain dependent dielectric with upper and lower sign of modulation; the presentations are made from 1014 Hz to 2.5 × 1014 Hz. During investigation, the results are found that the growth rate is non-uniform and non-sinusoidal with cyclotron frequency. Figure 8(a) shows a continuous and non-uniform growth-rate or a region of propagation with non-uniform pulse height from 1014 Hz to 2.5 × 1014 Hz, whereas Figure 8(b) shows a continuous and non- uniform growth-rate with large magnitude of non-uniform pulse height. The amplitude modulation and growth rate are enhanced 100 and 10 times by the diffusion of the CH. The profile shows only change in maxima and minima with pulse duration and cyclotron frequency. The curve has an oscillating character up and down with respect to zero planes which shows the pronounced conversion of the damping energy in different modes. The same observation has been observed for large pulsed duration as shown in Figure 9(a) and Figure 9(b). In this case, the duration of interaction varies up to 75 ns. Large numbers of pulse are generated during the propagation of high power pump wave with zero pulse width. The pulses are sinusoidal and the propagation is perfectly symmetrical. The carrier heating modifies the threshold amplitude modulation through diffusion as shown in Figure 9(a) and Figure 9(b). The pulse duration also modifies the growth rate by the transfer of momentum and energy with small pulse width or zero width. Figure 10(a) and Figure 10(b) show the growth rate with pulse duration and cyclotron frequency. The enhanced growth rates are observed in Figure 10(b) by the diffusion process from the carrier heating.

Below the pump frequency, the mass dependence threshold electric field controls the growth rate and the amplitude modulation. This figure illustrate the variation of growth rate with cyclotron frequency, which shows that growth rate with positive and negative are always in phase but it does not follows the sinusoidal nature: hence it shows the frequency and amplitude modulation from diffusion of hot carriers.

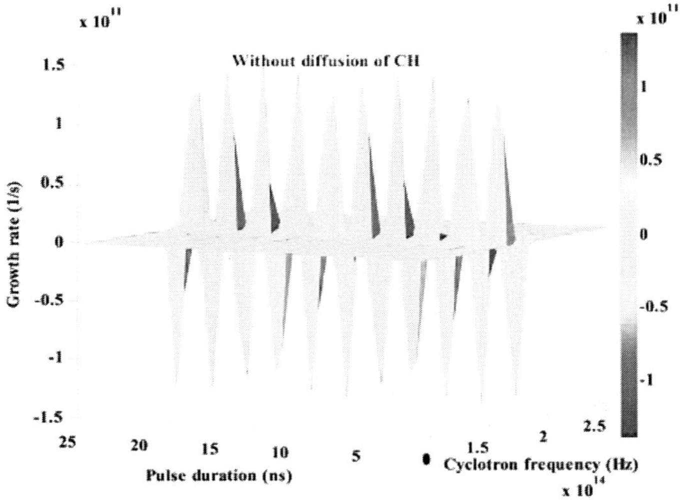

(a)

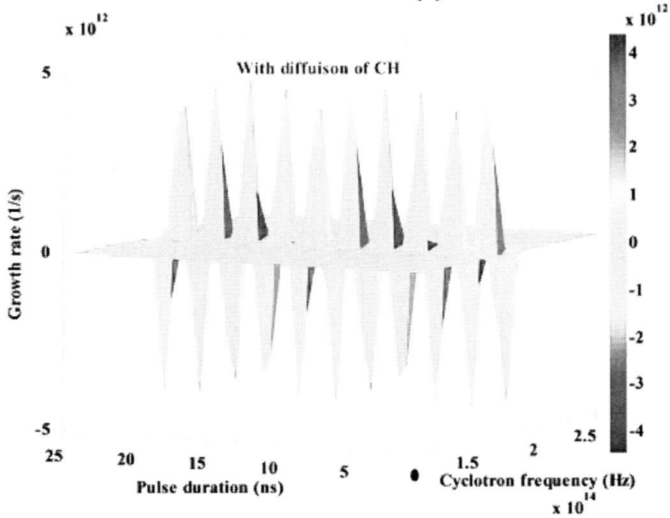

(b)

Figure 8. Dependence of growth rate from the pump wave with interaction length and cyclotron frequency varies from $\omega_0 < \omega_c > \omega_0$. Curve (a) without CH effect; (b) with CH effect.

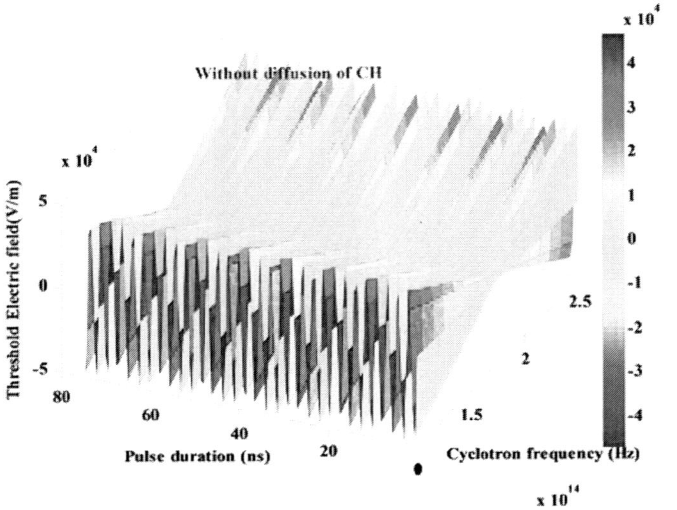

(a)

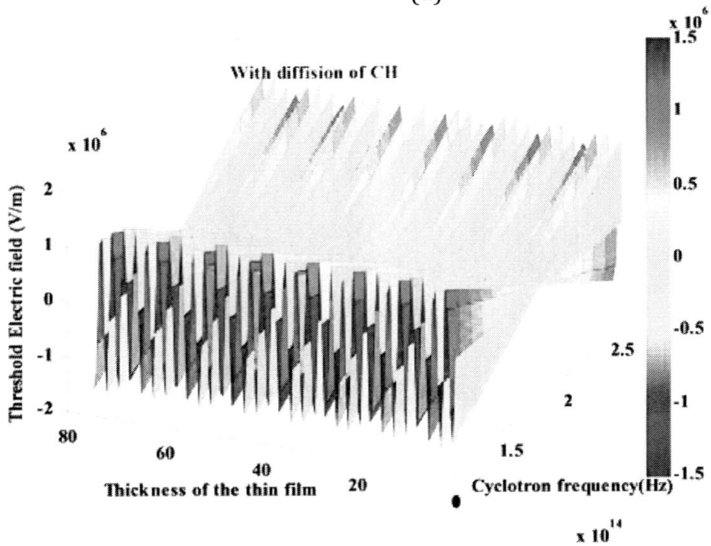

(b)

Figure 9. Dependence of growth rate from the pump wave with pulse duration and cyclotron frequency. Curve (a) without CH effect; (b) with CH effect.

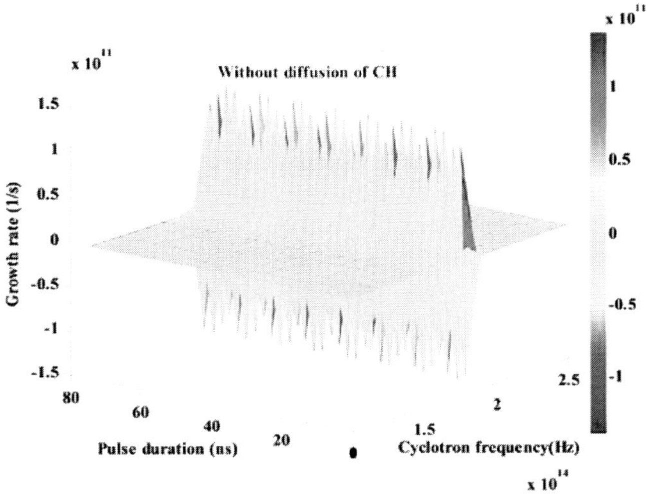

(a)

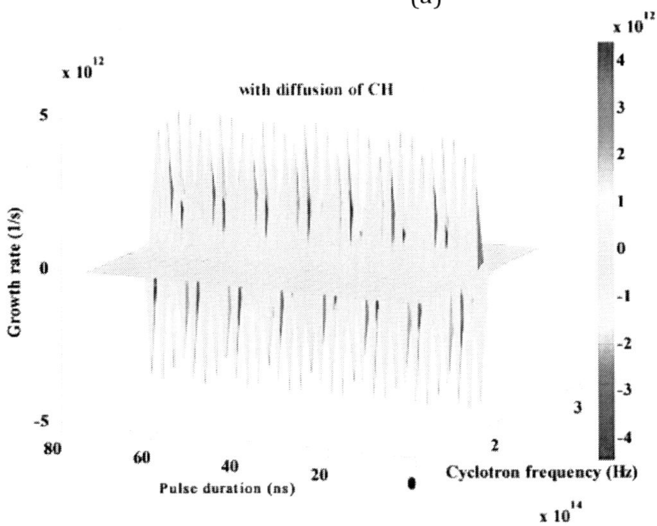

(b)

Figure 10. Dependence of growth rate from the pump wave with interaction length and cyclotron frequency varies from $\omega_0 < \omega_c > \omega_0$. Curve (a) without CH effect; (b) with CH effect.

5. CONCLUSION

The results of this paper suggest that the modulation and growth rate of an electromagnetic wave can be easily achieved in many ferroelectric materials such as BaTiO3. The acousto-helicon is excited by the modulation of high power helicon pump wave in a longitudinally magnetized strain dependent dielectric material. The CH effect always increases the magnitude of the modulation and growth rate either interacting with change in interaction length (z) or oscillating the applied electric field by changing the pulse duration. The diffusion of carrier heating just modifies only the magnitude by retaining the shape of profile either in threshold electric field or growth rate. Hence, hot carriers are always much more favorable to modulate the pump wave via the interaction of acousto-helicon wave. This interaction parameter of the plasma media is very limited between the carriers heating and the lattice via diffusion process and is applied over a wide range of parametric wave number k or k+. The sinusoidal nature is found only in threshold electric field; meanwhile this nature is completely lost in growth rate of frequency and amplitude modulation. Non-uniform nature of growth rate is observed during the change in cyclotron frequency from 1014 Hz to 2.5 × 1014 Hz.

ACKNOWLEDGEMENTS

The authors are very much thankful to Principal Govt. M.V.M College for encouragement.

REFERENCES

1. 1. Mamum, A.A. and Slimullah, M. (1991) Parametric Excitation of Alfven and Helicon Waves in a Magnetoactive Compensated Semiconductor by Microwave Radiation. Physical Review B, 44, 8685-8693.
2. Nimje, N., Dubey, S. and Ghosh, S. (2012) Amplitude Modulation and Demodulation of an Electrometric Wave in Magnetized Acoustio-Optic Diffusive Semiconductor Plasma: Hot Carrier Effects. Optics & Laser Technology, 44, 744-748.
3. Gupta, P.K. and Sen, P.K. (2001) The Role of Electrostriction on Parametric Dispersion and Amplification in Doped Piezoelectric Semiconductors. Non linear Optics, 26, 361-377.
4. Paiella, R., Martini, R., Capasso, F., Gmachl, C., Hwang, H.Y., Baillargeon, J.N., Sivco, D.L., Cho, A.Y., Whittaker, E.A. and Liu, H.C.

(2001) High-Frequency Modulation without the Relaxation Oscillation Resonance in Quantum Cascade Lasers. Applied Physics Letters, 79, 2526.
5. Lashmore-Davies, C.N. (1976) Modulated Instability of a Finite Amplitude Alfven Waves. Physics of Fluids, 19, 587.
6. Vazquez, R.A., Vachss, F.A., Neurgaonkar, R.R. and Ewbank, M.D. (1991) Large Photorefractive Coupling Coefficient in a Thin Cerium-Doped Strontium Barium Niobate Crystal. Journal of the Optical Socitey of America B, 8, 1932-1941.
7. Tervonen, E., Friberg, A.T. and Turunen, J. (1993) Acousto-Optic Conversion of Laser Beams into Flat-Top Beams. Journal of Modern Optics, 40, 625-641.
8. Vartharajah, P., Newell, A.C., Moloney, J.V. and Aceves, A.B. (1990) Transmission, Reflection and Trapping of Collimated Light Beams in Diffusive Kerr-Like Nonlinear Media. Physical Review A, 42, 1767-1774.
9. Ghosh, S. and Rishi, M.P. (2002) Acousto-Optic Modulation in Magnetized Diffusive Semiconductor. European Physical Journal D: Atomic Molecular, Optical and Plasma Physics, 19, 223-230.
10. Ghosh, S. and Saxena, R.B. (1985) Amplification of Acousto-Helicon Wave Due to Modulation of a High-Power Helicon Wave in Longitudinal Magnetized Cubic Piezoelectric Semiconducting Plasmas. Acustica, 58, 91-97.
11. Odulov, S.G. and Sturman, B.I. (1996) Coupling of Orthogonally Polarized Waves in Barium Titanate by Parametric Scattering. JEPT, 82, 1095-1101.
12. Conwell, E.M. (1967) High Field Transport in Semiconductor. Academic Press, New York, 159.
13. Ghosh, S. and Agarwal, V.K. (1982) Exicitation of Acouto-Helicon Waves Due to Modulation of a Laser Beam in Longitudinally Magnetised Semiconductors. Physica Status Solidi (b), 110 K37-40 70 A.
14. Steele, M.C. and Vural, B. (1969) Wave Interactions in Solid State Plasmas. Mc-Graw Hill, New York, 154.

6
CHAPTER

IMPROVING THE DIELECTRIC PROPERTIES OF HIGH DENSITY POLYETHYLENE BY INCORPORATING CLAY-NANOFILLER

Ossama E. Gouda[1], Sohair F. Mahmoud[2], Ahmed A. El-Gendy[3], Ahmed S. Haiba[2]

[1]*Electrical Department, Faculty of Engineering, Cairo University, Giza, Egypt*

[2]*High Voltage Metrology Lab, National Institute of Standards (NIS), Giza, Egypt*

[3]*Nanotechnology and Nanometrology Lab, National Institute of Standards (NIS), Giza, Egypt*

ABSTRACT

Polymer nanocomposites have been used for various important industrial applications. The prep- aration of high density polyethylene composed with Na-montmorillonite nanofiller using melt compounding method for different concentrations of clay-nanofiller of 0%, 2%, 6%, 10%, and 15% has been successfully done. The morphology of the obtained samples was optimized and characte- rized by scanning electron microscope showing the formation of the polymer nanocomposites. The thermal stability and dielectric properties were measured for the prepared samples. Thermal gravimetric analysis results show that thermal stability in polymer nanocomposites is more than that in the base polymer. It has been shown that the polymer nanocomposites exhibit some very different dielectric characteristics when compared to the base polymer. The dielectric breakdown strength is enhanced by the addition of clay-nanofiller. The dielectric constant (ε_r) and dissipation factor (Tan δ) have been studied in the frequency range 200 Hz to 2 MHz at room temperature in- dicating that enhancements have been occurred in ε_r and Tan δ by the addition of clay-nanofiller in the polymer material when compared with the pure material.

Keywords: Polymer Nanocomposites, High density polyethylene, Dielectric breakdown, Dielectric constant, Dissipation factor

1. INTRODUCTION

Polymers play an important role for many applications due to their unique properties which can be classified as heat sensitive, flexible, electrically insulating, amorphous, or semi-crystalline materials. For that reason, polymers are the most commonly used dielectrics because of their reliability, availability, ease of fabrications, and low cost. The selection of the proper dielectric polymer for a desired application depends on the requirements and operating conditions of the applied system [1] . The electrical properties of polymers can be improved by the addition of inorganic nano-fillers to the polymers forming new materials called polymer nanocomposites (PNC). Polymer nanocomposites are composite materials having several wt% of inorganic particles of nanometer di- mensions homogeneously dispersed into their polymer matrix. PNC with better dielectric and electrical insula- tion properties are slowly emerging as excellent functional materials for dielectrics and electrical insulation application and the term "nanodielectrics" for such materials is increasingly becoming popular. Although the technology of addition of fillers to polymers to enhance a particular dielectric property has been in existence for several decades [2] - [4] , the effect of filler size on the dielectric behaviour of the polymer composites has not been understood fully. It is with the advent of nanotechnology leading to the availability and commercialization of nanoparticles that polymer nanocomposite technology started to gain momentum. Polymer nanocomposites have been found to exhibit enhanced physical, thermal, mechanical, and dielectric properties when compared to the traditional polymer materials especially at low nano-filler concentrations (1% - 10%) [5] - [7] . However, it is only recently that the dielectric properties of such polymer nanocomposites were looked into and limited re- search results demonstrate very encouraging dielectric properties for these materials. Irrespective of the type of base polymer material (thermoplastic or thermoset), significant enhancements in several physical properties, like thermal conductivity (with conducting fillers) or dielectric properties like resistivity, permittivity, dissipation factor, dielectric strength, tracking and partial discharge resistant characteristics (with insulating fillers) were obser- ved when compared to similar properties in traditional neat polymers [8] - [10] . These observations were mainly attri- buted to the unique properties of nanoparticles and the large interfacial area in polymer nanocomposites [11] - [13] .

The present work focuses on the dielectric properties of polyethylene (PE) nanocomposites. Polyethylene is one of the thermoplastic polyolefin which is traditionally one of the most widely used polymer classes with ap- plications in structural, textile, and packaging industries, and their nanocomposites have found multiple applica- tions for the same uses. This paper shows the preparation and characterization of high density polyethylene com- posed with Na-montmorillonite clay-nanofiller (HDPE/clay) with different concentrations of clay-nanofiller as 0%, 2%, 6%, 10% and 15%. Then the dielectric properties, such as dielectric constant, dissipation factor, dielec- tric breakdown, and insulation resistance, of the prepared samples will be discussed and compared to the base polymer material.

2. EXPERIMENTAL WORK

2.1. Materials

HDPE with melt flow rate of 0.75 g/min and density of 960 kg/m^3 is chosen as the base polymer material for the current study. It was manufactured by the International Company for Manufacturing Plastic Products. Sodium montmorillonite clay K10 (MMT) was acquired from fluka chemika company. Hexadecyl Trimethyl Ammo- nium Bromide, modifier or surfactant material, was obtained from Merck KGaA, Darmstadt, Germany.

2.2. Modification of Clay

The preparation of polymer/clay nanocomposites with good dispersion of clay layers within the polymer matrix is not possible by physical mixing of polymer and clay particles. It is not easy to disperse nanolayers in most polymers due to the high face to face stacking of layers in agglomerated tactoids and their intrinsic hydrophilis- ity which make them incompatible with hydrophobic polymers. The intrinsic incompatibility of hydrophilic clay layers with hydrophobic polymer chains prevents the dispersion of clay nanolayers within polymer matrix and causes to the weak interfacial interactions. Modification of clay layers with hydrophobic agents is necessary in order to render the clay layers more compatible with polymer chains, and result in a larger interlayer spacing. In addition, modification process improves the strength of the interface between the inorganic clay and the polymer matrix. So, sodium montmorillonite (Na-MMT) clay was modified with the compatiblizer of Hexadecyl Trimethyl Ammonium Bromide [14] .

100 g of clay was dispersed into 1000 ml of methanol solvent and placed on hot plate with magnetic stirrer to allow continuous stirring for 2 hours. On the other hand, 100 g of hexadecyl trimethyl ammonium bromide was dissolved in 500 ml of methanol. Then the solution was added to clay dispersion. The stirring continued for 72 hours. After that, the modified clay was filtered and collected. Finally the filtrate was dried in a vacuum oven at 70°C for 6 hours [15] .

2.3. Preparation of HDPE/clay composites

The concentrations of modified clay-nanofiller were added as 0%, 2%, 6%, 10%, and 15% into the base polymer material. HDPE/clay nanocomposites were prepared by melt compounding method (master batch method) using twin screw extruder (TSE) at zones temperature 163°C, 167°C, and 167°C, for Zone 1, Zone 2, and Zone 3 re- spectively. The screw speed was maintained at 30 rpm. After extrusion, the dried pellets of nanocomposites were preheated using Morgan Press Injection unit at 160°C for 30 min and injected to produce test samples with dimensions 7.5 cm × 7.5 cm × 0.25 cm for dielectric measurements [15] . The prepared samples are referred to in this paper as HDPE 0% (pure material), HDPE 2%, HDPE 6%, HDPE 10%, and HDPE 15%.

2.4. Characterization of HDPE/clay composites

The prepared samples were characterized by the scanning electron microscope (SEM). SEM is a type of electron microscope that produces images of a sample by scanning it with a focused beam of electrons. The electrons in- teract with atoms in the sample, producing various signals that can be detected and that contain information about the sample's surface topography and composition. The scanning electron microscope images were carried out by using SEM, model Quanta 250 FEG (Field Emission Gun) attached with EDX unit (Energy Dispersive X-ray Analyses), with accelerating voltage 30 kV, magnification 14× up to 1,000,000×, and a resolution of 1 nm.

Thermal stability was measured by using thermo gravimetric analysis (TGA). TGA experiments were done by a shimadzu TA-50 thermal analyzer using scanning rate of 5°C/min under N2 with 20 ml/min flow rate, from room temperature to 600°C.

2.5. Dielectric Properties

Dielectric breakdown refers to a rapid reduction in the resistance of an electrical insulator when the voltage ap- plied across it exceeds the breakdown voltage. Dielectric breakdown measurements were performed using AC Dielectric Test Set. The samples were sandwiched between two electrodes and tested at room temperature under an ac voltage ramp of 750 V/sec. The ac voltage was increased with a rate of 750 V/Sec until breakdown occurred.

Dielectric constant is called relative permittivity which is a parameter that indicates the relative charge storage capability of dielectrics in the presence of an electric field. The used instrument is an Agilent E4980A LCR me- ter with dielectric sample holder. The equivalent parallel capacitance (C_p) was measured directly by the LCR meter, then the dielectric constant is calculated as shown below in the results section.

Dissipation factor is called loss tangent or Tan δ. It represents the energy loss in the dielectrics and it is pre- ferred to be smaller for insulation materials. It was measured directly by an Agilent E4980A LCR meter with di- electric sample holder in the frequency range 200 Hz to 2 MHz at room temperature.

Also, insulation resistance was measured directly by LCR meter at the same conditions.

3. RESULTS AND DISCUSSION

3.1. Scanning Electron Microscopy (SEM)

The morphology of the SEM images for HDPE with 2% clay, 6% clay, 10% clay, and 15% clay composites is shown in figures 1-4 respectively. Each sample has two images with different magnifications. All SEM images for all samples revealed that, clay was dispersed in polymer matrix very well and there wasn't any accumulation of clay-nanofiller in it. An important observation is that the thickness of clay content is still in nano-size range (1 - 100 nm). This means that the samples were successfully prepared.

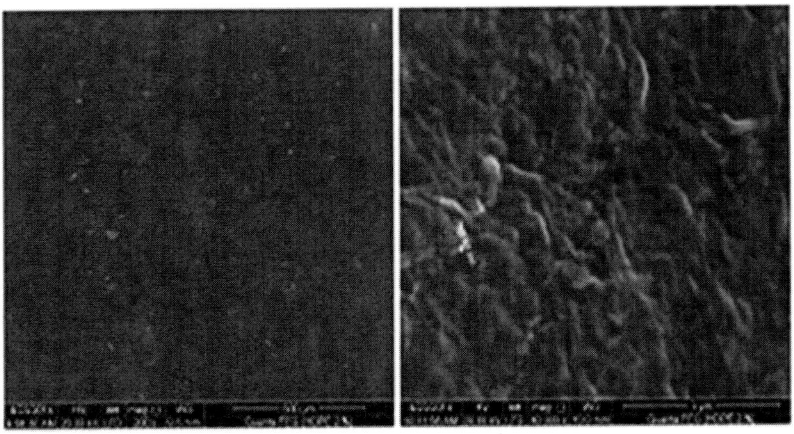

Figure 1. SEM images for HDPE 2% sample at (200× & 40,000×) magnifications.

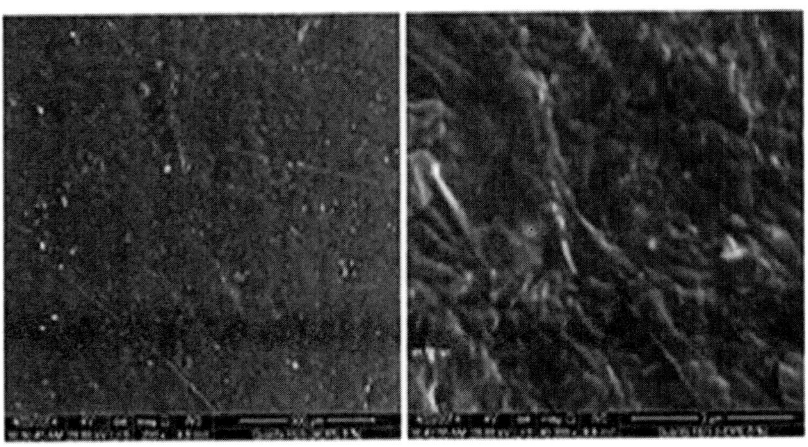

Figure 2. SEM images for HDPE 6% sample at (200× & 40,000×) magnifications.

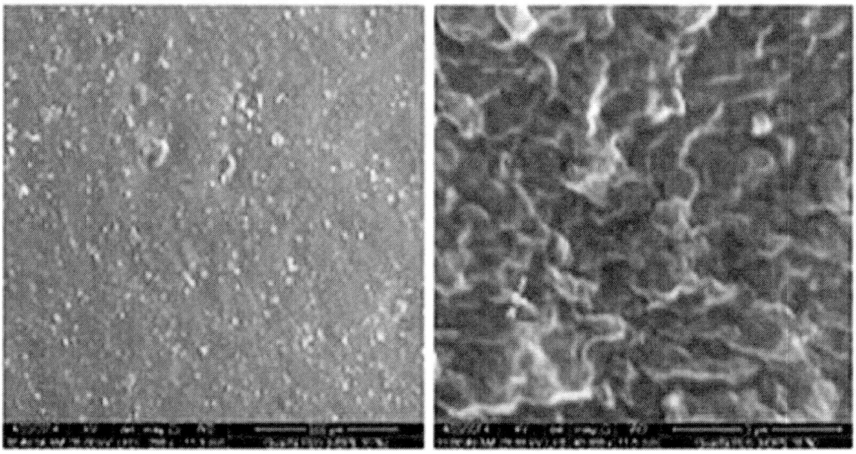

Figure 3. SEM images for HDPE 10% sample at (200× & 40,000×) magnifications.

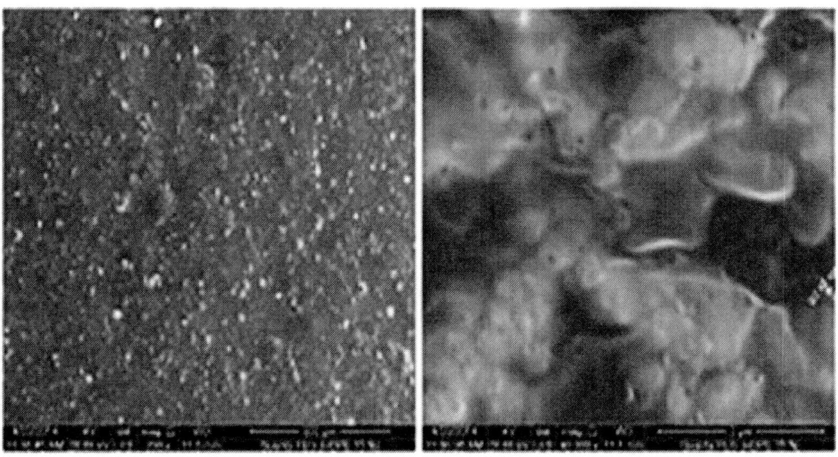

Figure 4. SEM images for HDPE 15% sample at (200× & 40,000×) magnifications.

3.2. Thermal Analysis

The thermal stability of the prepared samples was measured using thermogravimetric analyzer (TGA). In this technique, the weight loss of the material due to the formation of volatile compounds under degradation because of the heating and temperature rise is monitored.

The data available from TGA is tabulated in table 1 and graphed in figure 5 including $T_{10\%}$ (onset tempera- ture), the temperature at which 10% degradation from the sample occurs, $T_{50\%}$, the temperature at which 50% degradation occurs, T_{max}, the temperature at which maximum degradation occurs, and residual loss at 600°C.

According to TGA results as shown in figure 5, the incorporation of MMT to HDPE improved the thermal stability at higher degradation temperature ranges compared to pure HDPE. The temperature of the 10% degra- dation of HDPE 2%, HDPE 6% and HDPE 10% has been shifted to lower temperatures relative to HDPE 0%, while the 10% degradation temperature of HDPE 15% shifted to higher temperatures compared to HDPE 0%. The 50% and maximum degradation temperatures have been shifted to higher temperatures compared to HDPE 0%. This means that, thermal stability has been occurred with increasing the concentration of MMT composed to HDPE. The residual weight of the samples at 600°C increased with increasing the concentration of clay com- posed to HDPE. Thus, thermal stability of HDPE/clay has been improved compared to pure HDPE.

3.3. Dielectric Properties

3.3.1. Dielectric breakdown strength

The dielectric breakdown strength of the composites is analyzed using an AC dielectric test set at room tempera- ture. The test was repeated 5 times for each sample and the average value was recorded and plotted as shown in figure 6.

Figure 6 shows the behavior of the dielectric strength for HDPE/clay composites. Results show that the di- electric breakdown voltage increases with increasing the concentrations of clay-nanofiller to HDPE when compared to the pure material having the same dimensions until reaching to an optimum value (36.1 kV) at HDPE 6% then, the breakdown voltage starts to decrease at 10%, and 15% clay-nanofillers. Although the breakdown voltage decreases at 10% clay-nanofiller, its value is larger than the value of pure material. It is observed that the breakdown voltage value of the sample HDPE 15% is lower than that of pure material. As a result, the dielectric breakdown strength has been improved at all concentrations of clay-nanofiller except 15% clay concentration when compared to the unfilled material. The optimum enhancement occurred at HDPE 6%.

Table 1. TGA results for HDPE and HDPE/clay composites.

Samples	$T_{10\%}$ (°C)	$T_{50\%}$ (°C)	T_{max} (°C)	Residual Weight (mg) at 600°C
HDPE 0%	403.6	451.4	478.3	0.32
HDPE 2%	403.5	451.6	479.3	0.42
HDPE 6%	396.7	459.0	481.2	1.24
HDPE 10%	399.1	463.1	484.2	1.77
HDPE 15%	405.3	464.8	485.1	4.34

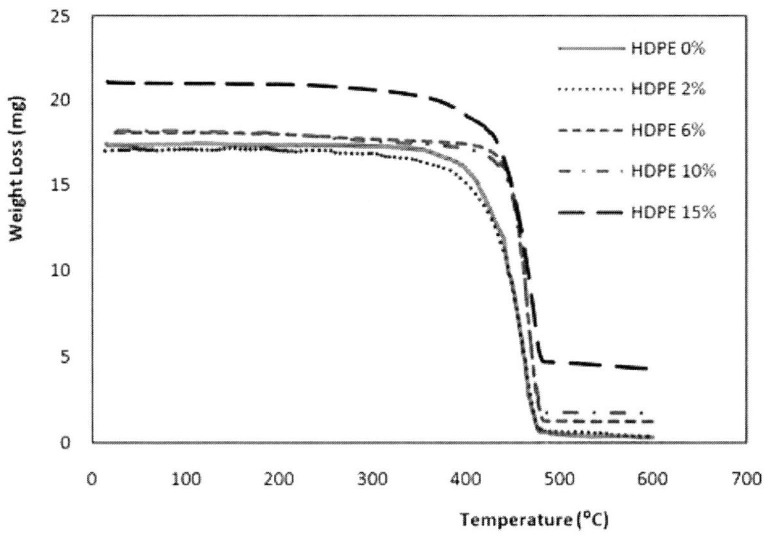

Figure 5. TGA curves for HDPE and HDPE/clay composites.

3.3.2. Dielectric constant (ε_r)

Measured quantity was the equivalent parallel capacitance (C_p) of the samples in the frequency range of 200 Hz to 2 MHz, then the dielectric constant (ε_r) was calculated by the following equations [16] and plotted as shown in figure 7.

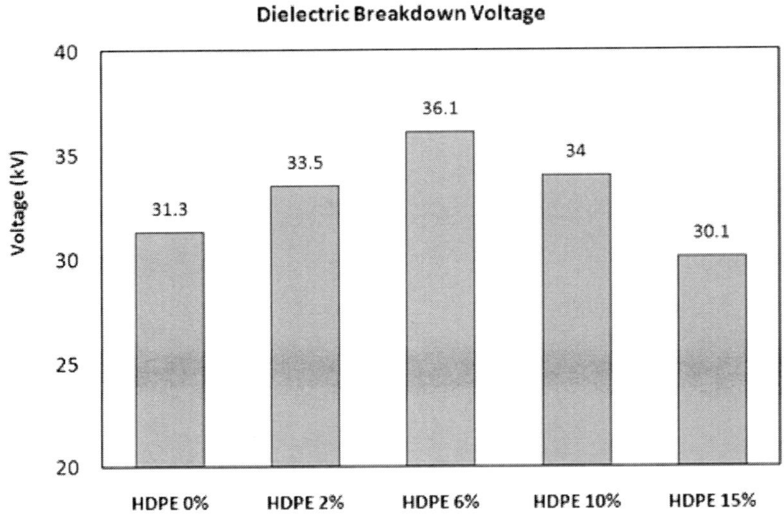

Figure 6. Dielectric breakdown strength measurement for HDPE/ clay composites.

$$C_p = \frac{\varepsilon_0 \varepsilon_r A}{t} \quad (1)$$

$$\varepsilon_r = C_p \frac{t}{\varepsilon_0 A} \quad (2)$$

where ε_0 = 8.854 × 10 - 12 F/m is the permittivity of free space, (A) is the area of electrodes, and (t) is the thickness of the samples.

Figure 7 shows the variation of the dielectric constant (ε_r) with frequency at room temperature for all samples. Observed differences were found in dielectric constant between pure HDPE and HDPE composites with different concentrations of clay-nanofiller. It is seen that, ε_r decreases with increasing frequency for all samples. An important observation is that dielectric constant decreases considerably with the addition of clay-nanofiller up to 6% filler concentration, and then it starts to increase at 10% and 15% filler concentrations. The value of ε_r at HDPE 10% is still lower than that of pure HDPE and its value at HDPE 15%, is higher than the pure material. HDPE 6% has the lowest dielectric constant and HDPE 15% has the highest dielectric constant. The increasing of ε_r at 10% and 15% filler concentrations may be due to the effect of ε_r of composites (inclusions

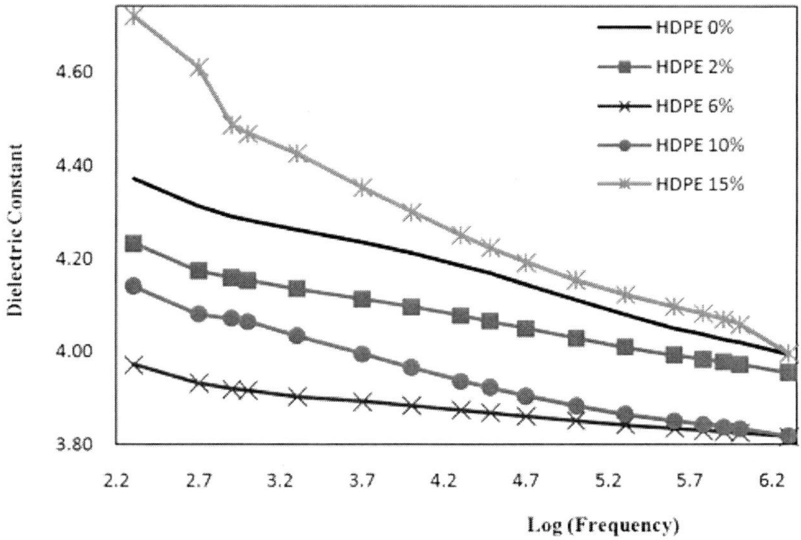

Figure 7. Frequency dependence of dielectric constant at room temperature.

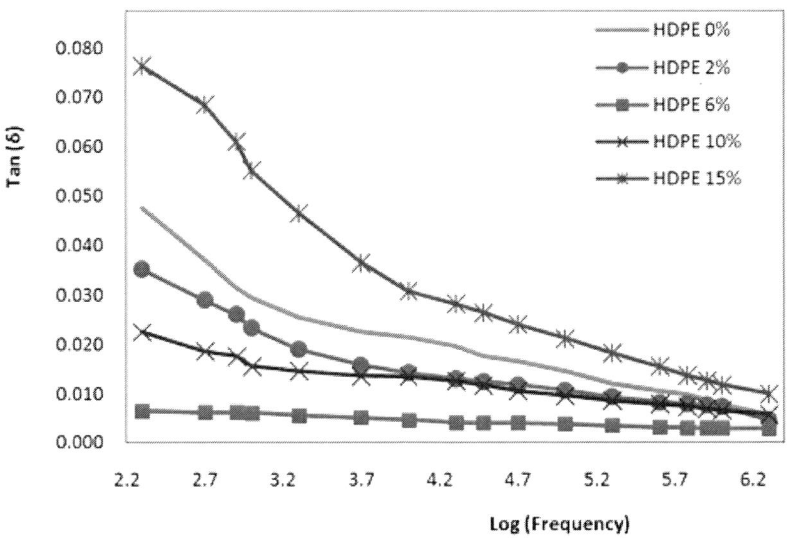

Figure 8. Frequency dependence of dissipation factor at room temperature.

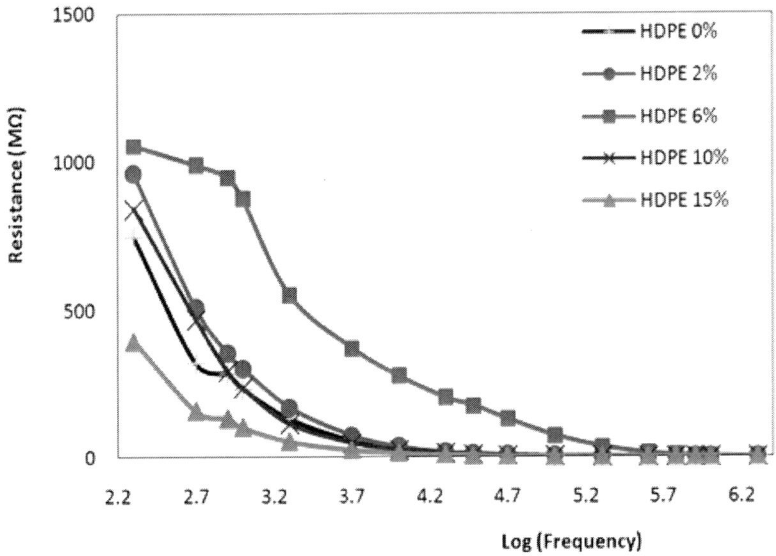

Figure 9. The insulation resistance variation with variable fre- quencies at room temperature.

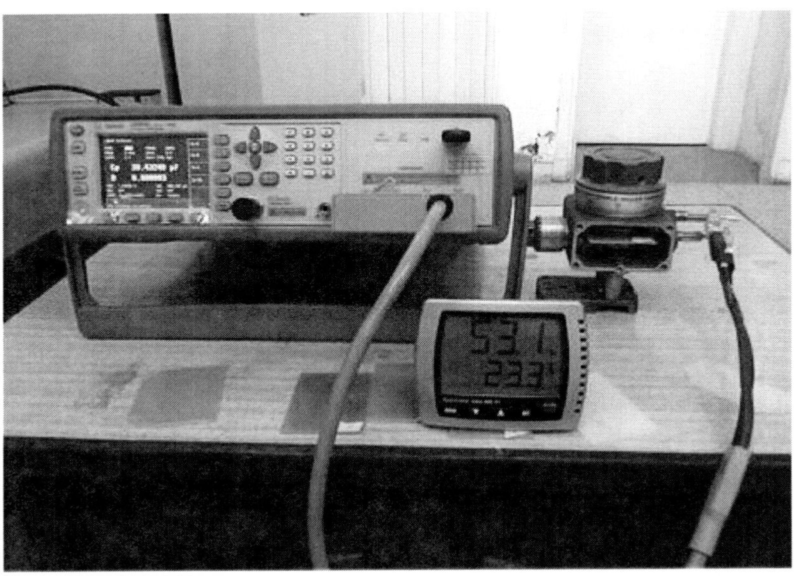

Figure 10. Agilent E4980A LCR meter with dielectric sample holder.

Figure 11. Dielectric test set for dielectric breakdown measure- ment.

+ matrix) on the resultant permittivity [17] [18] . This means that an enhancement occurred in dielectric constant at 2%, 6%, and 10% filler concentrations when these composites used as insulating materials.

3.3.3. Dissipation Factor (Tan δ)

Figure 8 shows the variation of the dissipation factor (Tan δ) with frequency at room temperature for all sam- ples. As shown in the figure, Tan δ decreases with increasing frequency for all samples. Also an important ob- servation is that Tan δ decreases with increasing the concentrations of clay-nanofiller incorporated in polymeric material up to 6% filler concentration, then it further increases at 10% and 15% filler concentrations. This may be due to the increasing of conductivity according to increasing of nano-filler concentration [17] [18] . HDPE 6% has the lowest dissipation factor and HDPE 15% has the highest dissipation factor. In addition, the values of Tan δ for HDPE 2%, HDPE 6% and HDPE 10% samples are less than that of pure material. On the other hand, the values of Tan δ for 15% filler concentrations are higher than that of pure material. This means that clay-nano- filler improves the dissipation factor for HDPE polymeric material when used for insulating purpose.

3.3.4. Insulation Resistance (R)

Figure 9 shows the variation of the insulation resistance (R) with frequency at room temperature for all samples. Marked differences were found in the insulation resistance between pure HDPE and HDPE composites with different concentrations of clay-nanofiller. It is seen that, the insulation resistance decreases with increasing fre- quency for all samples. An important observation is that the insulation resistance increases with the addition of clay-nanofiller up to 6% filler concentration, and then it starts to decrease at 10% and 15% filler concentrations. Although that, the insulation resistance value at 10% filler concentration is still higher than that of pure HDPE. On the other hand, the insulation resistance value at 15% filler concentrations is lower than that of pure sample. The decreasing of the insulation resistance at 10 % and 15% filler concentrations may be due to the increasing of conductivity of composites at high concentrations. This means that an enhancement occurred in the insulation re- sistance up to 10% filler concentration when compared to pure material.

Figure 10 and Figure 11 show the instruments which used for dielectric and breakdown measurements re- spectively.

4. CONCLUSION

HDPE/clay composites are prepared by melt compounding method (Master Batch method). Morphology struc- ture (clay dispersion in polymer matrix and the thickness of clay-nanofiller) of the prepared samples is investi- gated by SEM. SEM images show that clay content is well dispersed in the polymer matrix indicating samples are successfully prepared. Thermal stability and dielectric properties are investigated for the prepared samples. TGA results show that HDPE nanocomposites have thermal stability more than unfilled polymer material. Di- electric breakdown strength is improved by the addition of clay-nanofillers. Dielectric constant and dissipation factor are studied at room temperature in the frequency range 200 Hz to 2 MHz. The experimental results show that there is an enhancement in both ε_r and Tan δ due to the unique behavior of clay-nanofiller when incorporated into the polymer base matrix HDPE. Also insulation resistance has been improved by the addition of clay- nanofiller. From all results, it can be noticed that 6% filler concentration is the optimum clay content for HDPE/ clay system.

ACKNOWLEDGEMENTS

The authors would like to thank the Fire and Explosion Protection Lab at NIS for giving access to the TSE facil- ity used in this research work. The authors are also grateful to Prof. Dr. Mostafa, in Thermometry Lab at NIS, for his help in the TGA measurements.

REFERENCES

1. Du, B.X. and Liu, H.J. (2010) Effects of Atmospheric Pressure on Tracking Failure of Gamma-Ray Irradiated Polymer Insulating Materials. IEEE Transactions on Dielectrics and Electrical Insulation, 17, 541-547.
2. Carmona, F. (1989) Conducting Filled Polymers. Physica A, 157, 461-469. Bai, Y., Cheng, Z.-Y., Bharti, V., Xu, H.S. and Zhang, Q.M. (2000) High Dielectric-Constant Ceramic-Powder Polymer Composites. Applied Physics Letters, 76, 3804-3806.
3. Ueki, M.M. and Zanin, M. (1999) Influence of Additives on the Dielectric Strength of High-Density Polyethylene. IEEE Transactions on Dielectrics and Electrical Insulation, 6, 876-881
4. Messersmith, P.B. and Giannelis, E.P. (1994) Synthesis and Characterization of Layered Silicate-Epoxy Nanocompo- sites. Chemistry of Material, 6, 1719-1725.
5. Ray, S.S. and Okamoto, M. (2003) Polymer/Layered Silicate Nanocomposites: A Review from Preparation to Pro- cessing. Progress in Polymer Science, 28, 1539-1641.
6. Gensler, R., Groppel, P., Muhrer, V. and Muller, N. (2002) Applications of Nanoparticles in Polymers for Electronic and Electrical Engineering. Particle and Particle Systems Characterization, 19, 293-299.
7. Tanaka, T. (2005) Dielectric Nanocomposites with Insulating Properties. IEEE Transactions on Dielectrics and Electrical Insulation, 12, 914-928.
8. Cao, Y., Irwin, P.C. and Younsi, K. (2004) The Future of Nanodielectrics in the Electrical Power Industry. IEEE Transactions on Dielectrics and Electrical Insulation, 11, 797-807.

9. Imai, T., Sawa, F., Ozaki, T., Inoue, Y., Shimizu, T. and Tanaka, T. (2006) Comparison of Insulation Breakdown Pro- perties of Epoxy Nanocomposites under Homogeneous and Divergent Electric Fields. 2006 IEEE Conference on Elec- trical Insulation and Dielectric Phenomena, Kansas, 15-18 October 2006, 306-309. http://140.98.202.196/xpl/login.jsp?tp=&arnumber=4105431&url=http%3A%2F%2F140.98.202.196%2Fstamp%2Fstamp.jsp%3Ftp%3D%26arnumber%3D4105431

10. Lewis, T.J. (2004) Interfaces Are the Dominant Feature of Dielectrics at the Nanometric Level. IEEE Transactions on Dielectrics and Electrical Insulation, 11, 739-753.

11. Roy, M., Nelson, J.K., MacCrone, R.K. and Schadler, L.S. (2005) Polymer Nanocomposite Dielectrics—The Role of the Interface. IEEE Transactions on Dielectrics and Electrical Insulation, 12, 629-643.

12. Ajayan, P.M., Schadler, L.S. and Braun, P.V. (2003) Nanocomposite Science and Technology. John Wiley & Sons Inc., New York.

13. Liu, X.H. and Wu, Q.J. (2001) PP/Clay Nanocomposites Prepared by Grafting-Melt Intercalation. Elsevier Polymer, 42, 10013-10019.

14. Hassan, M.A., El-Sayed, I., Nour, M.A. and Mohamed, A.A. (2012) Flammability Properties of HDPE Nanocompo- sites Based on Modification of Na-MMT with Organo Silane and Ammonium Phosphate Mono and Di Basic. Elixir Applied Chemistry, 46, 8328-8333.

15. Li, L. (2011) Dielectric Properties of Aged Polymers and Nanocomposites. Ph.D. Thesis, Iowa State University, Iowa.

16. Singha, S. and Thomas, M.J. (2008) Permittivity and Tan Delta Characteristics of Epoxy Nanocomposites. IEEE Trans- actions on Dielectrics and Electrical Insulation, 15, 2-11.

17. Singha, S. and Thomas, M.J. (2008) Dielectric Properties of Epoxy Nanocomposites. IEEE Transactions on Dielectrics and Electrical Insulation, 15, 12-23.

7

CHAPTER

DIELECTRIC AND MICROWAVE PROPERTIES OF NATURAL RUBBER BASED NANOCOMPOSITES CONTAINING GRAPHENE

Omar A. Al-Hartomy[1,2], Ahmed Al-Ghamdi[2], Nikolay Dishovsky[3], Rossitsa Shtarkova[4], Vladimir Iliev[5], Ibrahim Mutlay[6], Farid El-Tantawy[7]

[1] *Department of Physics, Faculty of Science, University of Tabuk, Tabuk, Saudi Arabia;*
[2] *Department of Physics, Faculty of Science, King Abdulaziz University, Jeddah, Saudi Arabia;*
[3] *Department of Polymer Engineering, University of Chemical Technology and Metallurgy, Sofia, Bulgaria;*
[4] *Department of Chemistry, Technical University, Sofia, Bulgaria;*
[5] *College of Telecommunications and Posts, Sofia, Bulgaria;*
[6] *Hayzen Engineering Co., Ankara, Turkey;*
[7] *Department of Physics, Faculty of Science, Suez Canal University, Ismailia, Egypt.*

ABSTRACT

The development of carbon nanotubes based materials has been impeded by both their difficult dispersion in the polymer matrix and their high cost. The discovery of graphene and the subsequent development of graphene-based polymer nanocomposites is an important addition in the area of nanoscience and technology. In this study the influence of graphene nanoparticles (GNP) in concentrations from 2.0 to 10.0 phr on the dielectric (dielectric permittivity, dielectric loss angle tangent) and microwave (reflection coefficient, attenuation coefficient, shielding effectiveness) properties of nanocomposites on the basis of natural rubber has been investigated in the wide frequency range (1 - 12 GHz). The results achieved allow recommending graphene as a filler for natural rubber based composites to afford specific dielectric and microwave properties, especially when their loading with the much more expensive carbon nanotubes is not possible.

Keywords: Nanocomposites; Natural Rubber; Graphene; Dielectric and Microwave Properties

1. INTRODUCTION

Over the past five decades, industrial scale "composite materials" have been produced by adding different minerals and metals to thermosetting, thermoplastic and elastomeric polymers. As compared to the bulk polymers, these composites have shown moderate mechanical performance improvements in the mechanical properties such as Young's modulus, tensile strength, abrasion resistance and storage modulus. However, recent advances in nanoscale particle synthesis have definitely accelerated the growth of the composite industry. The capacity to synthesize and characterize atomic-level particles has produced a new generation of high-performance fillers and new class of materials—polymer nanocomposites.

Polymer nanocomposites are commonly defined as the combination of a polymer matrix and additives that have at least one dimension in the nanometer range. The additives can be one-dimensional, such as nanotubes and fibres, two-dimensional, which include layered clay minerals or graphene sheets, or three-dimensional, including spherical particles [1].

At present, nanocomposites employing carbon-based reinforcement materials are dominated by carbon nanotubes (CNTs) [2-4]. However, the development of CNTreinforced composites has been impeded by both their difficult dispersion in the polymer matrix and their high cost.

The discovery of graphene [5] and the subsequent development of graphene-based polymer nanocomposites is an important addition in the area of nanoscience and technology. Deservedly the Nobel Prize in Physics for 2010 was awarded to Andre Geim and Konstantin Novoselov from the University of Manchester for their work on a single free-standing atomic layer of carbon (graphene). Graphene is an allotrope of carbon, whose structure is one-atom-thick planar sheets of sp^2—bonded carbon atoms packed in a honeycomb lattice. It is the basic structural element of some carbon allotropes including graphite, charcoal, carbon nanotubes and fullerenes.

Compared to carbon nanotubes, as well as its high aspect ratio and low density graphene has attracted considerable attention because of its unique and outstanding mechanical, electrical and electronic properties [6]. In addition to good thermal conductivity, remarkable mechanical stiffness and high fracture strength, graphene has been supposed to be a semiconductor with zero gap which is quite different from conventional silicon semiconductors. In graphene, electrons shoot along with minimal resistance which may allow for low-power, fasterswitching transistor and become a candidate to replace silicon in the area of microchip electronics.

All these unique properties in a single nanomaterial have made physicists, chemists and material scientists exited about graphene's potential. The history, chemistry, preparation methods and possible applications of graphene are reviewed in [7-9]. Another new review focused on trends and frontiers in graphene-based polymer nanocomposites was published a year ago [10]. Recently, the trustees of Princeton University received a patent for graphene-elastomer nanocomposites where functionalized graphene sheets (FGS) had been dispersed in vulcanized natural rubber, styrene butadiene rubber, Psisoprene-Ps and PDMS [11]. The patented work could find a wide range of industrial applications, including food packaging, gasketing and automotive. The authors conclude that graphene-rubber nanocomposites possess qualities like those of carbon nanotube composites but are much cheaper to make.

The data about graphene influence on the microwave properties of the elastomeric composites are very scarce. Y. Chen et al. used functionalized graphene-epoxy composites as lightweight shielding materials for electromagnetic radiation [12]. I. M. De Rosa and coworkers have a wide expertise in the design of micro/nanocomposites based on carbon fibers and carbon nanotubes, for the realization of high performing radar absorbing screens, with tailored properties [13-15]. In a recent paper [16], the authors have accomplished a Salisbury screen, that consists of three layers. The second layer (the spacer) is a low-loss-tangent nanocomposite based on a Bisphenol-A based epoxy resin filled with GNPs at 0, 5% and 1% wt [16]. The real and imaginary parts of the complex effective permittivity within the 8 - 18 GHz range of the nanocomposite filled with GNPs have been shown. It has been observed that the real part of the effective permittivity is nearly constant.

There are no literature data about the dielectric and microwave properties of nanocomposites comprising a higher amount of graphene nanoparticles (GNP), e.g. over 1 phr. Neither there are available reports on investigations on those properties carried out in a wider frequency range, first of all at frequencies lower than 8 GHz. Therefore the aim of this work is to study the influence that graphene nanoparticles (in amounts of 2 to 10 phr) have on the dielectric (dielectric permittivity, dielectric loss angle tangent) as well as on the microwave properties (absorption and reflection of the electromagnetic waves, the effectiveness of the electromagnetic shielding) of natural rubber based composites in a significantly greater frequency range—from 1 to 12 GHz.

2. EXPERIMENTAL

2.1. Characterisation of the Graphene Used

Graphene as produced by Hayzen Engineering Co., Ankara, Turkey was used in our investigation. Graphene nanoplatelets (GNP) have a "platelet" morphology, meaning they have a very thin but wide aspect. Aspect ratios for this material can range into the thousands. Each particle consists of several sheets of graphene with an overall thickness of 50 nm and average plate diameter 40 micron.

2.2. Prepapation and Vulcanization of Rubber Compounds

Table 1 summarizes the formulation characteristics of the rubber compounds (in phr) used for the investigations.

The rubber compounds were prepared on an open tworoll laboratory mill (L/D 320 × 360 and friction 1, 27) by incorporating pre-characterized graphene nanoplatelets (GNP) into a natural rubber matrix at various loadings (Table 1). The speed of the slow roll was 25 min^{-1}. The experiments were repeated for verifying the statistical significance. The ready compounds in the form of sheets stayed 24 hours prior to their vulcanization.

The optimal vulcanization time was determined by the vulcanization isotherms, taken on an oscillating disc vulcameter MDR 2000 (Alpha Technologies) at 150°C according to ISO 3417:2002.

These composites were evaluated for their dielectric (dielectric permittivity, dielectric loss angle tangent) and microwave (reflection coefficient, attenuation coefficient, and shielding effectiveness) properties in the 1 - 12 GHz frequency range.

Table 1. Composition of the rubber compounds studied.

	NR 1	NR 2	NR 3	NR 4
Natural rubber	100	100	100	100
Foaming agent	8	8	8	8
Stearic acid	1	1	1	1
ZnO	4	4	4	4
Processing oil	10	10	10	10
GNP	0	2	6	10
MBTS	2	2	2	2
TMTD	1	1	1	1
IPPD 4020	1	1	1	1
Sulphur	2	2	2	2

3. MEASUREMENTS

3.1. Microwave Properties

3.1.1. Reflection and Attenuation

Measurements of reflection and attenuation were carried out using the measurement of output (adopted) power P_a in the output of a measuring line without losses, where samples of materials may be included. Because of the wide frequency measurement a coaxial line was used. Samples of the materials were shaped into discs with an external diameter D = 20.6 mm, equal to the outer diameter of the coaxial line and thickness of $\Delta \approx 2$ mm. The internal diameter depended on the relative dielectric permittivity of the material.

The sample reflected a part of the incident electromagnetic wave with power P_{in}. The rest of the wave with power P_p penetrated the material, so that the attenuation L depended on the coefficient of reflection $|\Gamma|$. Its module was determined by a reflect meter.

Thus the attenuation was determined by

$$L = 10\log\frac{P_a}{P_p}, dB \qquad (1)$$

where

$$P_p = P_{in.}\left(1 - |\Gamma|^2\right) \qquad (2)$$

The following scheme presents the equipment used for testing both parameters (Figure 1)

1) A set of generators for the whole range: HP686A and G4 - 79 to 82;

2) Coaxial section of the deck E2M Orion, with samples of material;

3) Power meter HP432A;

4) Scalar reflectance meter HP416A; R—Reflect meter, including:

—Two directional couplers Narda 4222.16;

—Two crystal detectors Narda 4503-N.

3.1.2. Shielding Effectiveness (S.E.)

This parameter is defined as the sum of the reflection losses R, dB and attenuation L, dB in the material.

It can be directly measured or calculated from the measured reflectance and attenuation in the material. In the first case, as measured: incident power on the sample

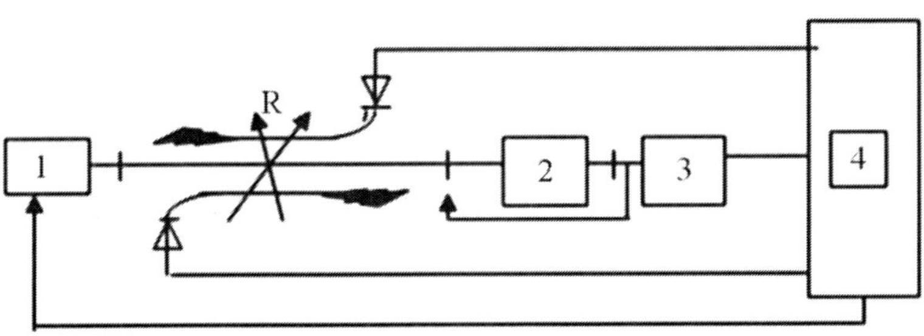

Figure 1. Scheme of the equipment for measuring the microwave properties.

P_{in} and adopted after the sample P_a (Figure 1), S.E. is determined by

$$\text{S.E.} = 10\log\frac{P_0}{P_a}, dB \tag{3}$$

In the second, if known reflection and absorption in the material, S.E. is determined, by a definition, as

$$\text{S.E.}, dB = -20\lg|\Gamma| + L, dB \tag{4}$$

where R, dB is the attenuation due to the reflection of power at the interfaces.

In the present work the shielding effectiveness was determined by Equation (4).

3.2. Dielectric Properties

3.2.1. Complex permittivity

The determination of complex permittivity was carried out by the resonance method, based on the cavity perturbation technique.

Having measured the resonance frequency of the empty cavity resonator f_r a measurement of the shift in resonance frequency f_ε was carried out in the presence of the sample material. Then the dielectric constant ε_r was calculated from the shift in resonance frequency, cavity and the sample cross sections S_r and S_ε, respectively

$$\varepsilon_r = 1 + \frac{S_r}{2S_\varepsilon} \cdot \frac{f_r - f_\varepsilon}{f_r} \, . \tag{5}$$

The sample was in the form of a disc with a diameter of 10 mm and about 2 mm thick. It was placed at the maximum electric field location of the cavity. Because the thickness of the sample was not equal to the height of the resonator, a dielectric occurred with an equivalent permittivity e_e at the place of its inclusion. The parameter was determined by (5) and instead e_r be saved e_e. Then e_r was determined by

$$\varepsilon_r \approx \varepsilon_e(k+1) - k, (\Delta \ll l) \, , \tag{6}$$

where $k = l/\Delta$ and l is the distance from the disc to the top of the resonator.

3.2.2. Loss Factor tanδ

The loss factor tanδ was calculated from the quality factor of the cavity with Q_ε and without sample Q_r

$$\tan\delta = \frac{1}{4\varepsilon_r}\frac{S_r}{S_\varepsilon}\left(\frac{1}{Q_\varepsilon}-\frac{1}{Q_r}\right) \quad .(7)$$

The measurement setup used several cavity resonators for the whole range, generators for the whole range, frequency meter and oscilloscope.

The following scheme presents the set used for measuring the dielectric properties (Figure 2):

1) Generators for the whole range: HP686A and G4 - 79 to 82;

2) Frequency meters: H 532A; FS - 54;

3) Cavity resonator;

4) Sample;

5) Oscilloscope EO 213.

4. RESULTS AND DISCUSSION

4.1. Dielectric Properties

Figures 3 and 4 present the dependence of relative dielectric permittivity (e_r) and dielectric loss angle tangent (tanδ) on frequency and filler concentration for the studied composites filled with graphene nanoparticles.

4.1.1. Complex Relative Dielectric Permittivity. Real Part

As seen from the plots relative dielectric permittivity increases with the increasing frequency at constant filler concentration as well as with the increasing filler concentration at constant frequency.

The real part of dielectric permittivity is the ratio of the capacity of an electric capacitor filled with the substance investigated to that of the same capacitor in vacuum, at a definite external field of frequency [17]. The dependence of the real part of the complex relative dielectric permittivity on the frequency e_r (f) is shown in Figure 3. As seen in the concentration interval studied dielectric permittivity values for the non-filled and filled composites are relatively close at lower frequencies (up to 7 GHz). At frequencies higher than 7 GHz the dielectric permittivity increases with the increasing filler concentration. Moreover, there is a tendency of a more pronounced difference in the values for the non-filled and filled composites. The values we obtained for the nonfilled composites are close to the ones for natural rubber at 1000 Hz reported in literature which are 2.40 - 2.70 [18]. As the figure shows our investigations also confirm the fact that the dielectric permittivity increases at higher frequencies. Of particular interest is the 9 - 12 GHz range wherein there is a relatively fast increase in the dielectric permittivity and its dependence on the GNP is the most prominent. Having in mind that the dielectric permittivity is related to the composites polarity [17], probably at frequencies higher than 9 GHz the polarization of the natural rubber matrix is hampered. Hence, the dielectric

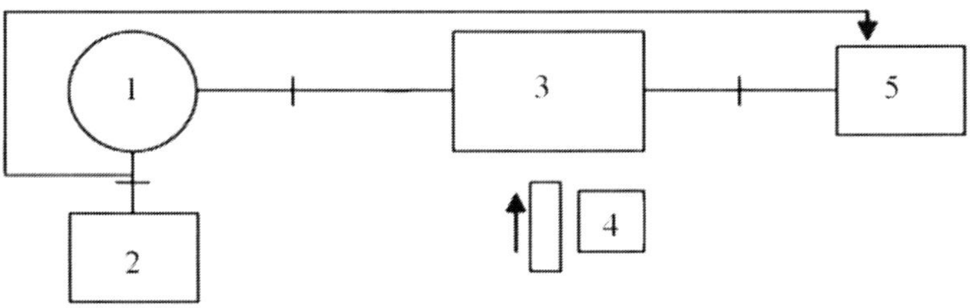

Figure 2. Scheme of the equipment for measuring the dielectric properties.

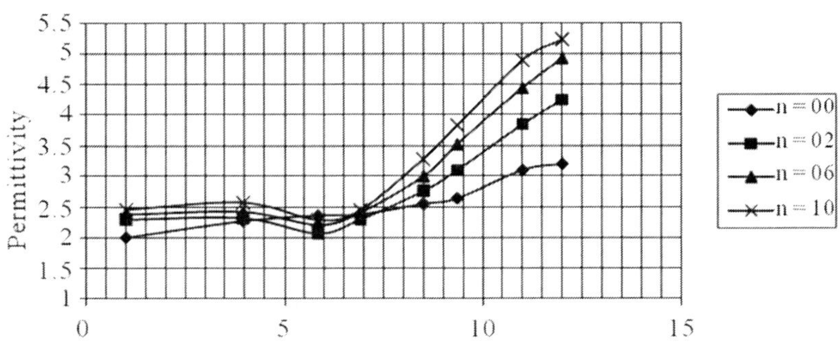

Figure 3. Frequency dependence of relative dielectric permittivity e_r at various filler content (n-phr of graphene).

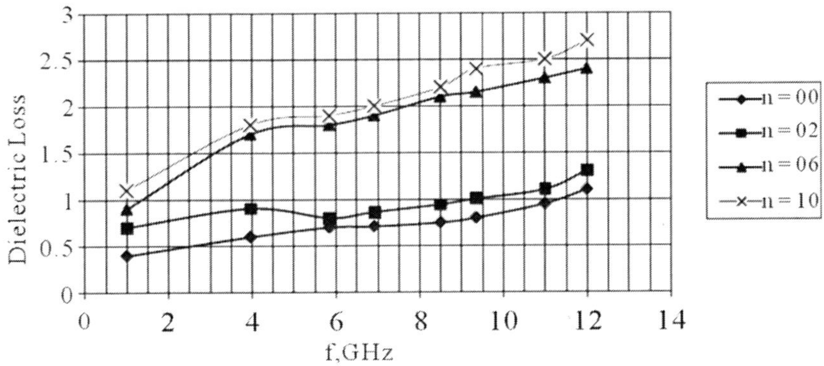

Figure 4. Frequency dependence of dielectric loss tanδ$_\varepsilon$ at various filler content (n-phr of graphene).

permittivity increases. On the other hand, the increase in e_r proves the lower polarization of the composite at higher frequencies.

The figure allows to calculate that the maximum difference within the range of 2 units is between the nonfilled composite and the one comprising 10 phr GNP at 12 GHz. Besides, at frequencies higher than 7 GHz the frequency dependence of dielectric permittivity for the filled composites is much closer to the linear one.

4.1.2. Complex Relative Dielectric Permittivity. Imaginary part (Dielectric loss)

Figure 4 shows the frequency dependence of the imaginary part of the relative complex dielectric permittivity, also known as the dielectric loss angle tangent (tanδ$_\varepsilon$) in the 4 - 12 GHz range. As expected, the increase in frequency and filler amount leads to an increase in tanδ$_\varepsilon$ values. The filler does not change the character of this dependence since it exists for the non-filled composite as well. Noteworthy is the fact that the addition of 2, 0 phr GNP has a relatively slight effect upon tanδ$_\varepsilon$ values of the composite which are close to the ones of the matrix. However, tanδ$_\varepsilon$ increases considerably at 6, 0 phr filler concentration. It is also seen that the frequency dependence of tanδ$_\varepsilon$ both for the non-filled composite and the filled ones has an almost linear character. In the case of composites comprising a higher filler amount the frequency dependence of this parameter is more marked.

4.2. Microwave Properties

Figures 5-7 summarize the microwave properties of the composites studied within the 1 - 12 GHz frequency range.

4.2.1. Coefficient of reflection

The coefficient of reflection is a complex number $\dot{\Gamma}$. For the aims of the present study it is enough to estimate its modulus |Γ| that is the reflected part of the electromagnetic wave falling upon the sample. That allows determining the absorption properties of the tested material. Figure 5 shows the frequency dependence of |Γ|. As seen the higher the frequency and filler amount the higher |Γ| is. The figure also reveals that |Γ| values do not exceed 0.60 within the entire frequency range. A connection should be made between the dielectric losses mentioned above (Figure 4) which smoothly increase, though in a relatively small range. That to a great extent explains the small changes in the attenuation coefficient. One should keep in mind the fact that the composite comprises a foaming agent making its structure porous what lowers its attenuation coefficient values. As a whole the attenuation coefficient slightly increases with the increasing frequency and filler amount. The increase is particularly in the 6 - 12 GHz region. In the 1 - 6 GHz region the effect of frequency and filler amount on the attenuation coefficient is less pronounced. The effect is in accordance with the one upon the real part of the complex dielectric permittivity (Figure 3).

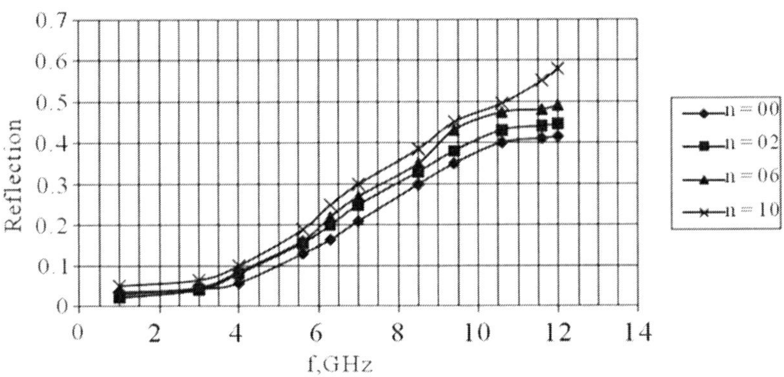

Figure 5. Frequency dependence of reflection |Γ| at various filler concentrations (n-phr of graphene).

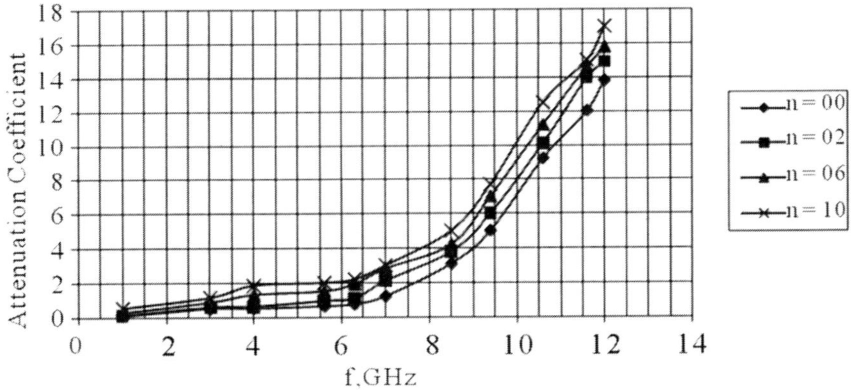

Figure 6. Frequency dependence of attenuation coefficient α of electromagnetic waves at various filler concentrations (nphr of graphene).

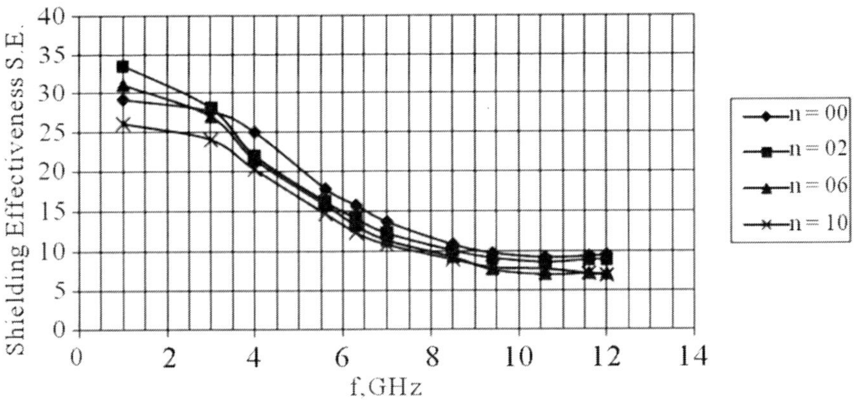

Figure 7. Frequency dependence of shielding effectiveness at various filler content (n-phr of graphene)

4.2.2. Attenuation Coefficient

The frequency dependence of the attenuation coefficient (α) could be studied in three frequency ranges: 1) 1 - 6 GHz; 2) 6 - 9 GHz and 3) 9 - 12 GHz (Figure 6). In the first range α £ 2.2 dB/cm; in the second 2.2 £ α £ 6 dB/cm and in the third 6 £ α £ 17. As the figures show in the first frequency range the attenuation remains almost unchanged, especially for the non-filled composites and for those with a small amount of filler.

There is a statistically significant attenuation increase with the increasing frequency and filler amount. This increase is considerable in the third frequency range especially for the filled composites. The effect is less pronounced for the non-filled ones. The comparison of the said dependence with those of the reflection coefficient and dielectric permittivity reveals a similarity that is obviously due to the same factor (the structure of grapheme and its specifics). It should be noted that the maximum attenuation of 17 dB/cm at 12 GHz is for a composite comprising 10 phr GNP.

4.2.3. Shielding Effectiveness

Figure 7 plots the frequency dependence of the electromagnetic shielding effectiveness (S.E.). At lower frequencies S.E. values are higher but decrease with the increasing frequency. In the (1 - 7) GHz the dependence has a character close to the linear one. There is also a range (7 - 12 GHz) of a relatively small effect of frequency and filler amount on S.E. values which remain in the 7 и 10 dB range. The initial values are due mainly to the return loss which are of interference nature. At the highest frequencies in the said range the attenuation in the sample is not great enough to compensate the increase in the reflection coefficient (Figure 5) and (Figure 6), hence the shielding effectiveness decreases.

5. CONCLUSIONS

1) The effect that graphene nanoparticles have upon the dielectric and microwave properties of natural rubber based composites filled at 2.0 up to 10.0 phr has been studied in the 1 - 12 GHz range.

2) It has been found that the dielectric permittivity increases slightly with the increasing frequency and filler amount. The effect is pronounced especially in the 9 - 12 GHz range. The tendency is the same in the case of dielectric loss angle tangent, although the impact of the filler amount is less marked, i.e. the imaginary part of the complex relative dielectric permittivity is less sensitive to alternations of the filler amount and frequency than the real part.

3) The reflection coefficient increases with the increasing filler amount and frequency. The effects are not well pronounced. The availability of a foaming agent contributes to the low reflection coefficient.

4) The attenuation coefficient has been found to be almost frequency and filler amount independent in the 1 - 6 GHz range, then in the 6 - 9 GHz range it increases slightly, while in the 9 - 12 GHz range its values increase

drastically with the increasing frequency and filler amount. The dependence is almost a linear one.

5) The frequency and filler amount have an effect upon the reflection and attenuation coefficients as well as upon the electromagnetic shielding effectiveness. S.E. values are in the 10 - 34 dB range. The attenuation is not great enough to compensate the increasing reflectance. Therefore as a whole the shielding effectiveness decreases gradually with the increasing frequency, especially at frequencies up to 7 GHz.

6) The effect of graphene nanoparticles upon the dielectric and microwave properties of the composites studied occurs mainly in the 6 - 12 GHz range at a minimal filler amount of 6 phr.

7) The results achieved allow recommending graphene as a filler for natural rubber based composites to afford specific dielectric and microwave properties, especially when their loading with the much more expensive carbon nanotubes is not possible.

6. ACKNOWLEDGEMENTS

The present research is a result of an international collaboration program between the University of Tabuk, Tabuk 71491, Kingdom of Saudi Arabia and the University of Chemical Technology and Metallurgy, Sofia, Bulgaria. The authors gratefully acknowledge the financial support from the University of Tabuk.

REFERENCES

1. S. Thomas and R. Stephen, "Rubber Nanocomposites Preparation, Properties and Applications," Wiley, Singapore, 2010.
2. P. M. Ajayan, "Single-Walled Carbon Nanotube-Polymer Composites: Strength and Weakness," Advanced Materials, Vol. 12, No.10, 2000, pp. 750-753.
3. E. T. Thostenson, "Advances in the science and technology of Carbon Nanotubes and Their Composites: A Review," Composites Science and Technology, Vol. 61, No. 13, 2001, pp. 1899-1912.

4. A. Krueger, "Carbon Materials and Nanotechnology," Wiley, New York, 2010.

5. K. S. Novoselov, "Electric Field Effect in Atomically Thin Carbon Films," Science, Vol. 306, No. 5696, 2004, pp. 666-669.

6. H. J. Salavagione, G. Martínez and G. Ellis, "GrapheneBased Polymer Nanocomposites" in: S. Mikhailov, Ed., Physics and Applications of Graphene—Experiments, InTech, Rijeka, 2011, pp. 169-192.

7. M. J. Allen, V. C. Tung and R. B. Kaner, "Honeycomb carbon: A Review of Graphene," Chemical Reviews, Vol. 110, No. 1, 2010, pp. 132-145.

8. Y. Zhu, S. Murali, W. Cai, X. Li, J. W. Suk, J. R. Potts and R. S. Ruoff, "Graphene and Graphene Oxide: Synthesis, Properties, and Applications," Advanced Materials, Vol. 22, No. 35, 2010, pp. 1-19.

9. O. C. Compton and S. T. Nguyen, "Graphene Oxide, Highly Reduced Graphene Oxide, and Graphene: Versatile Building Blocks for Carbon-Based Materials," Small, Vol. 6, No. 6, 2010, pp. 711-723.

10. P. Mukhopadhyay and R. K. Gupta, "Trend and Frontiers in Graphene-Based Polymer Nanocomposites," Plastics Engineering, 2011. www.4spe.org

11. R. K. Prud'Homme, B. Ozbas, I. Aksay, R. Register and D. Adamson, "Functional Graphene-Rubber Nanocomposites," US Patent No. 7745528, 2010.

12. J. Liang, Y. Huang, Y. Ma, Z. Liu, J. Cai, C. Zhang, H. Gao and Y. Chen, "Electromagnetic Interference Shielding of Graphene/Epoxy Composites," Carbon, Vol. 47, No. 3, 2009, pp. 922-925.

13. I. M. De Rosa, F. Sarasini, M. S. Sarto and A. Tamburrano, "EMC impact of Advanced Carbon Fiber/Carbon Nanotube Reinforced Composites for Next Generation Aerospace Applications," IEEE Transactions on Electromagnetic Compatibility, Vol. 50, No. 3, 2008, pp. 556- 563.

14. I. M. De Rosa, R. Mancinelli, F. Sarasini, M. S. Sarto and A. Tamburrano, " Electromagnetic Design and Realization of Innovative Fiber-Reinforced Broad-Band Absorbing Screens," IEEE Transaction on Electromagnetic Compatibility, Vol. 51, No. 3, 2009 pp. 700-707.

15. I. M. De Rosa, A. Dinescu, F. Sarasini, M. S. Sarto and A. Tamburrano, "Effect of Short Carbon Fibers and MWCNTs on Microwave Absorbing Properties of Polyester Composites Containing Nickel-Coated Carbon Fibers," Composites Science and Technology, Vol. 70, No. 1, 2010, pp. 102-109.

16. G. De Bellis, I. M. De Rosa, A. Dinescu, M. S. Sarto and A. Tamburrano, "Electromagnetic Absorbing Nanocomposites Including Carbon Fibers, Nanotubes and Graphene Nanoplatelets," Proceedings of the 2010 IEEE International Symposium on Electromagnetic Compatibility, Fort Lauderdale, 25-30 July 2010, pp. 202-207.

17. P. Banerjee and S. Biswas, "Dielectric Properties of EVA Rubber Composites at Microwave Frequencies," Journal of Microwave Power and Electromagnetic Energy, Vol. 45, No. 1, 2011, pp. 24-29.

18. A. Kornev, A. Bukanov and O. Sheverdiaev, "Technology of Elastomeric Materials," in Russian, Istek, Moscow, 2005.

8 CHAPTER

LOW DIELECTRIC MATERIALS FOR MICROELECTRONICS

He Seung Lee, Albert S. Lee, Kyung-Youl Baek and Seung Sang Hwang

[1] *Center for Materials Architecturing, Korea Institute of Science Technology, Seoul, South Korea*

1. INTRODUCTION

Over the past half century, low dielectric materials have beenintensively researched by ceramic and polymer scientists. However, these materials possess a vast myriad of electrical, thermal, chemical, and mechanical properties that are just as crucial as the name that classifies them. Therefore, in many cases, the applications of low dielectric constant materials are dictated by these other properties, and the choice of low dielectric material may have a tremendous effect on a device's performance and lifetime.

In the field of microelectronics, many of the early low dielectric materials have been satisfactory in covering the required properties. But as the microelectronics industry continuously boomed through the 21st century, more and more advanced processes and materials have been in demand. Since the invention of microprocessor, the number of active devices on a chip has been exponentially increasing, approximately doubling every year, famously forecast by Gordon Moore in 1965. All of this is driven by the need for optimal electrical and functional performance.

Figure 1 showsthe shrinking of the device dimensions over signal delay value. And while the total capacitance can be traded for resistance and vice versa by changing the geometry of the wire cross-section, theRCwill always increase for future nodes. In other words, in order to enhance performance, decreasing the device size, as well as decreasing the interconnecting wire distance, gate and interconnect signals delay is the main challenge for ceramic and polymer scientists to overcome. In another approach to solve this RC delay problem, researchers have already changed the aluminum line to Cu line, which has lower resistance. But due to limitations in metal lines being applicable for use, research of low dielectric materials are continually being pursued today. The main challenge for researchers in the microelectronic industry is not to develop materials with the lowest dielectricconstant, but to find materials that satisfy all of the electrical, thermal, chemical, and mechanical properties required for optimal device performance.

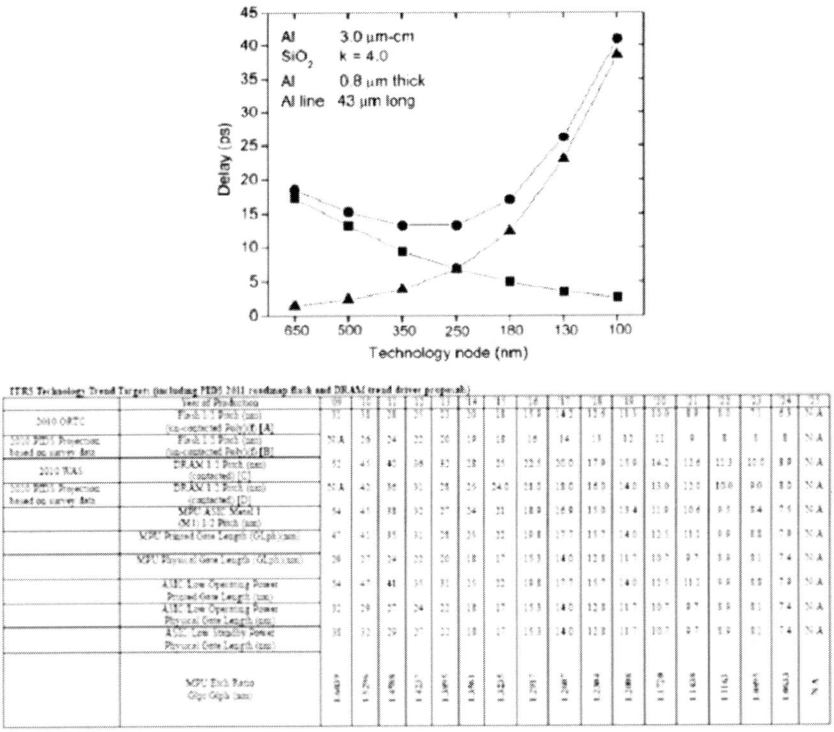

Figure 1. Calculated gate and interconnect dely as a function of technology node according to the National Technology Roadmap for Semiconductores(NTRS) in 1997 (top): █ gate delay; ▲ interconnect delay (Al and SiO_2); • sum of delays (Al and SiO_2) and ITRS technology trend targets (bottom)

2. DEFINITION OF DIELECTRIC CONSTANT

Dielectric constant k (also called relative permittivity εr) isthe ratio of the permittivity of a substance to that of freespace. A material containing polar components, such as polar chemical bonds, which arepresented as electric dipoles in Figure 2,has an elevated dielectric constant, in which the electrical dipoles alignunder an external electric field. This alignment of dipoles adds to the electric field. As a result, a capacitor witha dielectric medium of higher k will hold more electric chargeat the same applied voltage or, in other words, itscapacitance will be higher. The dipole formation is a result ofelectronic polarization (displacement of electrons), distortionpolarization (displacement of ions), or orientationpolarization (displacement of molecules) in an alternatingelectric field. These phenomena have characteristicdependencies on the frequency of the alternating electricfield, giving rise to a change in the real and imaginary part ofthe dielectric constant between the microwave, ultraviolet,and optical frequency range.

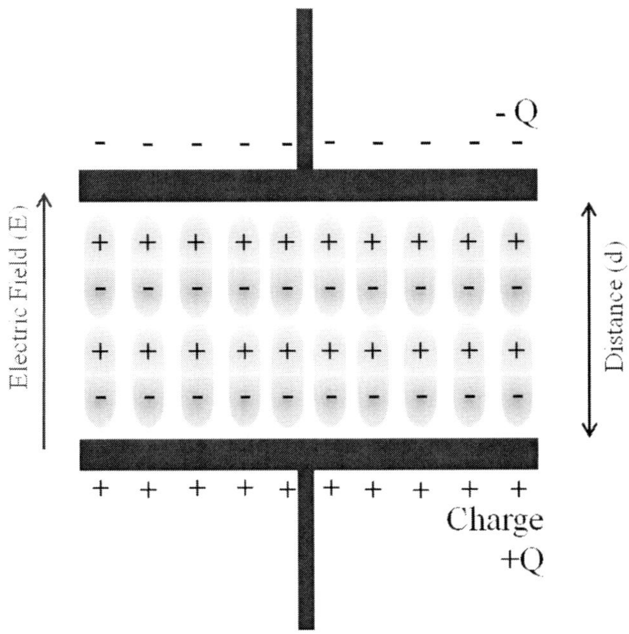

Figure 2. Schematic illustrationof a capacitor.

3. PROPERTY REQUIREMENTS OF LOW DIELECTRIC MATERIAL

Dielectric materials must meet stringent material property requirements for successful integration into the interconnect structures. These requirements are based on electrical properties, thermal stability, thermomechanicaland thermal stress properties, and chemical stability. The desired electrical properties can be outlined as low dielectric constant, low dielectric loss and leakage current, and high breakdown voltage. As RC delay and crosstalk are primarily determined by the dielectric constant, in a typical CVD SiO_2 film, the dielectric constant is around 4. And although many polymeric materials satisfy these electrical criteria, the dimensional stability, thermal and chemical stability, mechanical strength, and thermal conductivity of polymers are inferior to those of SiO_2.

Table 1. Property Requirements of Low-k Dielectrics

Electrical	Chemical	Mechanical	Thermal
Dielectirc constant Anisotropy Low dissipation Low leakage current Low charge trapping High electric-field strength High reliability	Chemical resistance Etch selectivity Low moisture uptake Low solubility in H_2O Low gas permeability High purity No metal corrosion Long storage life Enviromentally safe	Thickness uniformly Good adhesion Low stress High hardness Low shrinkage Crack resistance High tensile modulus	High thermal stability Low coefficient of thermal expansion Low thermal weight loss High thermal conductivity

In the fabrication of the multilevel structures, as many as 10 to 15 temperature treatments are repeated at elevated temperatures exceeding 400-425°C. This inherent processing of inter-dielectric (ILD) materials makes thermal stability a key prerequisite of low dielectric materials in microelectronics. Not only is the thermal stability in terms of degradation key, but the insensitivity to thermal history may be just as important. For example, changes in the crystallinity and/or crystalline phases during these thermal cycles may cause changes in the electrical and/or mechanical properties, making the material dependent on its thermal history. Other problems seldom seen in thermal processing include outgassing of volatile solvents and/or decomposition products which may cause poisoning, delamination, blistering, or cracking in the ILD.

Another thermomechanical concern of ILD materials is its coefficient of thermal expansion (CTE). The extensive thermal cycling of microelectronics may also cause stresses in the interconnect structure if there is a CTEmismatch between the ILD material and the metal or substrate. These stresses invariably cause delamination if adhesion is poor. And while adhesion promoters may be added to enhance wetting and chemical bonding at the interface between the ILD and substrate, this is mostly undesired from manufacturing point of view, as it adds unnecessary processing steps. Also, if the adhesion promoter thermally degrades, it may lead to adhesion failures or create a leakage path.

Adhesion is determined by chemical bonding at the metal/ILD interface and the mechanical interaction between the metal and ILD. Thus, ideal ILDs should have good mechanical properties such as a large Young's modulus (E), tensile strength, and elongation-at-break. Andalthough it is uncertain what constitutes sufficient mechanical strength for successful integration into a manufacturable process, the elongation-at-break should be as large as possible to sustain the deformation and impart crack resistance, even at elevated temperatures. Also, a high modulus retention at elevated temperatures, E(T), is required for the ILD to maintain its structural integrity and dimensional stability during subsequent processing steps. Related to E(T) is the glass transition temperature, Tg. Since exceeding the Tg causes a large decrease in the modulus and yield stress in amorphous, non-crosslinked polymers, a Tg greater or equal to the highest processing temperature is desired. For example, residual compressive stresses in capping layers can cause buckling and delamination of the capping films due to the compliance of an ILD above its Tg [1,2]. Buckling has also been observed in capping layers deposited below the ILD's Tg if the capping film is highly compressive [3].

Other processing concerns include chemical resistance to the solvents and etchants commonly used during chip fabrication, chemical interaction with the metal lines causing corrosion, and moisture uptake. Moisture is a primary concern because even trace amounts can have a detrimental impact on the dielectric constant. The ILDs should also be free of trace metal contaminants, have long shelf-lives, and, preferably, not require refrigeration. Metal contamination, which can compromise the device and provide a leakage path between lines, is often a problem for polymers synthesized using metal catalysts. Other processing requirements include the ability to pattern and etch the film, etch selectivity to resists, good thickness uniformity, gap-fill in submicron trenches, and planarization.

The long-term reliability of chips fabricated using low-k materials must also be evaluated. Electromigration and stress voiding are primary failure mechanisms in integrated circuits [4-6] and these are reliability concerns when replacing SiO_2 with an alternative ILD that has thermal and mechanical properties inferior to those of SiO_2.

4. DESIGN OF LOW DIELECTRIC MATERIAL

There are two strategies for designing a low dielectric material: decreasingdipole strength or the number of dipoles (Fig. 4) or a combination of both. In the first strategy, materials with chemical bonds of lower polarizability than Si-O or lower density would be used. Today, the microelectronics industry has already moved to certain low-k materials, where somesilica Si-O bonds have been replaced with less polar Si-F orSi-C bonds. A more elementary reduction of the polarizability can be attained by utilizing all nonpolar bonds, such as C-C or C-H, as in the case of organic polymers.

Table 2. Electronic polarizability [7] and bond enthalpieds [8]

Bond	Polarizability (Å³)	Average bond energy (Kcal/mole)
C - C	0.531	83
C - F	0.555	116
C - O	0.584	84
C - H	0.652	99
O - H	0.706	102
C=O	1.020	176
C=C	1.643	146
C≡C	2.036	200
C≡N	2.239	213

[a] Reference [7]
[b] Reference [8]

The second strategy involves decreasing the number of dipoles within the ILD material by effectively decreasing the density of a material. This can be achieved by increasingthe free volume through rearranging the material structure orintroducing porosity. Porosity can be constitutive orsubtractive. Constitutive porosity refers to the self-organizationof a material. After manufacturing, such amaterial is porous without any additional treatment.Constitutive porosity is relatively low (usually less than 15%)and pore sizes are ~ 1 nm in diameter. According toInternational Union of Pure and Applied Chemistry (IUPAC)classification[9], pores less than 2 nm are denoted 'micropores'.Subtractive porosity involves selective removal of part ofthe material. This can be achieved via an artificially addedingredient (e.g. a thermally degradable substance called a'porogen', which is removed by annealing to leave behindpores) or by selective etching (e.g. Si-O bonds in SiOCHmaterials removed by HF).

5. UTILIZATION OF LOW DIELECTRIC MATERIALS IN MICROELECTRONICS

A particularly difficult challenge for low dielectric materials development has been to obtain the combination of low dielectric constant and good thermal and mechanical stability. Generally, the types of chemical structures that imbue structural stability are those having strong individual bonds and a high density of such bonds. However, the strongest bonds often are the most polarizable, and increasing the bond density gives a similar increase in polarization. For example, the rigidity and thermal stability of SiO_2 is in part due to the dense (2.2–2.4 g/cc) chemical network. Unfortunately, the high bond and material density in SiO2 lead to a large atomic polarizability, and therefore a high dielectric constant. Organic polymeric materials often have a lower dielectric constant due to the lower material density (<1.0 g/cc) and lower individual bond polarizabilities.

In this part, the relationship between molecular structure and low dielectric properties is discussed with consideration of factors such chemical bond, density, and polarizability.

5.1. Linear structure

Linear structured materials have been actively researched for various microelectronic applications. In the early stages of microelectronics development, IBM implemented a polyimide-based material in microchips based on its good thermal, mechanical, chemical, and electrical properties. However, as required properties have become stricter because of narrowing interconnect line distance, polyimide-based materials have been unable to satisfy device performance with the main reason due to its high water absorption. Despite its superior properties, it became apparent that a linear polymeric structure was unfeasible for application as more high performance devices were being demanded.

However, linear polymeric structures have given polymer scientists invaluable clues into the possible molecular content of low dielectric materials. According to the definition of a dielectric, the material density has a direct relationship with respect to its dielectric constant. Linear polymersoccupy a free volume, derived from large steric hindrance comparedto single small molecules. For this reason, linear structured materials such as organic polymers, polyethylene and polypropylene show quite low density (0.8~0.9), and thus low dielectric value (2.1~2.6). Unfortunately, these organic polymers suffer fromcritical disadvantages such as thermal instability such as low glass transition temperature and low degradation temperature.

Therefore, many scientists turned to polymeric materials having an aromaticmoiety. This chemical structure showed enhanced thermal properties and was expected to have a low density due its rigid molecular structure. The high polarizability of these materials due to their relatively high dipole moment was expected to compensate for the inherently large free volume. Some of the various aromatic, linear polymers are outlined below.

5.1.1. Polyimides (PIs)

Excellent thermomechanical propertiescan be obtained byincorporating a very stiff polymer. The classicexample of a stiff polymer chain is aromatic polyimides, whichhave a rigid backbone due to the many aryl and imide rings along the chain. These structural characteristics give rise to excellent mechanical and thermal properties in the form high modulus (8–10 GPa) and high Tg (350 to 400°C)[10]. However,the rigid chain structure causes the PI chains to align preferentially parallel to thesubstrate, especially when deposited as thin films, which results in anisotropic properties[11-18]. For example, while the out-of-plane k valueof BPDA-PDAis 3.1, the more important in-plane value is >3.5 [14].

The thermomechanical properties are likewise anisotropic. For instance, theCTE of thin films of rigid PIs is often <10 ppm/°C in the plane of the film,but can be more than ten times larger in the out-of-plane direction [14]. Anotherdrawback to PIs is that they absorb water effectively owing to the carbonylgroups, which raises the dielectric constant further. The release of thiswater duringprocessing can cause blistering of overlying layers [19].

Some of the drawbacks mentioned above can be ameliorated by tailoring thechemical structure of the PI. The k value and water adsorption can be lowered byincorporating fluorine into the material, while the anisotropy can be reduced byintroducing single bonds between rings, making the chain less rigid. For example,PMDA-TFMOB-6FDA-PDA, which utilizes both of these design strategies, hasan out-of-plane k=2.64 [20] and absorbs less moisture than unfluorinated PIs suchas BPDA-PDA [10]. However, the in-plane k value is still >3.0, and the wateruptake, although reduced, is significant enough to cause blistering in overlyinglayers during integration [19].

5.1.2. Poly (aryl ethers) (PAE)

The utilization of spin-on PAE materials results from attempts to balance the dielectric and thermomechanicalproperties. The aryl rings in these materials providebetter thermomechanical properties than do PIs, but the flexible aryl linkagesallow bending of the chains, which results in a more isotropic material than is obtainedfor PIs.

Additionally, the lack of polar groups, such as carbonyl, results ina lower k value and lower water uptake than the PIs. Fluorinated versions of PAEshad a k value of2.4 [21]. However, because of concerns about fluorine corrosion,the fluorine was removed from later versions of the material. The nonfluorinatedPAEs typically have a k of 2.8–2.9, whereas typical values for the modulus andCTE are 2.0 GPa and 50-60 ppm/°C, respectively. Resistance to thermal decompositioncan be quite good for PAEs as weight losses of only <2%over 8 h at 425°Chave been reported.One drawback of uncrosslinked PAEs is that they have a relatively low Tg of<275°C, which is lower than many of the thermal treatment temperatures of microelectronic devices.

5.1.3 Polynorbornene

Polynorbornene [22] is a pure hydrocarbon polymer without anypolar or polarizable groups. Known for their high thermal stability among organic polymers (Tg ~365°C)and low dielectric constant [23] (~2.2), polynorbornenes are soluble in common organic solvents despite its rigid backbond due to the randomly coiled nature of the polymerchains and lack of polar interactions.

This combination of properties makes polynorbornene an interesting candidate for ILD/IMD use. However, polynorbornenesexhibit insufficientadhesion to substrates with polar surfaces such as Si, oxidesor metals and its rigid backbone results in a rather brittlematerial. To overcome these shortcomings, a copolymer withalkoxysilyl and aikyi side groups at the norbornane rings inthe backbone of polynorbornene was developed byBFGoodrich (Avatre[]. [23,24] The alkoxysilyl groups enhanceadhesion to surfaces with hydroxyl groups and increase therelative mobility of the polymer chains, and hence theelongation at break of thin films [23]. However, the dielectricconstant increases from 2.2 of the unsubstitituted polymer to2.67 with an aikoxysilyl content of 20% [23].

Copolymers fromalkoxysilyl norbornene and alkylnorbornene derivatives showimproved elongation-at-break and reduced dielectricconstant [23] of E = 2.56. The glass transition temperatures andthermal stabilities of these materials are reduced comparedto non-functionalized polynorbornene [25,26]

5.1.4. Polytetrafluoroethylene (PTFE)

Proposals to use fluorinated organic materials like PTFE are aimed toward minimizingthe dielectric constant using the bonds of lowest polarizability. PTFE, which consists of singly bonded carbon chains saturated withfluorine atoms, has one of the lowest k values (<1.9) of any nonporous material,and is normally deposited by spin-on films [27]. One drawback of PTFE is that the flexible and uncrosslinked chain structure limits the thermomechanical stabilityof the material. For example, one PTFE material evaluated in our laboratory wasfound to have a low yield stress (12 MPa), low elastic modulus (0.5 GPa), lowsoftening temperature (<250°C), and high thermal expansion coefficient (CTE)(>100 ppm/°C). Together these factors can cause buckling or wrinkling of thefilm during process integration. A second issue for PTFE, which is a concern forall fluorine-containing materials, is the potential release of fluorine atoms that cancause corrosion of metals or other reliability problems in the interconnect structure.

5.1.5. Polysilsesquioxane

While research of linear, rigid, organic polymers have centered on lowering the density and enhancing thermal and mechanical properties [28,29], many of these materials tend to have elevated dielectric constants and decreased processability because the main frame of these polymers are composed of aromatics, double, and triple bonds, which can be readily polarized or have weak thermal stability.

Polysilsesquioxanes $(RSiO_{3/2})_n$ comprise a class of polymers that exhibit unique physical properties different from those of purely organic or inorganic compounds [30]. The various structures of polysilsesquioxanes, including ladder-like polysilsesquioxanes with double stranded backbones, cage-type polyhedral oligomeric silsesquioxanes (POSS), and sol-gel processed random branched structures have shown many inherent advantageous properties such as high thermal stability, low dielectric constant, good mechanical properties, and chemical resistance. Such properties have made polysilsesquioxanes a material of interest for polymer and ceramic scientists in the microelectronics industry.

In particular, polymethylsilsesquioxanes are of particular interest, as these materials exhibit low loss at high temperatures exceeding 500°C. However, to approach the favorable properties of polysilsesquioxanes in a reproducible manner, a regular structure with high molecular weight, such as ladder-structures, is to be favorable. This rigid ladder backbone with high molecular weight would support lower dielectric constants stemming from increase of inter-molecular space and high mechanical strength, as well as minimizing shrinkage during the ILD process. This material had 6.3 GPa of modulus and 2.7 of dielectric constant without curing process. [31]

5.2. Branched structures

In polymer chemistry, polymer branching induces a lower profile of material density without significant changes of in chemical properties. Because of this, many researchershave investigated polymer geometries such as graft and hyper branched structure for application as low dielectric material.

Branched polymers are advantageous in that through chemical modification of the side or end groups can give unique functionality that could not be realized by composites of two or three different materials. Also, the chemical bonding between the two components may offset one or more deficient property of the singular component.

5.2.1. Graft Polymers

Graft copolymers are a special type of branched copolymer in which the side chains are structurally distinct from the main chain. The Figure (3) depicts a special case where the main chain and side chains are composed of distinct homopolymers.

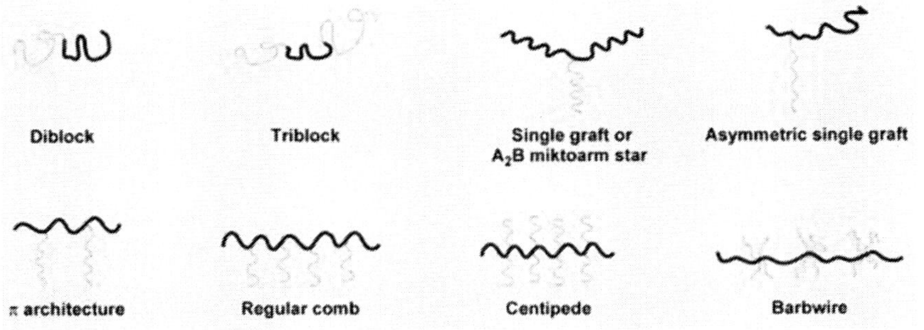

Figure 3. Special cases of grafted polymer

Kung-Hwa wei et al. reported a polyimide grafted polyhedral oligomeric silsesquioxane (POSS). They presented the dielectric constants and densitiesof the POSS/polyimide nanocomposites (figure 4). The dielectricconstants of the POSS/polyimide nanocomposites decreased as the amount of POSS increased. The maximumreduction in the dielectric constant of POSS/polyimidenanocomposites was found to be about 29%, compared to 16 mol%POSS/polyimide to pure polyimide (*k*) 2.32 vs 3.26. However, these graft polymers exhibited a slightly lower glass transition temperature about 40°C and increased thermal expansion efficient (CTE) from 31.9 ppm/K to 57.1 ppm/K after the grafting of POSS. [32]

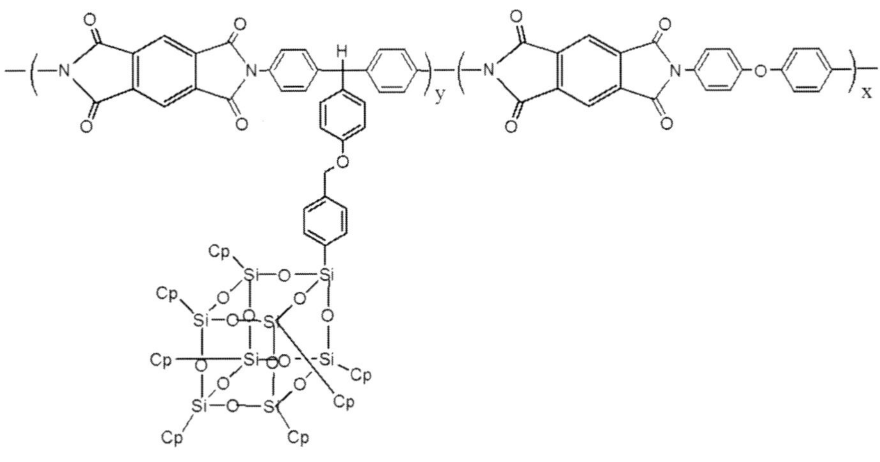

Figure 4. POSS/polyimide nanocomposites by grafted method

Another researcher studied about polyimide with grafted POSS structure. By introducing a polymerizable methyl methacryl functional groups to POSS and subsequent free-radical graft polymerization to an ozone treated polyimide, POSS grafted polyimide structures were obtained. Copolymerswith dielectric constants approaching 2.2could be achieved in the PI-*g*-PMA-POSS film containing23.5 mol % MA-POSS. In this approach, POSS content could be easily tuned by the grafting ratio of MA-POSS.[33]

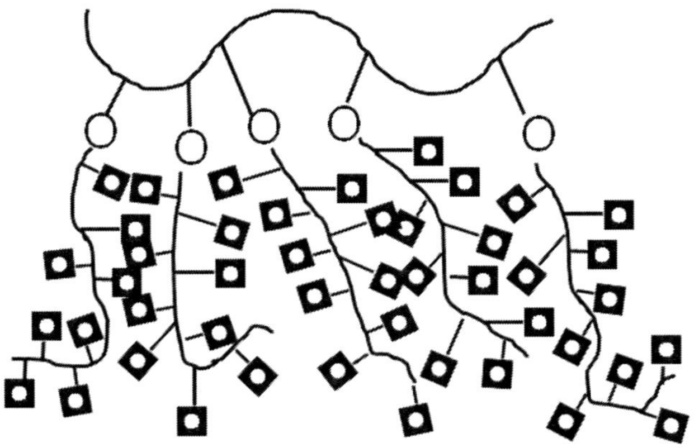

Figure 5. Synthesis of PI-*g*-PMA-POSS for low dielectrics

5.2.2. Hyperbranched Polymer

Hyperbranched polymersare denselybranched structures with large number of reactive groups. They are polymerized from monomers with mixed reactivities, commonly denoted A2B or A3Bmonomers, thus giving branched structures with exponential growth, in both end-groupfunctionalities and molecular weights.

One property often mentioned of hyperbranched polymers is the non-Newtonian relationship between viscosities and molecularweight, where hyperbranched polymers showed*low viscosities* at *high molecularweights.* For coating applications, this should be highly interesting in terms of microelectronics, where they may be used as an aid in critical patterning applications for back-end-of-line (BEOL) inter-level dielectric (ILD) materials.

Jitendra et al. showed that dense hyperbranchedcarbosiloxane (HBCSO) thin films have better mechanicalproperties than traditional organosilicates.[34] These materialsare obtained by sol-gel processing of methane-bridgedhyperbranched polycarbosilanes (HBPCSs), with the incorporatedmethane bridges being reminiscent of the systemsdescribed above (Figure 6). For example, Young's moduliof 17-22 GPa are obtained for films with dielectric constantsranging from 2.6 to 3.1. These materials have excellentelectrical properties, breakdown voltages higher than 5 MV/cm, and leakage currents $<10^{-8}$ A/cm^2 measured at 2 MV.

It was also shown that the HBPCS structure is of considerableimportance in determining the properties of the thin filmsgenerated after sol-gel processing.

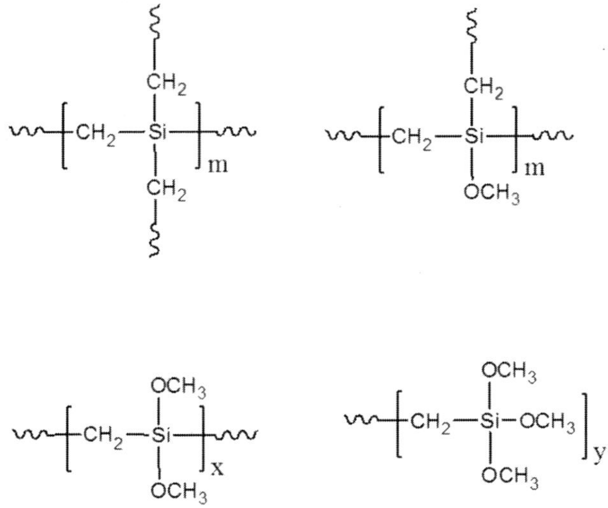

Figure 6. Chemical repeat units found in HBPCS precursors.

5.3. Network structure of low dielectric materials

In defining the different types of network structure materials for low dielectric constant applications in microelectronics, two classifications may exist. One is organic networks based on elemental carbon including amorphous carbon (diamond-like-carbon(DLC)) [35,36] and interpenetrating polymer network (IPN). The other is inorganic networks based on silicon oxide bonds such as amorphous SiO_2 and mesoporous crystalline silicon oxide.

Intuitively,network structures have excellent thermal stabilities and mechanical properties, and chemical resistance, but have relatively high density which is a factor in elevated dielectric constants. Because of this, dielectric constant and mechanical properties should be carefully controlled and careful consideration of its crystallinestructure, as crystalline structures tend to be easily polarized.

5.3.1. Organic Network

When compared with linear structured polymers, network polymers have significant advantages in thermal stability with increasing glass transition temperature and complying CTEs. Dielectric materials for microelectronics needed to have high Tg temperature up to 400°C and endurance of repeated thermal cycling, creating the thermal mismatches which can leadto flow, delamination, adhesive failure, etc. Examples or organic network materials are discussed below.

5.3.1.1. Diamond-like carbon [DLC]

Amorphous diamond-like carbon[DLC], which can be prepared by chemical vapor deposition [CVD] method, [35,36] are metastable materials composedof sp2, sp3, and even sp1 hybridized carbon atoms withhydrogen concentrations, *CH*, ranging from 1% to 50%, withthe composition being primarily determined by the natureof the precursor and the corresponding deposition conditions.

These sp^3 bonds can occur not only with crystals - in other words, in solids with long-range order - but also in amorphous solids where the atoms are in a random arrangement. In this case, there will only be bonding between a few individual atoms and not in a long-range order extending over a large number of atoms. The bond types have a considerable influence on the material properties of amorphous carbon films. If the sp^2 type is predominant the film will be softer, if the sp^3 type is predominant the film will be harder.

Under the right conditions, it is possible to deposit DLCfilms with compressive stress, spanning values from 200-800MPa, and dielectric constants approaching 2.7.[36, 37]

5.3.1.2. SiLK and BCB resins

A very promising class of networkpolymers has been developed byDow under the name 'Silk'. The formulations presumablyconsist of a mixture of monomeric and/or oligomericaromatic starting compounds, which contain ortho-bisethinylor -phenylethinyl groups [38]. The materials exhibit k values on the order of 2.6-2.7, with decomposition temperatures in excessof 500°C, no softening up to 490°C, good gap fill propertiesdown to below 0.1 um, a maximum water uptake of 0.25%,and a coefficient of thermal expansion 4s of 66 ppm/K.

Benzocyclobutene (BCB) resins were developed by Dow inthe 1980's [39], with a silicon-containing derivative formicroelectronics applications, and are commercially availableunder the name 'Cyclotene'. BCB resins with imide structurescan be extremely tough, and the dielectric constant of curedfilms from this monomer is 2.6-2.7, with thermal stability upto 375°C and water adsorption of only 0.2% [40,42].

5.3.2. Inorganic network

Inorganic networks mainly consist of ceramics or amorphous silica. In various materials for microelectronics, the silicon oxides play a major role due to its low polarizability, superior thermal and mechanical properties. In addition, tunable microspores can be made to reduce the dielectric constant through control of microstructure under special conditions.

5.3.2.1. Ordered Mesoporous Materials

Microporous zeolite thin films were first investigated by Yan and co-workers.[43] These films offer good thermal stability(i.e. no pore collapse or unidirectionalshrinkage) and inter-particle mechanicalstrength. They can be prepared via a simplespin-on method [43,44] or by in situ growth.[44,45] With simple spin-on methods a dispersionof small zeolite particles are prepared andcast onto a surface. The porosity originates fromthe interparticle porosity within the zeolitenanoparticles and the intra-particle porosityowing to the packing of the near-sphericalnanoparticles in thin film format.

5.3.2.2. Network polysilsesquioxane

Some of the most promising materials for dielectric materials are poly(silsesquioxanes). Mostcommon are polymethylsilsesquioxane (MSQ), e.g. AccuspinT-18 from Allied Signal [46], or poly(hydridosiLsesquioxane)(HSQ), e.g. FOx from Dow Corning [47]. Synthesis of these silsesquioxanes(MSQ and HSQ) have traditionally been through the sol-gel method, as its utility in being able to obtain highly cross-linked structures through acidic and or basic conditions has been well documented [48,49,50].Dielectric constant values of around 2.6can be achieved for HSQ and MSQ. But while MSQ exhibitsthis dielectric constant after curing at temperatures up to450°C, HSQ must be cured at temperatures lower than210°C [51]. Curing of HSQ at temperatures of 250°C or aboveresults in dielectric constant around 3 or even higher

32°[51,52]. Gap fill and planarization properties are also acceptable and because of their chemical structure, which isclosely related to SiO_2,polymethylsilsesquioxanes are also compatiblewith existing lithographic procedures.

Efforts to further decrease the dielectric constant without decreasing mechanical strength, POSS skeletons have been introduced in MSQ. To suppress the phase separation, incompletely condensed methyl functionalized POSS precursors have been used to form chemical bonds with oligomeric sol precursors. These incompletely condensed POSS moieties functioned as coupling agents while expanding the free volume of the final sol after curing which was accomplished to 4 GPa of modulus and 2.3 of dielectric constant. [53](figure 7).

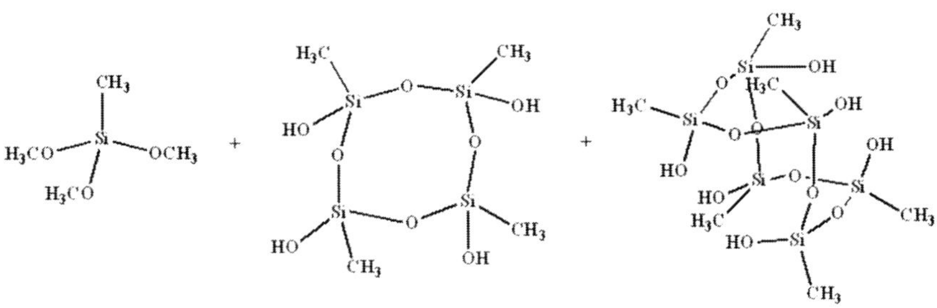

Figure 7. Introduction of POSS moiety by sol-gel method

5.4. Porous network polymer by subtraction of porogen

Numerous methods of introducing subtractive porosityinto spin-on deposited materials exist.Conventional methods of substractive porosity involve the addition of molecular or supramolecular particles called 'porogens'withtailored thermal stability to the dielectric precursor[54]. The stability of these particles issuch that they are not affected by the coating drying step, andthey are removed by pyrolysis during final film sintering orcure at temperatures typically in the range from 300 to400 °C. An example of amaterial for which the pore size and porosity, or the pore sizeand porogen load can be controlled independently. [55]However, it should be noted that the use of porogens should only be applied to dense materials having a k less than 2.5 and modulus over 5 GPa in order for the final material to satisfy the required mechanical property.

In organic materials, the SilK matrix has been the only known material to provide the thermal and mechanical properties at temperature up to 500°C for use in combination with porogens. C.E. Mohler et al. [56] reportedon porous SiLK dielectric film properties such as pore volume, porosity, size distribution, andshowed a 2.2 dielectric constant at 30% load of porogens.

In comparison with organic porous dielectric material, inorganic porous dielectric materials have been more rigorously investigated because of their superior mechanical properties. Representative studies have used polymethylsilsesquioxane (PMSQ) as matrix for the addition of various porogens such as the block copolymers, poly(styrerene-block-acrylic acid)[57], macromolecules of cyclodextrin[58], poly(caprolactone[59], and calix[4]arene[60].

Many of these studies with porogens have reported materials that have excellent mechanical and electrical properties, but lack in other practical aspects for application in microelectronics. When porogens are introduced into a matrix, critical problems may occur, such as thermal degradation products acting as a poison or contaminant within the matrix or interfacial adhesion problems. Therefore, use of porogens has yet to remain a difficult process for practical applications in microelectronics.

6. CONCLUSIONS

The search for materials with low dielectric constant in the microelectronics industry has and will continue feverishly into the future as the demand of faster processing speeds increases. Reduction of the dielectric constant of a material can beaccomplished by selecting chemical bonds with lowpolarizability and introducing porosity. Integration of suchmaterials into microelectronic circuits, however, poses anumber of challenges, as the materials must meet strictrequirements in terms of properties and reliability.The introduction of low-k materials in microelectronicsresearch and development is a good example of howindustrial needs drive new fundamental and applied researchtopics in science. Examples include pore structurecharacterization, deposition of thin films on poroussubstrates, mechanical properties of porous films, andconduction mechanisms in these materials.The substantial efforts made by materials and ICresearchers to integrate the low-k films and continuehistorical device performance improvements have contributedto, and are still leading to, innovative fundamental andapplied science.

ACKNOWLEDGEMENT

This work was financially supported by a grant from the Fundamental R&D Program for Core Technology of Materials funded by the Ministry of Knowledge Economy, Republic of Korea and Partially by a grant from Center for materials architecturing of Korea Institute of Science and Technology (KIST)

REFERENCE

1. G. W. Ray, 1998Mater. Res. Soc. Symp. Proc.511:199
2. R. Fox, J. R. Pellerin, 1997Res. Rep. Austin TX: SEMATECH
3. J. P. Hummel, 1995In Advanced Multilevel Metallization Materials Properties Issuesfor Copper Integration, ed. CS Schuckert,6:547.Wilmington, DE: DuPont Symp.Polyimides in Microelectronics
4. P. S. Ho, T. Kwok, 1989Rep. Prog. Phys.52:301
5. C. Hu-K, K. P. Rodbell, T. D. Sullivan, K. Y. Lee, D. P. Bouldin, 1995IBM J. Res. Dev. 39:465
6. S. R. Wilson, C. J. Tracy, eds, 1993Handbook of Multilevel Metallization for Integrated Circuits. Park Ridge, NJ: Noyes
7. K. J. Miller, H. B. Hollinger, J. Grebowicz, B. Wunderlich, 1990Macromolecules 23:3855
8. S. H. Pine, 1987Organic Chemistry. New York: McGraw-Hill. 5th ed.
9. J. Rouquerol, et al.Appl. Pure, Chem, 1994
10. B. C. Auman, 1995Mater. Res. Soc. Symp.Proc. San Francisco. 381:19
11. Y. Kang-S, 1994Microstructure and strengthening mechanisms in aluminum thin films on polyimide film. Ms thesis. Univ. Texas, Austin
12. J. Lee-K, 1998Structure-property correlation of polyimide thin films on line structure. PhD thesis. Univ. Texas, Austin
13. S. E. Molis, 1989In Polyimides: Materials,Chemistry and Characterization, ed. Amsterdam: Elsevier

14. M. Ree, K. J. Chen, D. P. Kirby, 199, J. Appl, Appl.Phys. 72:2014

15. S. T. Chen, H. H. Wagner, 1993J. Electron.Mater. 22:797

16. L. Lin, S. A. Bastrup, 1994J. Appl. PolymerSci. E 54:553

17. 1992 JD. Boese, H. Lee, D. Y. Yoon, J. F. Rabolt, 19, J. Polymer, Polymer. Sci, Phys. 30:1321

18. S. S. Hardaker, S. Moghazy, C. Y. Cha, R. J. Samuels, 199, J. Polymer, Polymer. Sci, r Phys. 31:1951

19. J. T. Wetzel, Y. T. Lii, S. M. Filipiak, B. Y. Nguyen, E. Travis, O., et al.1995Mater. Res. Soc.Symp. Proc. San Francisco. 381:217

20. G. B. De Maggio, W. E. Frieze, D. W. Gidley, M. Zhu, H. A. Hristov, A. F. Yee, 1997Phys.Rev. Lett. 78:1524

21. N. Hendricks, H., K. S. Y. Lau, A. R. Smith, W. B. Wan, 1995Mater. Res. Soc. Symp. San Proc. Francisco. 381:59

22. Heitz, W. 1995. Pure Appl.Chem. 67:1951

23. N. R. Grove, et al 1997Mater Res Soc Symp Proc 476:3

24. N. R. Grove, et al.1997Proceedings of the 6th International Conference on Multichip Modules, Institute of Electrical and Etecronics Engineers, New York

25. L. H. Treiche, et al.1999Low dielectric constant materiaLs for interlayer dielectrics. In: Nalwa, H.S. (Ed.) Low-k and high-k materials, Academic Press,Boston

26. N. Grove, et al.1997Proceedings of the 6th International Conference on Polyimides and other Low-k Dielectrics, McAfee, N J

27. C. T. Rosenmayer, J. W. Bartz, J. Hammes, 1997Mater. Res. Soc. Symp. Proc. San Francisco. 476:231

28. Voit,B.2000Journal of Polymer Science Part a-Polymer Chemistry, 38:2505

29. A. S. Mathews, I. Kim, C. S. Ha, 2007Macromolecular Research, 15:114

30. R. H. Baney, M. Itoh, A. Sakakibara, T. Suzukit, 1995Chem. Rev. 95:1409

31. Hwang, S.S., et al. Macromolecular Research, in press

32. C. Leu-M, Y. Chang-T, K. Wei-H, 2003Macromolecules, 36: 9122

33. Y. Chen, L. Chen, H. Nie, E. T. Kang, 200, Journal of Applied Polymer Science, 99:2226

34. J. S. Rathore, L. V. Interrante, G. Dubois, 2008AdV. Funct. Mater. 18: 4022.

35. A. Grill, In Dielectric Films for AdVanced Microelectronics; Baklanov, M., Maex, K., Green, M., Eds.; Wiley: New York, 2007

36. A. Grill, V. Patel, K. L. Saenger, C. Jahnes, S. A. Cohen, A. G. Schrott, D. C. Edelstein, J. R. Paraszczak, 1997Mater. Res. Soc. Symp. Proc., 443:155.

37. A. Grill, 2001Diamond Relat. Mater. 10: 234.

38. D. Babb, et elWodd Patent WO 97/10193, The Dow Chemical Company,March 20, 1997

39. R. A. Kirchhoff, K. J. Bruza, 1994Adv Polym Sci 117:1

40. Yang G.R. etal1997Mater Res Soc Symp Proc 476 :161

41. R. J. Gutmann, etal 1995Mater Res Soc Symp Proc 381:177

42. Ls. M. E. Mil, etal, 1997Microelectron Engng 33: 327

43. H. K. Hunt, C. M. Lew, M. Sun, Y. Yan, M. E. Davis, 2010Micro. And Meso. Mater., 128:12

44. Z. Wang, H. Wang, A. Mitra, L. Huang, Y. Yan, 2001Adv. Mater., 13:746.

45. S. Li, C. Demmelmaier, M. Itkis, Z. Liu, R. C. Haddon, Y. Yan, 2003Chem.Mater., 15:2687.

46. Hacker, N.P. et al1997Mater Res Soc Symp Proc 476:25

47. Bremmer, J.N. et al1997Mater Res Soc Symp Proc 476:37

48. D. Lee, et al. 200, 2008Macromolecular Research, 16353359

49. S. E. Rankin, et al.2000Macromolecules, 33:7639

50. Mackenzie,J.D.1988Journal of Non-Crystalline Solids, 100:162

51. S. M. Kim, et, al1998Mater Res Soc Symp Proc 511:39

52. D. Tobben, et al.1997Mater Res Soc Symp Proc 443:195

53. e. A. S. , H. S. Lee, S. Hwang, S.preparing to publish

54. Y. F. G. Lu, Z. Cao, R. P. Kale, S. Prabakar, G. P. Lopez, C. J. Brinker, 1999Chem. Mater. 11:1223

55. Baklanov, M. R. et al., Proceedings of the Advanced Metallization Conference (Materials Research Society, Pittsburgh, PA, 2002)

56. E. B. Mohler, G. Landes, F. Meyers, B. J. Kern, K. B. Ouellette, et, al.2003AIP Conf. Proc. 683:562

57. Y. Chang, C. Chen-Y, W. Chen-C, 2004Journal of Polymer Science: Part B: Polymer Physics, 42:4466

58. Y. Lyu-Y, J. Yim-H, Y. Byun, J. M. Kim, J. Jeon-K, 2006Thin Solid Films 496:526

59. J. Hyeon-Lee, Y. Y. Lyu, M. S. Lee, J. Hahn-H, J. H. Rhee, S. K. Mah, J. Yim-H, S. Y. Kim, 2004Macromol. Mater. Eng., 289:164

60. R. S. Vallery, M. Liu, D. W. Gidley, J. Yim-H, 2011Microporous and Mesoporous Materials 143:419

9

CHAPTER

THIN FILMS WITH HIGH SURFACE ROUGHNESS: THICKNESS AND DIELECTRIC FUNCTION ANALYSIS USING SPECTROSCOPIC ELLIPSOMETRY

Daniel Lehmann[1] , Falko Seidel[1], and Dietrich RT Zahn

[1] *Semiconductor Physics, Technische Universität Chemnitz*

ABSTRACT

An optical surface roughness model is presented, which allows a reliable determination of the dielectric function of thin films with high surface roughnesses of more than 10 nm peak to valley distance by means of spectroscopic ellipsometry. Starting from histogram evaluation of atomic force microscopy (AFM) topography measurements a specific roughness layer (RL) model was developed for an organic thin film grown in vacuum which is well suited as an example. Theoretical description based on counting statistics allows generalizing the RL model developed to be used for all non-conducting materials. Finally, a direct input of root mean square (RMS) values found by AFM measurements into the proposed model is presented, which is important for complex ellipsometric evaluation models where a reduction of the amount of unknown parameters can be crucial. Exemplarily, the evaluation of a N,N'-dimethoxyethyl-3,4,9,10-perylene-tetracarboxylic-diimide (DiMethoxyethyl-PTCDI) film is presented, which exhibits a very high surface roughness, *i.e.* showing no homogeneous film at all.

Keywords: Ellipsometry Roughness Thickness AFM RMS Dielectric function PTCDI

Electronic supplementary material

The online version of this article (doi:10.1186/2193-1801-3-82) contains supplementary material, which is available to authorized users.

Background

Spectroscopic ellipsometry (SE) is a commonly used tool to determine the thickness of thin films and investigate their dielectric function in a non-destructive way. During SE measurements the reflection properties of the whole sample are recorded. A subsequent evaluation by modeling the measured data is necessary to extract properties of individual layers. This procedure requires some knowledge of the general sample structure: the dielectric function of the substrate should be known with high precision and also the order of layers (*e.g.* substrate/oxide/film 1/film 2/roughness) is a necessary input parameter. The more parameter values are unknown in the beginning of the evaluation, the more complex is the evaluation and eventually no solution might be found if some unknown parameters are highly correlated. A possible solution is a step by step analysis of the sample during the preparation, *i.e.* by recording SE spectra after growth of each individual layer. As roughness is a side effect of layer growth itself, this procedure cannot be used for the determination of surface roughness. Another way of reducing the number of unknown parameters is the inclusion of results from other techniques as, for example, UV–vis spectroscopy. Here, we report on the inclusion of AFM topography measurements to support the evaluation of SE spectra of thin film samples with high surface roughness. We then generalize the RL model to be useful for any evaluation of surface roughness of thin films grown on substrates. Finally, it is shown how the root mean square (RMS) determined by AFM measurements can directly be used as input parameter during SE evaluation.

METHODS

The common way to handle a rough film in ellipsometric modeling is to divide it into two parts: one part containing the homogeneous film and the other part representing a non-homogeneous form referred to as "roughness layer" (Figure 1a).

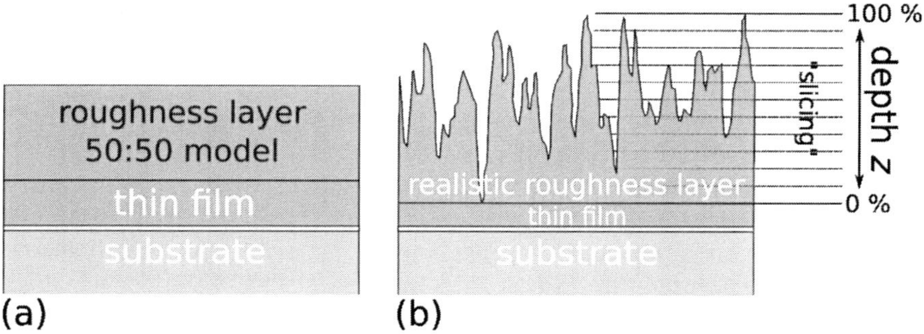

Figure 1. ROUGHNESS LAYER MODEL COMPARISON. **(A)** Common ellipsometry evaluation model: the RL is treated as an effective medium approximation (EMA) of 50% material and 50% void. **(B)** Realistic roughness layer derived from a cross section through an AFM picture of a DiMethoxyethyl-PTCDI film (Figure 2a). By z-slicing the RL into infinitesimal thin slices and treating each slice with an individual EMA material ratio, a more precise RL model can be derived.

The first part is simulated with the physically meaningful dielectric function belonging to the material. The determination of this dielectric function is usually the aim of the evaluation once the layer thicknesses have been determined. The second part is typically simulated using an effective medium approximation (EMA), a mixture of 50% material – having the dielectric function of the first part of the simulated layer – and of 50% void space with the dielectric function of vacuum (further referred to as "50:50 model") (Gordan et al.2006a,b; Louis et al.2007; Volintiru et al.2008). However, this estimation of the mixture ratio is not precise. The realistic roughness topography of grown layers has a profile comparable to mountains: if one virtually cuts a volume segment into slices of different height, at the height of the valleys the amount of material in each slice is high. At the height of the mountain tips, however, the amount of material is almost zero in each slice (Figure 1b). Hence, the material to void ratio is not constant throughout the roughness part of the layer, but changes as a function of distance to the virtual interface between the RL and the homogeneous part. A similar approach was already discussed by Aspnes et al. (1979), but with various geometrically constructed roughness profiles instead of AFM measured profiles.

The consequence of using a simple 50:50 model is an incorrect dielectric function of the roughness part and, hence, an incorrect dielectric function of the homogeneous part as both parts are correlated during the evaluation. These inaccuracies are small for transparent thick films with a small surface roughness. However, if the thickness of the RL has the same order of magnitude as the homogeneous film thickness, a reliable determination of the dielectric function is not possible.

AFM is a method to determine the topography of a sample surface. Commonly, an RMS value is given as value for the surface roughness. For determination details see Méndez-Vilas et al. (2002). However, much more information can be deduced for creating a sophisticated roughness model for SE. Exemplarily, an extreme case should be discussed: the dielectric function of an organic thin film of DiMethoxyethyl-PTCDI with a very high surface roughness is determined. A more detailed view on this semiconducting molecule itself and the implication of high surface roughness on electrical behavior when used in organic field-effect transistors can be found elsewhere (Lehmann and Zahn2009). The thin films were prepared by organic molecular beam deposition under high vacuum conditions ($p_{base} < 5 \times 10^{-7}$ mbar) on Silicon/SiO_2 (1.4 nm) substrates. The nominal thickness, respectively the deposited mass, during growth was monitored by a quartz micro balance (QMB). A nominal thickness of 30 nm, based on known QMB ratios for other PTCDI derivatives forming smooth films, was the deposition target. The SE measurement was performed 47 days after the film deposition using a Woollam VASE Ellipsometer, measuring at three incidence angles (65°, 70°, 75°) in a spectral range from 0.74 eV to 4.50 eV. The depolarization was measured additionally using an M2000 T-Solar Ellipsometer from Woollam in the same spectral range. Figure 2a shows an AFM picture of DiMethoxyethyl-PTCDI. As comparison a second picture of another molecule, N,N'-diphenyl-3,4,9,10-perylene-tetracarboxylic-diimide (DiPhenyl-PTCDI, nominally 30 nm thick), is also shown in Figure 2b. The comparison later demonstrates the error tolerance of the method with regard to a wrong input.

RESULTS AND DISCUSSION

For the method presented, the presumption has to be made that the Fresnel relations are valid. This is true for wavelengths much larger than the RL thickness and structure periodicity. Hence, the light scattering has to be weak and depolarization can be ignored. In the presented case the RL as well as the periodicity are in the range of 100 nm. As the film and RL thicknesses are determined in the transparent region below 1.4 eV ($\lambda \geq 886$ nm) we consider the Fresnel relations valid here. Figure 3 shows that the depolarization is low even for higher energies.

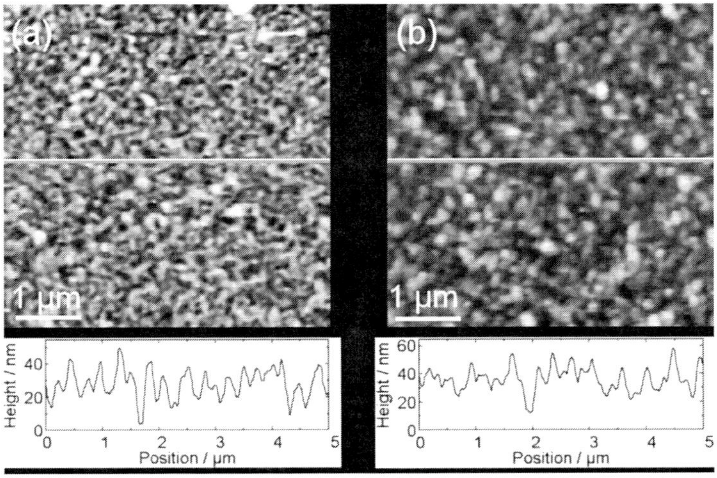

Figure 2. AFM MEASURED TOPOGRAPHIES. (A) Height profile of DiMethoxyethyl-PTCDI: grayscale from 0 nm (black) to 58 nm (white), **(B)** height profile of DiPhenyl-PTCDI for comparison: grayscale from 0 nm (black) to 94 nm (white). The roughness parameter, *i.e.* the material/void function, is deduced by integrating over a grayscale histogram of each AFM picture.

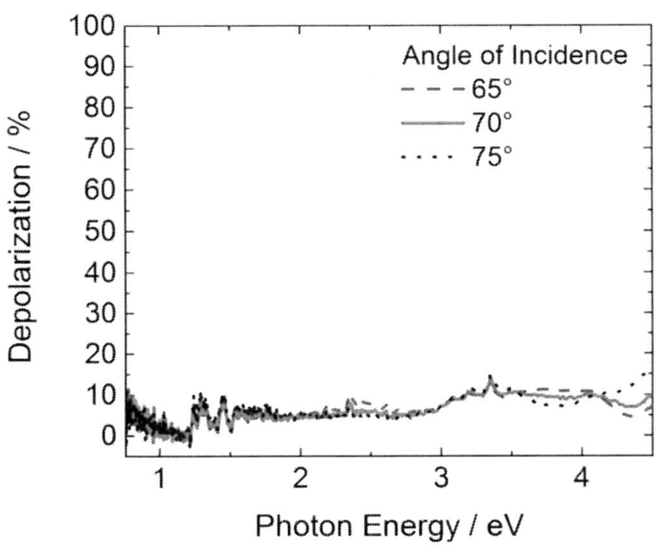

Figure 3. Depolarization of the dimethoxyethyl-ptcdi sample during ellipsometry measurement.

Slicing the measured volume segment into equal height slices as described above can be performed by taking a histogram of the grayscale AFM pictures. Each grey level corresponds to a specific height slice. The amount of pixels with the same grey level determines the amount of material in the corresponding height slice. An integration over all grey levels leads to a material/void ratio as a function of the height position within the RL (Figure 4).

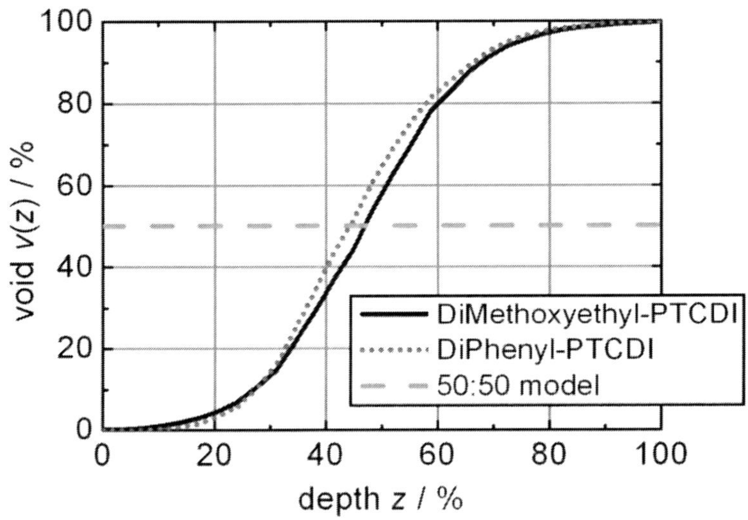

Figure 4 . MATERIAL/VOID RATIO COMPARISON. DiMethoxyethyl-PTCDI and DiPhenyl-PTCDI ratios are deduced from the AFM topography measurements of Figure 1.

The resulting material/void function can be fitted by the Boltzmann equation

$$v(z) = \frac{A-B}{1+e^{(z-C)/D}} + B, \tag{1}$$

where $v(z)$ is the percentage of voids as a function of the dimensionless depth z, starting from the homogeneous film with $v(z = 0\%) = 0\%$. The depth is also given as percentage as the thickness of the film is variable during fitting while SE modeling is performed. The resulting parameters are compared in Table 1; they are similar for both films.

Table 1. Boltzmann equation parameters determined by fitting the afm derived material/void ratios

	A/ %	*B*/ %	*C*/ %	*D*/ %
DiMethoxyethyl-PTCDI	−1.5	100.3	46.6	9.7
DiPhenyl-PTCDI	−2.8	99.7	44.0	9.3

The Boltzmann equation with the determined parameters is transferred into an SE RL model of the layer utilizing a Bruggeman EMA (Bruggeman1935) in the WVASE32 software (J.A. Woollam Co.). The homogeneous layer and RL thicknesses were determined using a Cauchy model in the range of 0.76 eV to 1.40 eV. Table 2 compares the results of the deduced material/void functions (layer order: Si/SiO_2/homogeneous film/AFM derived RL) to the results of the 50:50 model (layer order: Si/SiO_2/homogeneous film/50:50 model RL).

Table 2. Thicknesses d of a homogeneous DiMethoxyethyl-PTCDI film below the surface roughness and the corresponding roughness layer heights d r determined for a film with the use of different roughness profiles

Roughness profile of	*d*/nm	d_r/nm
DiMethoxyethyl-PTCDI (AFM)	0	107.6 ± 0.3
DiPhenyl-PTCDI (AFM)	0	104.2 ± 0.3
Proposed model with $\sigma = 13.5\%$	0	94.8 ± 0.3
50:50 model	27 ± 76	39 ± 70

The results from the material/void functions are very similar and predict both a Volmer-Weber type film growth with no homogeneous film below. The 50:50 model yields only uncertain values where a small change of starting parameters results in significantly different thicknesses. The peak valley distance – identical to the rough layer thickness in the material/void function model – is predicted to be larger than the AFM measured height profile. This can be explained with the limited resolution of the AFM tip:

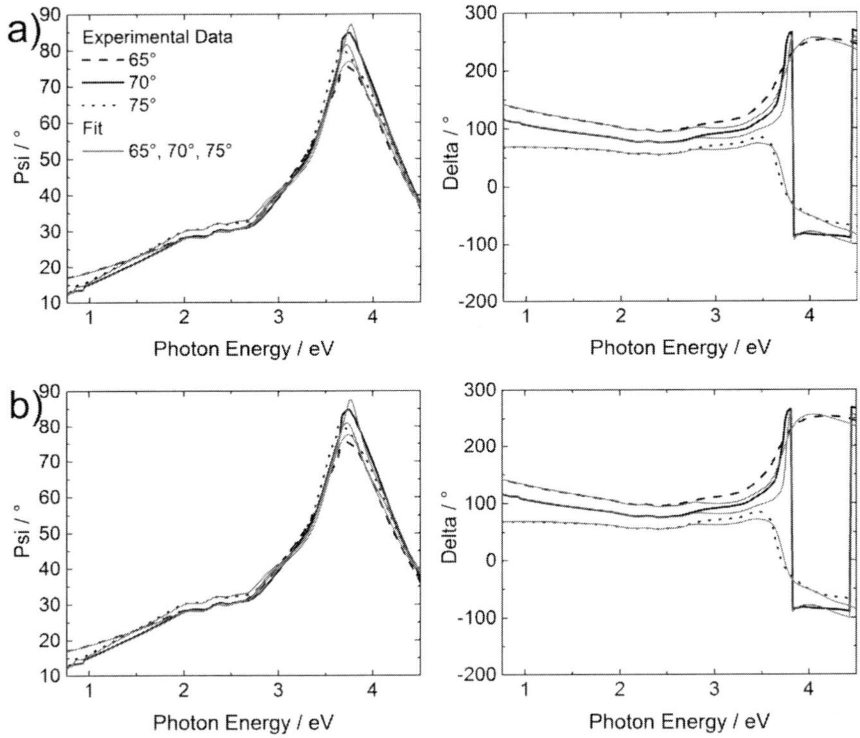

Figure 5. THE ELLIPSOMETRIC ANGLES **ψ** AND **Δ** AND THE CORRESPONDING FITS WITH GENERAL OSCILLATOR MODEL. **(A)** Fit of the 50:50 model, **(B)** fit of the AFM derived roughness profile model.

with a diameter of at least 10 nm only peaks and valleys of the same scale and above can be resolved.

Even though the model was derived from AFM pictures the depth z was included as a percentage and not as a thickness value in order to be generally applicable. Therefore, the shape of the material/void function does not depend on the actual thickness. This is especially supported by the results of the DiPhenyl-PTCDI topography measurements: although the AFM height profile exhibits a peak-to-valley distance almost twice as large as for the DiMethoxyethyl-PTCDI, the derived material/void function yields very comparable results during thickness determination in SE.

The determined thicknesses are used as input during the determination of the dielectric function. Therefore, a general oscillator model was utilized. Figure 5 shows the fits to ψ and Δ for the 50:50 model and the presented

AFM derived model. Figure 6 compares the results. Even though the quality of the fit does not differ significantly, the 50:50 model overestimates the dielectric function. Additionally, the error bar has to be considered to be as large as for the thickness determination: this result is not reliable. The material/void function model results for the dielectric functions deviate only slightly. The dielectric function of the AFM derived material/void function of DiMethoxyethyl-PTCDI itself is the most reliable one.

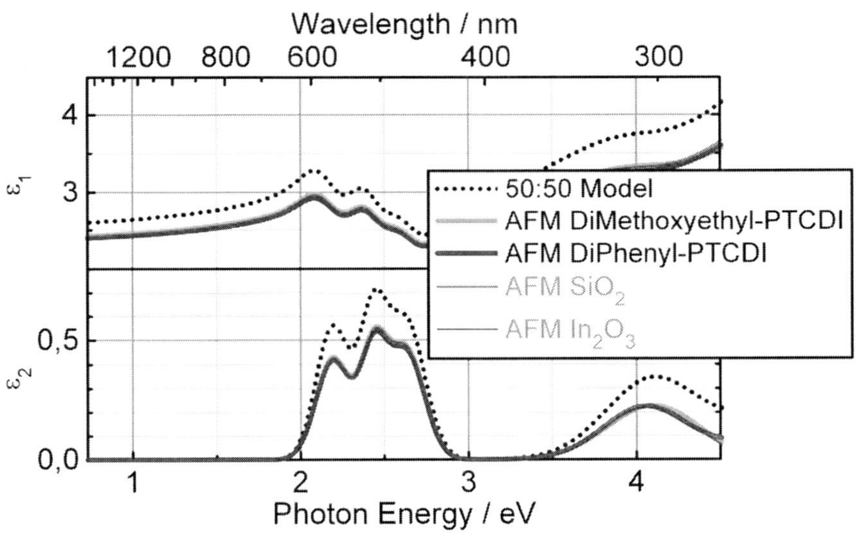

Figure 6. Dielectric functions determined for a dimethoxyethyl-ptcdi rough thin film using different roughness profiles. Dotted curve: dielectric function determined using the conventional 50:50 EMA model. Orange line: dielectric function determined using the actual AFM profile of the film (parameters in Table 1). Other lines: dielectric functions for the same DiMethoxyethyl-PTCDI film, but using the AFM roughness profiles of completely different films. The AFM derived roughness profiles yield almost identical results independent of their manifold origin, whereas the common 50:50 model overestimates the dielectric function.

Not shown in detail are further stress tests of the material/void function model: AFM pictures of thin films with a large range of thicknesses and from various sources where converted to material/void functions as presented above. Exemplarily, results from SiO_2 (1.23 nm peak to valley distance)[a] and In_2O_3 (200 nm peak to valley distance) (Dhananjay et al.2008) are also shown in Figure 6. These intentionally wrong inputs yield dielectric functions of DiMethoxyethyl-PTCDI with even less deviation from the actual DiMethoxyethyl-PTCDI material/void function compared to the DiPhenyl-PTCDI material/void function.

Based on these findings we propose a general RL model to be used instead of the 50:50 model for grown thin films:

The deposition of a large number of particles on a flat surface with random distribution of particles inside the deposition beam will yield a normal distribution, if converted to a histogram and integrated as described above, as long as surface diffusion is neglected:

$$N(Z=k) = \frac{1}{\sqrt{2\pi\lambda}} \exp\left(-\frac{(k-\lambda)^2}{2\lambda}\right). \tag{2}$$

The value λ is the expected value and k is the number of particles sticking to the surface with the probability $N(Z = k)$ at a specific area element having the size of each particle. The expected value equals the variance $E(Z) = Var(Z) = \lambda$. The mean square error (MSE) of this experiment is defined as

$$MSE(Z) = \frac{1}{m}\sum_{i=1}^{m} [k_i - \lambda]^2. \tag{3}$$

where m is the number of particles which can find space on the surface, if only one particle is allowed at each area segment. The corresponding RMS function can be defined as

$$RMS(Z) = \sqrt{MSE(Z)}. \tag{4}$$

Each particle consumes a fraction of area of the surface. Once this area is occupied, the next monolayer starts to grow on top at this specific position. Therefore, k is also equal to the number of monolayers, and, hence, $N(Z = k)$ is the probability to find k monolayers at a specific position on the substrate. Then, λ is the expected number of monolayers. For a discrete distribution as in equation (2) the variance can be calculated as follows:

$$Var(Z) \equiv \sum_{i=1}^{m} [Z_i - E(Z)]^2 p_i. \tag{5}$$

Here, $p_i = m_i/m$ is the probability to find a particle at the position i on the surface (m_i being the number of particles at each position i and m the total number of particles). For a monolayer one particle is considered on each position i and, hence, this ratio becomes $p_i = 1/m$. Then, equation (5) equals equation (3). Therefore, we can $S(Z) = \sqrt{Var(Z)}$. Since the particle size is small compared to the expected value a continuous distribution can be used instead of equation (2), a Gaussian distribution with $\lambda = C = \sigma^2$:

$$G(z) = \frac{1}{\sigma\sqrt{2\pi}} \exp\left(-\frac{(z-C)^2}{2\sigma^2}\right). \tag{6}$$

where z is the dimensionless height of the film. The resulting material/void function is

$$\begin{aligned} \nu(z) &= \int dz' G(z) = \frac{1}{\sigma\sqrt{2\pi}} \int dz' \exp\left(-\frac{(z'-C)^2}{2\sigma^2}\right) \\ &= c + \frac{1}{2} erf\left(\frac{z-C}{\sigma\sqrt{2}}\right). \end{aligned} \tag{7}$$

Additionally, in the boundary condition for $z \rightarrow \infty$ only voids are remaining. This leads to $c = 1/2$. If C is defined as 50%, the material/void function equals 50% at a relative depth of $z = 50\%$. This is in good agreement with $v(z)$ found from the evaluation of the AFM measurements. This also ensures that the mean amount of material in the roughness film found by the ellipsometry model equals 50% and, hence, the thickness of the $v(z)$-RL remains comparable with thicknesses determined by the common 50:50 model. Additionally, the nominal thickness of the RL can easily be defined as $d_{r,nom} = d_r/2$, which is important for growth rate control using a quartz microbalance.

The factor σ determines the slope of $v(z)$. A value of $\sigma = 13.5\%$ matches all the empirically found material/void functions best. The resulting material/void function

$$\nu(z) = 50\ \% \left[1 + erf\left(\frac{z-50\ \%}{13.5\ \%\sqrt{2}}\right)\right] \tag{8}$$

can be included into the ellipsometry software yielding with very little deviation the same dielectric function as the Boltzmann equation discussed above derived from the various materials.

If the ellipsometry software does not allow the use of the error function a similar $v(z)$ function can also be defined based on the Boltzmann equation (1) with $A = 0\%$ void at the bottom of the roughness film, $B = 100\%$ void at the top of the roughness film, $C = 50\%$ material at the relative depth of $z = 50\%$, and $D = 8.5\%$ as empirically found distribution parameter. The negligible difference to $v(z)$ based on the cumulative distribution is shown in Figure 7.

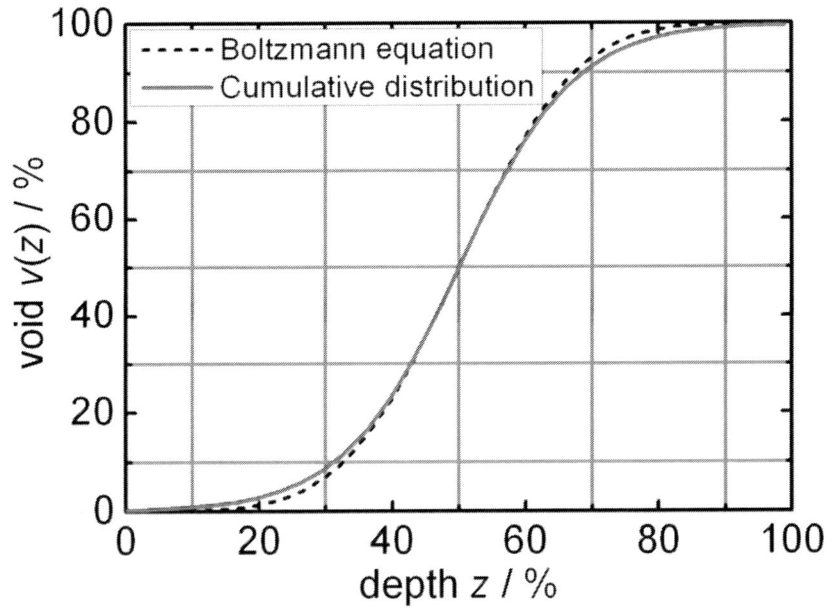

Figure 7. COMPARISON BETWEEN PROPOSED ROUGHNESS MODEL MATERIAL/VOID FUNCTIONS v (z). The functions formularized with the Boltzmann equation and with the cumulative distribution approach yield comparable results with negligible difference with respect to the evaluation of dielectric functions.

Even though the equations introduce multiple new parameters, there is no additional fit parameter during SE evaluation. Only the total thickness of the RL will be fitted as with the 50:50 model before. The values $\sigma = 13.5\%$ or $D = 8.5\%$, respectively, have also been used very successfully for other material systems, *i.e.* ZnO and Si (will be published elsewhere). There, AFM pictures were unavailable and these values had to be used as starting values. Even without freeing these parameters for fitting, the results were very satisfying. Without having a final proof, these values appear to be reliable starting values in general. Applying the model with these values, the thickness of the DiMethoxyethyl-PTCDI RL is determined to be (94.8 ± 0.3) nm, also without homogeneous film and with the same dielectric function.

RMS values determined by AFM measurements can be transferred into a RL thickness with $v(z)$ of equation (8) using

$$RMS = \sigma \cdot d_r = \sqrt{Var(z)}, \tag{9}$$

$$d_r = \frac{RMS}{13.5\ \%}. \tag{10}$$

It has to be pointed out that equation (10) only holds for an ideal AFM tip with a diameter of less than the smallest structure of the rough film. In the example of DiMethoxyethyl-PTCDI the thickness of 68.4 nm derived from the RMS value of 9.24 nm deviates from the SE determined film thickness of 94.8 nm by a factor of 1.39. The factor can be seen as a correction factor and might deviate between different AFM tips.

The RL model will not only hold for organic thin films as described exemplarily for DiMethoxyethyl-PTCDI, but as well for all organic and inorganic films grown from a particle beam, where surface diffusion and plasmonic effects can be neglected and the counting statistical approach presented is applicable.

The RL model presented might as well be used for interfaces between two materials, if the underlying material already exhibited surface roughness before. Consequently, *void* has to be substituted by the material grown on top. The formalism remains the same.

Franta and Ohlídal (2005) discussed the advantages of Rayleigh-Rice theory (RRT) compared to the EMA used here, especially for laterally large irregularities (*i.e.* islands). Following their discussion, we also recommend to use RRT over EMA for our RL model, if this option is available. However, to our knowledge today's commercially available software does not offer a sophisticated way for implementing RRT.

CONCLUSIONS

In summary a sophisticated roughness layer model for spectroscopic ellipsometry is developed starting from empirical evaluation of AFM measurements. The model is generalized utilizing statistical considerations to be useful for a broad range of materials exhibiting high surface roughness. The model allows a more precise determination of the dielectric function as well as a more reliable determination of film thicknesses of highly rough materials.

ENDNOTE

[a]unpublished AFM investigation on SiO_2 thin films by O. Zywitzki, Fraunhofer FEP Dresden.

ABBREVIATIONS

AFM: Atomic force microscopy

DiMethoxyethyl-PTCDI: N,N'-dimethoxyethyl-3,4,9,10-perylene-tetracarboxylic-diimide

DiPhenyl-PTCDI: N,N'-diphenyl-3,4,9,10-perylene-tetracarboxylic-diimide

EMA: Effective medium approximation

MSE: Mean square error

QMB: Quartz micro balance

RL: Roughness layer

RMS: Root mean square

RRT: Rayleigh-Rice theory

SE: Spectroscopic ellipsometry.

ACKNOWLEDGEMENTS

The authors are grateful to Franziska Lüttich for the AFM measurements of DiMethoxyethyl-PTCDI and DiPhenyl-PTCDI. This work was financially supported by the Scientific Priority Program 1121: *Organic field-effect transistors: Structural and dynamical properties* and the Bundesministerium für Bildung und Forschung (project 03IS2011 "nanett"). The publication costs of this article were funded by the German Research Foundation/DFG (Geschäftszeichen INST 270/219-1) and the Chemnitz University of Technology in the funding programme Open Access Publishing.

AUTHORS' CONTRIBUTIONS

DL prepared the DiMethoxyethyl-PTCDI and DiPhenyl-PTCDI thin films, measured and evaluated the ellipsometry data, evaluated the AFM measurements and developed, based on them, the ellipsometry roughness layer model. FS implemented the theoretical description starting from normal distribution, advancing the model to the use of the error function and contributing a direct translation factor for AFM measured RMS values into roughness layer thickness. DRTZ provided analysis instruments and revised the manuscript. All authors read and approved the final manuscript.

REFERENCES

1. Aspnes DE, Theeten JB, Hottier F: Investigation of effective-medium models of microscopic surface roughness by spectroscopic ellipsometry. Phys Rev B 1979, 20: 3292.
2. Bruggeman DAG: Berechnung verschiedener physikalischer Konstanten von heterogenen Substanzen. I. Dielektrizitätskonstanten und Leitfähigkeiten der Mischkörper aus isotropen Substanzen. Ann Phys 1935, 416: 636-679.
3. Dhananjay , Cheng S-S, Yang C-Y, Ou C-W, Chuang Y-C, Wu MC, Chu C-W: Dependence of channel thickness on the performance of In_2O_3 thin film transistors. J Phys D Appl Phys 2008, 41: 092006.
4. Franta D, Ohlídal I: Comparison of effective medium approximation and Rayleigh-Rice theory concerning ellipsometric characterization of rough surfaces. Opt Commun 2005, 248: 459.
5. Gordan OD, Sakurai T, Friedrich M, Akimoto K, Zahn DRT: Ellipsometric study of an organic template effect: H_2Pc/PTCDA. Org Electron 2006, 7: 521-527.

6. Gordan OD, Hermann S, Friedrich M, Zahn DRT: Optical properties of the interfaces in organic/organic multilayered heterostructures. J Phys IV France 2006, 132: 73-76.

7. Lehmann D, Zahn DRT: The electrical and dielectrical behavior of n-conducting perylene tetracarboxylic diimide derivatives. Appl Phys A 2009, 95: 203-207.

8. Louis JS, Lehmann D, Friedrich M, Zahn DRT: Study of dependence of molecular orientation and optical properties of zinc phthalocyanine grown under two different pressure conditions. J Appl Phys 2007, 101: 013503.

9. Méndez-Vilas A, Nuevo MJ, González-Martín ML, Labajos-Broncano L: Quantitative surface roughness determination of materials by AFM: some limitations. Mater Sci Forum 2002, 408–412: 245.

10. Volintiru I, Creatore M, van de Sanden MCM: In situ spectroscopic ellipsometry growth studies on the Al-doped ZnO films deposited by remote plasma-enhanced metalorganic chemical vapor deposition. J Appl Phys 2008, 103: 033704.

10 CHAPTER

DIELECTRIC PROPERTIES OF POROUS SILICON FOR USE AS A SUBSTRATE FOR THE ON-CHIP INTEGRATION OF MILLIMETER-WAVE DEVICES IN THE FREQUENCY RANGE 140 TO 210 GHZ

Panagiotis Sarafis and Androula Galiouna Nassiopoulou*

[1]*NCSR Demokritos, INN, Terma Patriarchou Grigoriou*

ABSTRACT

In this work, the dielectric properties of porous Si for its use as a local substrate material for the integration on the Si wafer of millimeter-wave devices were investigated in the frequency range 140 to 210 GHz. Broadband electrical characterization of coplanar waveguide transmission lines (CPW TLines), formed on the porous Si layer, was used in this respect. It was shown that the dielectric parameters of porous Si (dielectric permittivity and loss tangent) in the above frequency range have values similar to those obtained at lower frequencies (1 to 40 GHz). More specifically, for the samples used, the obtained values were approximately 3.12 ± 0.05 and 0.023 ± 0.005, respectively. Finally, a comparison was made between the performance of the CPW TLines on a 150-μm-thick porous Si layer and on three other radiofrequency (RF) substrates, namely, on trap-rich high-resistivity Si (trap-rich HR Si), on a standard complementary metal-oxide-semiconductor (CMOS) Si wafer (p-type, resistivity 1 to 10 Ω.cm) and on quartz.

Keywords: Porous Si Nanoscale semiconductors RF characterization Dielectric permittivity Millimeter-wave passive devices

BACKGROUND

The co-integration of radiofrequency (RF) and millimeter-wave passive devices with complementary metal-oxide-semiconductor (CMOS) circuitry is quite challenging due to the low resistivity of the CMOS Si substrate that introduces important losses during electromagnetic wave propagation (eddy currents into the substrate). Another drawback towards this integration is the high permittivity of Si ($\varepsilon_{r,Si}$ = 11.7) that causes an increase in crosstalk between lines, a decrease in antenna efficiency, and a reduction of the frequency of operation of the inductors. A viable solution recently investigated towards this integration is the formation of a local substrate with the appropriate dielectric properties on the Si wafer, on which the RF and millimeter-wave devices will be integrated. Such a substrate is a thick porous Si layer with high porosity, which can be optimized for best device performance by choosing the appropriate layer thickness, in order to minimize electromagnetic propagation losses into Si, and the appropriate low values of the dielectric permittivity, ε_r, and loss tangent. These last values are tunable by changing the material structure and morphology [1–6].

Porous Si structure (pore size, inter-pore distance) and morphology affect all its macroscopic properties (electrical, mechanical, optical, etc.) [7]. An intensive effort was made in the literature to correlate the electrical properties of the material with its structural parameters [8–12]. In view of the application of porous Si for the on-chip integration of RF and millimeter-wave devices, its dielectric properties (dielectric permittivity and loss tangent) as a function of frequency should be known, in order to be used by the device designer for an accurate prediction of device operation. In addition, since the dielectric properties of the material depend strongly on its structure and morphology [13], it is desirable to have an experimental method to extract the dielectric parameters of the specific material used in each application.

In this work, we will first discuss the existing models that correlate the structural properties of porous Si (porosity and morphology) with its dielectric properties and we will compare them with results obtained by a broadband extraction method, based on the measurement of the S-parameters of coplanar waveguide transmission lines (CPW TLines) integrated on the porous Si substrate. By combining these measurements with electromagnetic simulations, the dielectric permittivity and loss tangent of the substrate (porous Si) can be obtained. This method has been previously used by the authors to extract the dielectric parameters of porous Si in the frequency range 1 to 40 GHz [13, 14]. In this work, measurements are extended to the frequency range 140 to 210 GHz. Finally, by comparing the performance of CPW TLines on porous Si and three other substrates used in RF, namely, a trap-rich high-resistivity (HR) Si substrate [15–17], a standard CMOS Si wafer (p-type, resistivity 1 to 10 Ω.cm), and a quartz substrate, we demonstrate the superiority of porous Si as a local substrate for RF and millimeter-wave on-chip device integration.

METHODS

Fabrication of CPW TLines on porous Si layers

Thick PSi layers (thickness ~150 μm) were formed by anodization of highly doped p-type Si ($\rho = 1$ to 5 mΩ.cm). The electrolytic solution was a mixture of HF and ethanol (3 EtOH(99.9%)/2 HF(50%) v.v.) and the anodization current density was $J = 20$ mA/cm^2. The resulting layer had a porosity of 76% and a dendritic structure as presented in Figure 1. The porous Si layer was capped with 500 nm SiO_2 in order to stabilize it over time and achieve better planarization of the porous Si surface for further processing. On top of PSi, covered by SiO_2, a set of coplanar waveguide transmission lines (CPW TLines), made of 1-μm-thick patterned Al, was integrated (see Figure 2).

For comparison, identical CPW TLines were also fabricated on three other substrates, as follows: the first was the state-of-the-art trap-rich high-resistivity (HR) Si RF substrate [15]. This substrate was an n-type HR-Si wafer with nominal resistivity higher than 10 kΩ.cm, covered by a bilayer of a 500-nm-thick trap-rich poly-Si layer, deposited by low-pressure chemical vapor deposition (LPCVD) at 625°C, and a-500 nm-thick TEOS SiO_2 layer. The trap-rich layer is used to minimize the parasitic surface conduction within the Si layer underneath the silicon oxide by trapping the parasitic charges and thus restoring the initial high resistivity of the Si substrate [17]. The second substrate was a 380-μm-thick standard Si wafer used in CMOS-integrated circuits (ICs) (p-type, resistivity 1 to 10 Ω.cm). Finally, the last substrate was a 500-μm-thick quartz substrate, which is one of the off-chip RF substrates with almost negligible losses. This last substrate was used for comparison with the three other Si-based substrates.

Figure 1. SEM IMAGE OF HIGHLY POROUS SI. SEM image of highly porous Si formed on p^+ Si with resistivity 1 to 5 mΩ.cm. It depicts the vertical pores with dendrite structure of the material. Pore size is between 9 and 12 nm.

RF measurements and de-embedding

The S-parameters of the CPW TLines were measured in the 140-to-210-GHz range with an HP 8510B vector network analyzer (VNA) from Agilent (Santa Clara, CA, USA), combined with a millimeter-wave VNA extension module by Oleson Microwave Labs (Morgan Hill, CA, USA). All the measurements were calibrated using the Line-Reflect-Reflect-Match (LRRM) algorithm of the WinCal software from Cascade Microtech (Beaverton, OR, USA). A de-embedding procedure is always necessary in order to decouple the device response from the parasitics due to the contacts and pads. The method followed was the two-line method, using the measured S-parameters of two lines with different length (8 mm and 500 µm) [18]. The characteristic impedance, the propagation constant, and the effective permittivity of the lines were extracted and used to derive the dielectric parameters of the PSi substrate [13, 14].

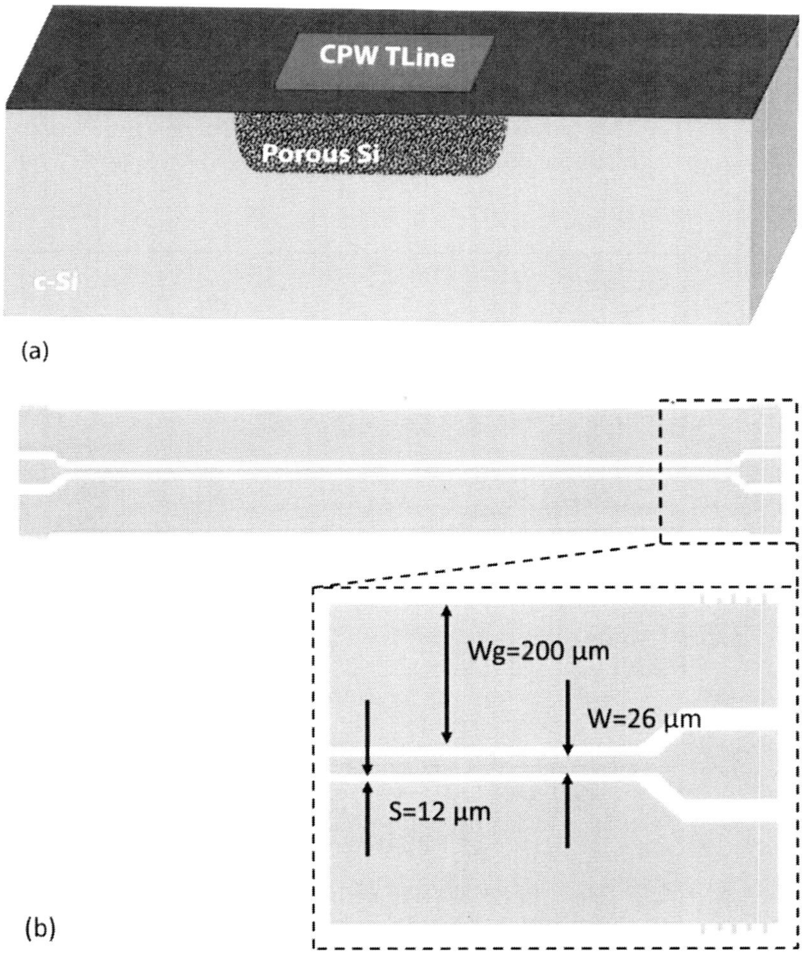

Figure 2. SCHEMATIC REPRESENTATION OF LOCAL POROUS SI LAYER ON SI WAFER AND GEOMETRY OF CPW TLINE. **(A)** Schematic representation of the locally formed porous Si layer on the Si wafer, on which the CPW TLine is integrated. **(B)** Topology of the CPW TLine with respective dimensions.

The de-embedding and the extraction method were first tested for the quartz substrate (fused silica), which is known to have a constant dielectric permittivity of 3.82 throughout the whole frequency range 1 to 210 GHz [19, 20]. The extraction method is described in detail in [13]. The obtained results are depicted in Figure 3 for the frequency ranges 1 to 40 GHz and 140 to 210 GHz. We can see that the curves show continuity between the two frequency ranges and the extracted values of the permittivity are 3.82 for frequencies in the range 1 to 40 GHz and 3.71 to 3.79 for frequencies in the range 140 to 210 GHz. These results are very close to the literature value of quartz permittivity (3.82) and give confidence that the de-embedding and the parameter extraction methods are valid. They were thus used to characterize the porous Si layer in the above frequency ranges.

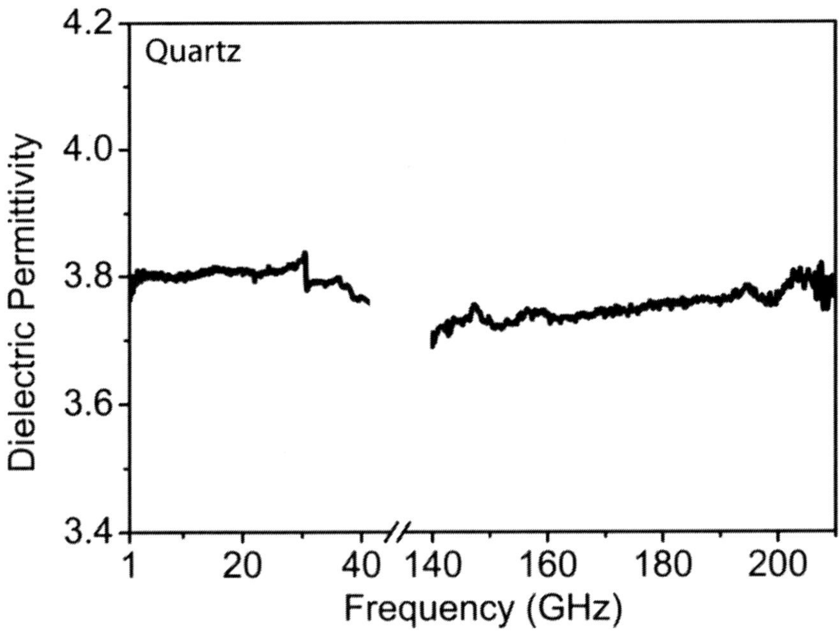

Figure 3. DIELECTRIC PERMITTIVITY OF QUARTZ AS A FUNCTION OF FREQUENCY IN FREQUENCY RANGES 1 TO 40 GHZ AND 140 TO 210 GHZ. The extracted dielectric permittivity of quartz as a function of frequency using the extraction method described in the text is depicted. A constant value of approximately 3.8 is obtained for the frequency range 1 to 40 GHz and on average 3.76 for the frequency range 140 to 210 GHz. The obtained values are very close to the nominal value of quartz permittivity in the whole frequency range under discussion (3.82).

Microscopic models for determining PSi dielectric properties

Porous Si structure and morphology depend on the electrochemical conditions used for its formation as well as on the starting wafer resistivity. Its dielectric properties are highly dependent on its structure and morphology. There are several works in the literature that correlate the material structure with its dielectric properties. According to [9, 21, 22], the ac electrical transport of porous Si follows two mechanisms. The first is limited by the length of the carrier random walk through the fractal structure of the material and is valid in the very low frequency range, while at higher frequencies, the random path is shorter and the hopping length stops to be the critical factor. In that case, conduction is mainly determined by the distance between inhomogeneous areas [22]. The dielectric permittivity of porous Si (ε_{PSi}) describes the polarization of the atoms and the impurities inside the material. As it is shown in [22], ε_{PSi} depends on frequency only for frequencies <100 Hz. For higher frequencies, its value is saturated and remains constant up to at least 100 kHz. This value is also independent of temperature.

Effective medium approximation

Instead of considering carrier transport at the nanoscale, it is easier to correlate the dielectric properties of the material with macroscopic parameters, as for example the material porosity. An effective medium approximation (EMA) is used in this respect. This approximation is valid when the wavelength of the signal is much larger than the typical dimensions of pores and nanostructures composing the material. The most common models used in the literature to correlate the permittivity of non-oxidized porous Si with its porosity are the following:

Vegard's approximation[1]

Vegard's approximation is a simple mixing model correlating the dielectric permittivity with porosity (P) through the relation:

$$\varepsilon_{PSi} = P\varepsilon_{air} + (1 - P)\varepsilon_{Si}. \tag{1}$$

where ε_{PSi} is the permittivity of porous Si, ε_{air} is the permittivity of air, and ε_{Si} is the permittivity of Si and P is the porosity.

Maxwell-Garnett's approximation[23]

This is valid for systems in which the filling fraction f (where $f = 1 - P$) of the porous material is far smaller than the porosity (P) [23]. The following expression is obtained:

$$\frac{\varepsilon_{PSi} - \varepsilon_{air}}{\varepsilon_{PSi} + 2\varepsilon_{air}} = (1 - P)\frac{\varepsilon_{Si} - \varepsilon_{air}}{\varepsilon_{Si} + 2\varepsilon_{air}} \tag{2}$$

Bruggeman's approximation[23].

This is applied to structures where the filling fraction is comparable to the porosity [23]. The following expression relates the dielectric permittivity with the porosity:

$$P\frac{\varepsilon_{PSi} - \varepsilon_{air}}{\varepsilon_{PSi} + 2\varepsilon_{air}} = (1 - P)\frac{\varepsilon_{Si} - \varepsilon_{air}}{\varepsilon_{Si} + 2\varepsilon_{air}} \tag{3}$$

Bergman's approximation[24]

It introduces the spectral density function *g(n,P)* to take into account the nanotopology of the material. The following expression is obtained:

$$\varepsilon_{PSi} = \varepsilon_{air}\left(1 - f\int_0^1 \frac{g(n.f)}{\frac{\varepsilon_{air}}{\varepsilon_{air} - \varepsilon_{Si}} - n}dn\right) \tag{4}$$

From all the above models, Vegard's approximation is the simplest one. The most commonly used model is the Bruggeman's model [11, 25]. Both the Vegard's model for non-oxidized Si and the Maxwell-Garnett's model have been proven to be insufficient to explain the results of several experiments [13, 24, 26]. An improved version of the Vegard's model incorporates also the SiO_2 native oxide surrounding the Si nanostructures composing the material [27]. Better agreement between the model and experimental results is obtained in this case. The oxidation of the Si skeleton leads to a decreased permittivity of the material [11, 27]. This is because the oxidation not only changes material composition, but also leads to reduction of material porosity. Finally, the Bergman's approximation predicts quite well the dielectric behavior of PSi in the optical frequencies. The spectral density function *g(n,f)* that describes the micro-topology of the material has to be extracted in this respect [12].

Dielectric parameter extraction using broadb and electrical measurements

The models of Vegard, Maxwell-Garnett, and Bruggeman, as presented above, relate ε_{PSi} with material porosity. However, they were insufficient to explain the experimental results of several groups [13, 26, 28]. This can be attributed to the complexity of the PSi structure and morphology, which differs from one sample to another, even if the macroscopic porosity is the same. It is also quite difficult to find a representative *g(n,f)* function that accurately describes the specific porous Si structure and morphology in each case, making the Bergman's model difficult to use. It is thus interesting to have a method to characterize the specific PSi material used in each experiment by simply measuring the electromagnetic (EM) response of a device integrated on it. This is the approach we use in this work. By integrating CPW TLines on top of porous Si and measuring their S-parameters, we extract porous Si dielectric parameters by combining the experimental results with electromagnetic simulations and conformal mapping calculations. This method has been described in detail in [13, 14], and the results have been proven to be in very good agreement with full-wave EM simulations [14].

In Figure 4 the extracted dielectric permittivity of three PSi layers with 70%, 76%, and 84% porosity using the above method are depicted in full black circles. The PSi layers were fabricated on a p^+-type Si wafer with resistivity 1 to 5 mΩ.cm and had a surface area of 4 cm^2. Identical transmission lines were integrated on all three samples (see Figure 2b). The obtained results were compared with those obtained using Vegard's, Maxwell-Garnett's and Bruggeman's models for PSi by applying formulas (1) to (3) given above. From Figure 4, it can be seen that the values of the extracted permittivity using broadband electrical measurements of the specific CPW TLines are between those obtained with the Bruggeman's and Vegard's models for non-oxidized PSi. On the other hand, by using the more elaborated Vegard's law described in [27], which takes into account the presence of a native oxide shell surrounding the Si nanostructures (in our case, we considered a native oxide thickness of 1.5 nm and a Si skeleton thickness of 10 nm), better agreement is achieved between our experimental results and the calculated ones.

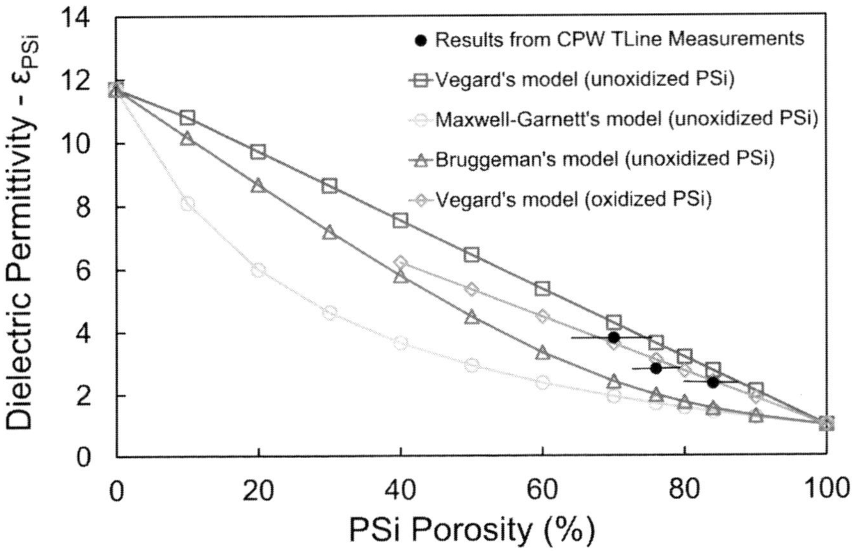

Figure 4. DIELECTRIC PERMITTIVITY OF POROUS SI AS A FUNCTION OF POROSITY. Full black dots: extracted values of the dielectric permittivity ε_{PSi} of porous Si from measurements of CPW TLines. Open squares: results using Vegard's model for unoxidized porous Si. Open circles: results using Maxwell-Garnett's model for unoxidized porous Si. Open triangles: results using Bruggeman's model for unoxidized Si. Open rhombi: results using Vegard's model for oxidized porous Si.

RESULTS AND DISCUSSION

Porous Si dielectric parameters in the frequency range 140 to 210 GHz

Using broadband electrical measurements combined with simulations, the dielectric parameters of PSi in the frequency range 140 to 210 GHz were extracted. The obtained results are presented in Figure 5 in comparison with the extracted parameters for the frequency range 1 to 40 GHz. At low frequencies (1 to 40 GHz), there is an initial slight monotonic decrease of ε_{PSi} from 3.19 to 3.12 and it then stabilizes around this value (Figure 5a). In the high-frequency range (140 to 210 GHz), ε_{PSi} oscillates around the values of 3.1 and 3.2, within a maximum deviation of 0.1. Similarly, the value of the loss tangent is between 0.031 and 0.023 in the range 5 to 40 GHz (see Figure 5b), while it stays constant at 0.023 in the range 140 to 210 GHz, with a maximum deviation of 0.005. In the 1-to-5-GHz range, the results are not reliable due to the contact resistance between the RF probes and the pads and were omitted. This effect becomes negligible at higher frequencies.

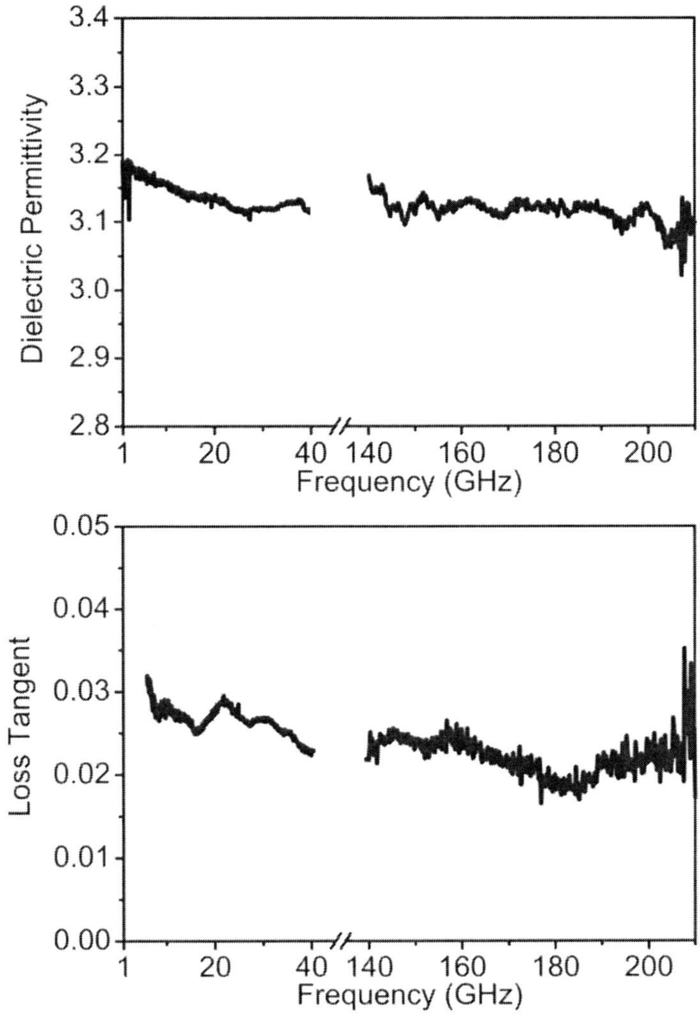

Figure 5. PSI DIELECTRIC PERMITTIVITY AND LOSS TANGENT IN FREQUENCY RANGES 1 TO 40 GHZ AND 140 TO 210 GHZ. The curves depict PSi dielectric permittivity **(A)** and loss tangent **(B)**, extracted from broadband electrical measurements combined with simulations of CPW TLines integrated on the PSi substrate for the frequency ranges 1 to 40 GHz and 140 to 200 GHz.

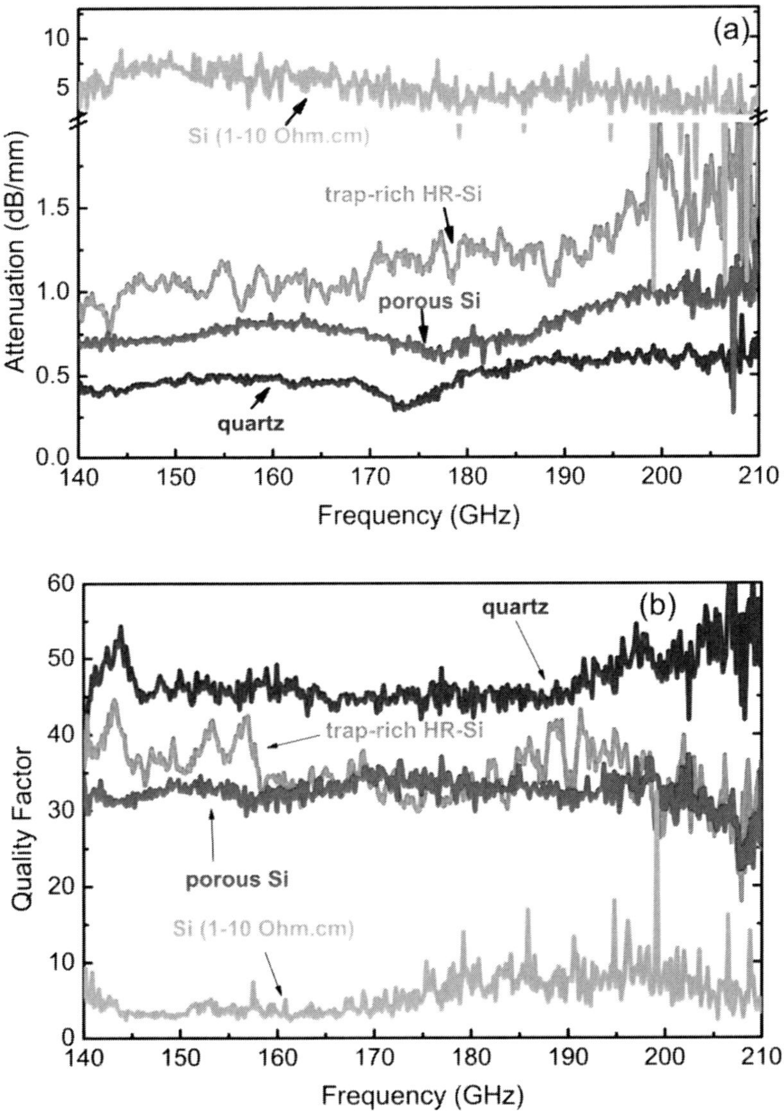

Figure 6. ATTENUATION (A) AND QUALITY FACTOR (B) OF CPW TLINES ON PSI COMPARED WITH THREE OTHER SUBSTRATES. Comparison of signal attenuation and quality factor of CPW TLines on PSi (blue lines) compared to that of similar CPW TLines on trap-rich HR Si (green lines), quartz (dark red lines) and low-resistivity CMOS Si (orange lines) in the frequency range 140 to 210 GHz.

In overall, from the above, we can deduce that the dielectric permittivity of porous Si is almost constant in the studied frequency ranges. It also shows a continuity of the two curves, suggesting the same constant value in the frequency range 40 to 140 GHz. The loss tangent shows a slight decrease with frequency, while again there is continuity between the low- and high-frequency curves.

Comparison of PSi with other RF and millimeter-wave substrates

In order to demonstrate the high performance of porous Si for use as a substrate for RF and millimeter-wave devices, a comparison was made between this substrate and three other substrates used in the same respect. Identical CPW TLines were integrated on the four different substrates, their S-parameters were measured, and the propagation constant for each line was extracted. Figure 6 shows the extracted values of signal attenuation (a) and quality factor (b) for the CPW TLines on the four different substrates. We deduce that the lines on the three substrates, trap-rich HR Si, PSi, and quartz, have better performance than those on the low-resistivity CMOS Si. More specifically, trap-rich HR-Si reduces losses from 4.8 to 1.6 dB/mm at 210 GHz, while PSi leads to a further decrease of the attenuation loss of 1.2 dB/mm at 210 GHz. Both the above substrates show similar performance with quartz, which is a non-Si, off-chip substrate.

The observed reduction of signal attenuation a and the increase of the quality factor Q of the CPW TLine on PSi versus bulk Si is attributed to the reduction of the material loss tangent and dielectric permittivity through nanostructuring. As shown previously by the authors, the achieved low permittivity of porous Si at high porosities shows advantages in many RF and millimeter-wave devices, namely, high-characteristic impedance of the CPW TLines [5], inductors operating at higher frequencies [29, 30] and antennas with reduced surface waves induced into the substrate can be obtained.

CONCLUSIONS

The dielectric properties of a 150-μm-thick PSi layer with 76% porosity were presented for the first time for the frequency range 140 to 210 GHz, using broadband electrical measurements combined with electromagnetic simulations. It was found that the dielectric permittivity is almost constant in the above frequency range, having approximately the same value as at lower frequencies. The loss tangent is also almost constant with frequency. Finally, a comparison between the performance of CPW TLines on PSi, trap-rich HR Si, quartz, and standard low-resistivity CMOS Si was made in the above frequency range. An almost equal performance was obtained between the trap-rich HR Si, PSi, and quartz. At 210 GHz, porous Si showed an attenuation as low as 1 dB/mm and the quality factor was ~30. This performance is added to the other advantages of PSi compared to other Si-based substrates, e.g., its compatibility with the low-resistivity CMOS substrate (permitting co-integration of CMOS logic with RF and millimeter-wave devices on the same substrate) and its low achievable permittivity). All the above make PSi an excellent local substrate on the Si wafer for RF and millimeter-wave device integration on the Si chip, paving the way towards the digital/RF analog system-on-chip (SoC) of the future.

ACKNOWLEDGEMENTS

The trap-rich high-resistivity Si wafers were provided by UCL Belgium (Jean-Pierre Raskin), while measurements in the frequency range 140 to 210 GHz of the CPW TLines were conducted in the facilities of VTT, Helsinki, Finland (arranged by A. Markus) during a visit of P. Sarafis to VTT. This work was supported by the EU Network of Excellence 'Nanofunction' through the

AUTHORS' CONTRIBUTIONS

PS made the experiments and drafted the paper, while AGN supervised the work and revised the paper. Both authors read and approved the final manuscript.

REFERENCES

1. Kim H-S, Xie Y-H, DeVincentis M, Itoh T, Jenkins K a: Unoxidized porous Si as an isolation material for mixed-signal integrated circuit applications. J Appl Phys 2003, 93: 4226. 10.1063/1.1555700
2. Welty R, Park S, Asbeck PM, Dancil K-PS, Sailor MJ: Porous silicon technology for RF integrated circuit applications. In 1998 Top. Meet. Silicon Monolith. Integr. Circuits RF Syst. Dig. Pap. (Cat. No.98EX271). IEEE; 1998:160–163.

3. Gautier G, Leduc P: Porous silicon for electrical isolation in radio frequency devices: a review. Appl Phys Rev 2014, 1: 011101. 10.1063/1.4833575

4. Capelle M, Billoué J, Poveda P, Gautier G: RF performances of inductors integrated on localized p + -type porous silicon regions. Nanoscale Res Lett 2012, 7: 523. 10.1186/1556-276X-7-523

5. Issa H, Ferrari P, Hourdakis E, Nassiopoulou AG: On-chip high-performance millimeter-wave transmission lines on locally grown porous silicon areas. IEEE Trans Electron Devices 2011, 58: 3720–3724.

6. Capelle M, Billoué J, Poveda P, Gautier G: Study of porous silicon substrates for the monolithic integration of radiofrequency circuits. Int J Microw Wirel Technol 2013, 6: 39–43.

7. "Properties of porous silicon" Emis Datareviews. Ser. No18, IEE, an INSPEC Publ.UK, edited by L.T.Canham. 1997. 1997.

8. Di Francia G, La Ferrara V, Maddalena P, Ninno D, Odierna LP, Cataudella V: AC conductivity of porous silicon: a fractal and surface transport mechanism. Il Nuovo Cimento D 1996, 18: 1187–1196. 10.1007/BF02464696

9. Lehmann V, Hofmann F, Möller F, Grüning U: Resistivity of porous silicon: a surface effect. Thin Solid Films 1995, 255: 20–22. 10.1016/0040-6090(94)05624-M

10. Boarino L, Borini S, Amato G: Electrical properties of mesoporous silicon: from a surface effect to coulomb blockade and more. J Electrochem Soc 2009, 156: K223. 10.1149/1.3232202

11. Astrova E, Tolmachev V: Effective refractive index and composition of oxidized porous silicon films. Mater Sci Eng B 2000, 69–70: 142–148.

12. Theiβ W: The dielectric function of porous silicon - how to obtain it and how to use it. Thin Solid Films 1996, 276: 7–12. 10.1016/0040-6090(95)08036-8

13. Sarafis P, Hourdakis E, Nassiopoulou AG: Dielectric permittivity of Porous Si for use as substrate material in Si-integrated RF devices. IEEE Trans Electron Devices 2013, 60: 1436–1443.

14. Sarafis P, Hourdakis E, Nassiopoulou AG: Porous Si dielectric parameter extraction for use in RF passive device integration :

Measurements and simulations. 43rd Eur. Solid-State Device Res. Conf. Bucharest 2013, 99–102.

15. Neve CR, Ben Alia K, Malaquin C, Allibert F, Desbonnets E, Bertrand I, Van Den Daele W, Raskin J-P: RF and linear performance of commercial 200 mm trap-rich HR-SOI wafers for SoC applications. In 2013 IEEE 13th Top. Meet. Silicon Monolith. Integr. Circuits RF Syst. Austin, TX, USA: IEEE; 2013:15–17.

16. Ben AK, Neve CR, Gharsallah A, Raskin J: RF performance of SOI CMOS technology on commercial 200-mm enhanced signal integrity high resistivity SOI substrate. IEEE Trans Electron Devices 2014, 61: 722–728.

17. Lederer D, Raskin J-P: Effective resistivity of fully-processed SOI substrates. Solid State Electron 2005, 49: 491–496. 10.1016/j.sse.2004.12.003

18. Mangan AM, Voinigescu SP, Tazlauanu M: De-embedding transmission line measurements for accurate modeling of IC designs. IEEE Trans Electron Devices 2006, 53: 235–241.

19. Lamb JW: Miscellaneous data on materials for millimetre and submillimetre optics. Int J Infrared Millimeter Waves 1996, 17: 1997–2034. 10.1007/BF02069487

20. Cheng H-J, Whitaker JF, Weller TM, Katehi LPB: Terahertz-bandwidth characteristics of coplanar transmission lines on low permittivity substrates. IEEE Trans Microw Theory Tech 1994, 42: 2399–2406. 10.1109/22.339773

21. Ben-Chorin M, Möller F, Koch F: Nonlinear electrical transport in porous silicon. Phys Rev B 1994, 49: 2981–2984. 10.1103/PhysRevB.49.2981

22. Ben-Chorin M, Möller F, Koch F, Schirmacher W, Eberhard M: Hopping transport on a fractal: ac conductivity of porous silicon. Phys Rev B 1995, 51: 2199–2213.

23. Campos AM, Torres J, Giraldo JJ: Porous silicon dielectric function modeling from effective medium theories. Surf Rev Lett 2002, 09: 1631–1635. 10.1142/S0218625X02004116

24. Theiβ W, Henkel S, Arntzen M: Connecting microscopic and macroscopic properties of porous media: choosing appropriate effective medium concepts. Thin Solid Films 1995, 255: 177–180. 10.1016/0040-6090(94)05649-X

25. Khardani M, Bouaïcha M, Bessaïs B: Bruggeman effective medium approach for modelling optical properties of porous silicon: comparison with experiment. Phys Status Solidi 2007, 4: 1986–1990. 10.1002/pssc.200674420

26. Ramani S, Cheville A, Escorcia Garcia J, Agarwal V: Conductivity of free-standing porous silicon layers using Terahertz differential time-domain spectroscopy. Phys Status Solidi 2007, 4: 2111–2115. 10.1002/pssc.200674393

27. Theodoropoulou M, Pagonis DN, Nassiopoulou AG, Krontiras CA, Georga SN: Dielectric characterization of macroporous thick silicon films in the frequency range 1 Hz-1 MHz. Phys Status Solidi 2008, 5: 3597–3600. 10.1002/pssc.200780153

28. Menard S, Fevre A, Capelle M, Defforge T, Billoue J, Gautier G: Dielectric behaviour of porous silicon grown from p-type substrates. In Int. Conf. Porous Semicond. - Sci. Technol, 0. Benidorm-Alicante; 2014:122–123.

29. Sarafis P, Hourdakis E, Nassiopoulou AG, Roda Neve C, Ben Ali K, Raskin J-P: Advanced Si-based substrates for RF passive integration: comparison between local porous Si layer technology and trap-rich high resistivity Si. Solid State Electron 2013, 87: 27–33.

30. Capelle M, Billoue J, Poveda P, Gautier G: N-type porous silicon substrates for integrated RF inductors. IEEE Trans Electron Devices 2011, 58: 4111–4114.

11 CHAPTER

NATURAL LIGHTING SYSTEMS BASED ON DIELECTRIC PRISMATIC FILM

Daniel Vázquez-Moliní[1], Antonio Álvarez Fernández-Balbuena[1] and Berta García-Fernández[1]

[1] Dept. of Optics, School of Optics, University Complutense of Madrid, Spain

1. INTRODUCTION

Daylight provides high quality lighting, reduces energy use and has numerous beneficial physical and psychological effects on people. Furthermore, natural lighting has many benefits in creating indoor spaces, such as energy saving and better quality of vision; two facts that improve environments and thus, productivity.

The light pipe is a device that can transfer natural light from a building´s roof into the depths of the building, this straight construction consist of a reflective closed walled structure (P.D. Swift & G.B. Smith, 1995). Daylight guidance has been one of the mayor areas of innovation in interior lighting in recent years, with the development of light pipes daylight and electric light are simultaneously delivered into a building where they are combined and distributed via luminaries. As a result, the overall wattage of artificial light is reduced and the consumption of electricity decreases (Mayhoub, et al., 2010). The commonest light pipes are reflective mirror guides which use high reflectance aluminium, also fiber optic guides are widely used for illumination purposes. Optical design with new materials like dielectric ones, with regard to their reflection, transmission and absorption is as important as its geometry study.

In recent works, Vazquez-Moliní et al. introduced an illumination system called ADASY® integrated into a building's façade that consists of a horizontal light guide inside the building. ADASY® comprises a collection system, a light guide, and daylight luminaries (Vazquez-Moliní et al, 2009). Prior developments in solar lighting systems based on micro-replicated light film had been studied showing that prism light guides performance varies with the length of the guide, maintenance conditions, the collecting system, the luminaries, and the direction from which light is directed (Whitehead, L. A., 1982). The objective of previous investigations of the group had been the study and development of efficient dielectric prismatic hollow light pipes that direct natural light into interior spaces applied in office buildings One of the proposed system is a daylight illumination system by vertical transparent prismatic light guide for an office building; this model consist in a hollow tube internally coated with thin polycarbonate prismatic film. In this model, two different prismatic sheets have been used; the light guiding system works with 90º prismatic film and the extraction system is composed of 70º prismatic film perpendicular to the previous one that works extracting the light outside the guide with a specific angular distribution. This design allows us to obtain a transparent simple and beautiful pipe which is integrated in building design; in addition, the spatial distribution of the extraction sheets can be adapted to the requirements of each space. (Alvarez Fernandez-Balbuena A. et al, 2010).

Light color quality is an important issue to evaluate in natural lighting systems. High reflectance aluminum lighting guides are giving bad light quality because the spectral reflectance of the aluminium, changes the color characteristics of the output light at the end of the guide (Vázquez-Moliní et al, 2007). When light guides are made of a dielectric prismatic film, the influence of the spectral reflectance is minimized due to the total internal reflection produced in the surface of the prismatic film, absorptance is not usually considered significant in the literature when the sheets are thin. Color Rendering Index and Correlated Color Temperature are important parameters in order to evaluate lighting quality in Museums, office buildings and production centers to get the normative approval (García-Fernández et al, 2011).

A skylight is a technology for obtaining natural light into a building. Skylights are an opening in a roof that is covered with translucent or transparent material and that is designed to admit light provided a connection to the outdoor environment to occupants. Nowadays the skylight technology is widely used in outstanding buildings (Dubois, 2003). Almost all of the skylights used are just an opening in the ceiling that ensures watertight but without significant optics, just a diffuser on it to prevent direct sunlight.

The Compound Parabolic Concentrators (CPC) is a non-focusing light funnel with specular reflecting surface of a parabolic shape designed to give the maximum concentration ratio for a given acceptance angle θmax. CPCs are relevant for solar energy caption because they achieve good concentration for many acceptance angles. There are many applications where the Compound Parabolic Concentrator is used not only for concentration but also for collimation (in reverse mode) like in natural lighting, thermal collector, LED's optic, car light and optical fiber coupling (Winston, 1975). Winston et al. had explored a growing field that is applicable to areas where the collection, concentration, transport and distribution of light is important. Systems for natural light caption offer very important advantages when use some kind of CPC optics (Winston et al., 2005). CPC passive optics made of dielectric film allows the operation of the collimating system during long periods of time without need to fit its orientation since a CPC in reverse mode is capable of redirecting all the light entering in 2π in the designed angle (Alvarez Fernandez-Balbuena et al., 2009).

2. PRISMATIC FILM

A prismatic film is a thin plastic that works with the optical principle of total internal reflection (TIR) through the prism structure. It can be used in several applications for replacing metal guides with best performance. The film geometry has one flat surface and the other one is a textured surface consisting of an array of linear right angle prisms inclined at 45 degrees to the flat surface. This configuration has a light angular acceptance cone determined by the refractive index of the prismatic film material, approximately a 30º semiangle cone. If this angular criterion is not met, the light will instead be rejected out of the light pipe through the prismatic material (OLF).

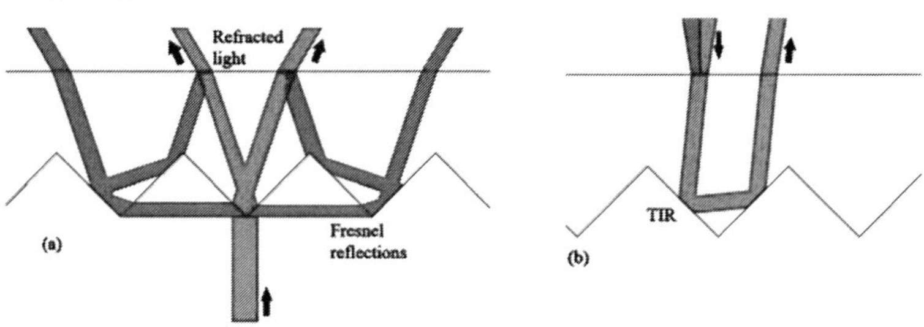

Figure 1. a) Example of refraction and (b) total internal reflection in prismatic film.

Figure 1 (a) is an example of light entering in a prismatic film through prism apex side, in this case light is divided in two main arms (refracted light). The incidents rays are drawn in red lines and the Fresnel refractions are plotted with blue lines. Most of the rays directly emerge from the prism as displayed in red lines; rest of light than is guided in the prismatic structure is Fresnel light with low flux energy. In figure 1 (b) the prismatic film produces TIR, in this case the light is returned in the direction from which it came, this principle is used for guiding light with a prismatic film.

The degree to which the film's prisms deviate from perfect prisms also affects the efficiency of the total internal reflection process, and therefore, the effectiveness of the film in transporting and distributing light. These imperfections include 90º corners which are not precise, surfaces which are not optically flat or which deviate from the correct angle and optical inhomogeneities in the material (Remillard et al.1992). Absorption is due to bulk absorptivity of the material used to produce the film and transmission can be used to advantage of the application in light distribution. With the typical losses due to absorption and transmission, the reflectance efficiency has been estimated as approaching 99% (Keipp, 1994). Precision micromachining, polymer processing and certain other manufacturing technologies like microreplication have made possible the development of an optimized prismatic film (Wang, 2009). The structure of prism film under different magnifications is shown in Figure 2.

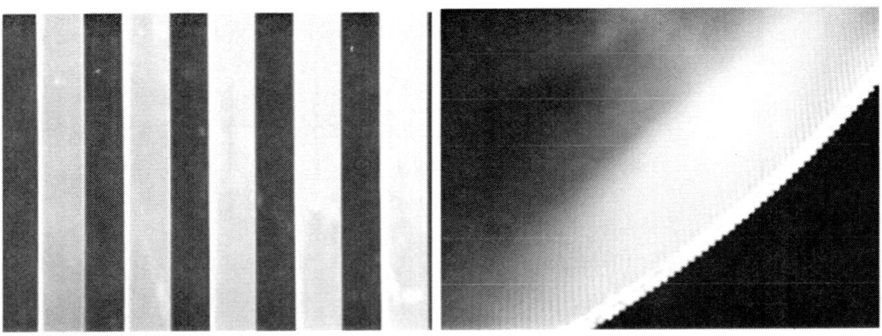

Figure 2. Prismatic film used to distribute light. a) Surface area detail of prisms structure (25X), b) prismatic film (2.5X).

2.1. Image processing algorithms

This section deal with a brief description of the recognition procedure algorithm to analyse the prism profile structure. First, a diagram with a description of the recognition procedure to extract the inclination of the lines from the image and to calculate the radio of the peak prism is given (Fig.3), secondly, a more detailed description will follow.

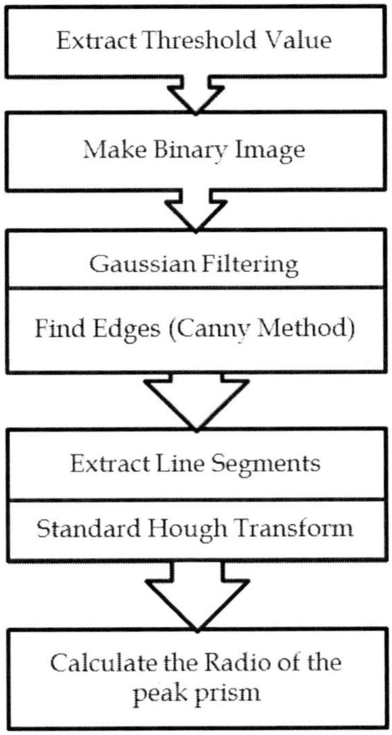

Figure 3. Structure of the recognition procedure and extraction process performed.

In order to investigate the light behavior in the system the effect of accuracy of prismatic structures and peak defects has been examined; computer analyses such as changes of prism angle and plane shape were carried out. By analyzing the image processing prism structure and by using morphological operations measurement of the inclination angle with high accuracy can be achieved. To execute the processing, the prismatic film profile showed in figure 4 is employed (the optical microscope used to obtain the prism image is Motic SMZ-143 equipped with a digital camera Moticam 2000).

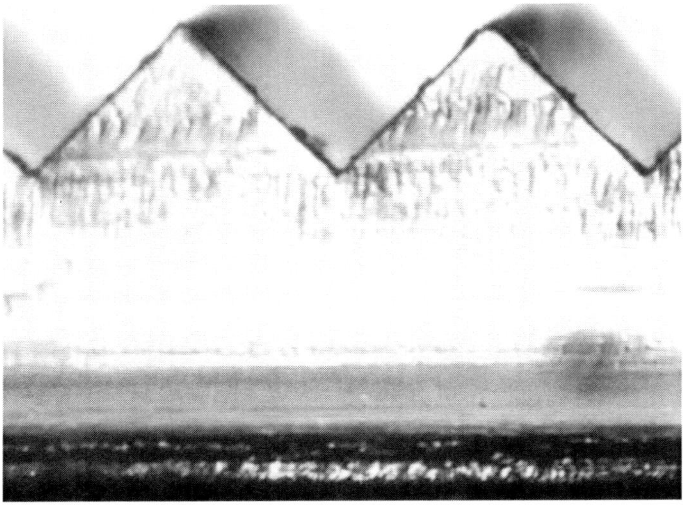

Figure 4. Prism structure used to analyse the parameters of the prism (57X)

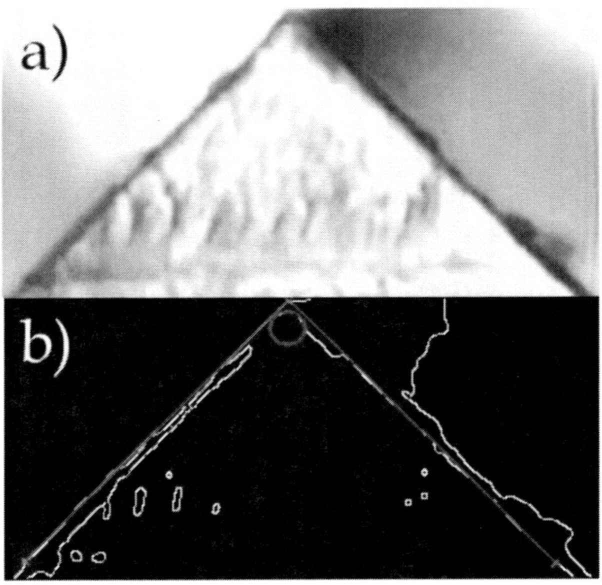

Figure 5. a) Cropped image before the digital processing used to obtain the dimensions of micro prism structures. (b) Edge detected images resulting from the Canny Method (white), the line after Hough transforms (red), and the estimation to calculate the radius of the prism´s peak (blue).

Firstly, a threshold is applied to the input image in order to make it binary. The threshold value is determined from a grey-level histogram of the image, later the edge is separated from the background (Fig.5.b).

The edge map is used to determine the existence of lines around the regions. This edge description is obtained from the operator Canny (Canny, 1986) this operator is consider as an optimal edge detector to find boundaries between poorly defined objects as well as hard edges. The Canny method finds edges by looking for local maxima of the gradient of the image; the gradient is calculated using the derivative of a Gaussian filter. The method uses two thresholds, to detect strong and weak edges, and includes the weak edges in the output only if they are connected to strong edges.

Later, we use the Hough transform (Fig.5. (b) and Fig.6) to detect the parameters that controls the accuracy of the right angle at the vertex of the prism (Hough, 1962).

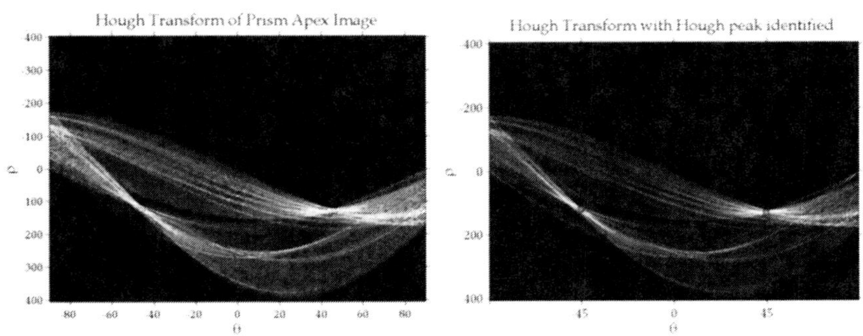

Figure 6. a) The Hough transform of the prism image. (b)The red square shows the peaks of data in the Hough matrix. ρ is the distance from the center and θ the angle at which the sum of intensities in the image peaks, it is thus the slope of the line along with the position.

Originally, the Hough Transform was proposed to extract straight lines in the particle tracks recognizing procedure. Nowadays, the Hough transform is a technique which is used to insulate features of a particular shape within an image. In this case, the Hough transform is used to identify the parameters of the line and it uses the parametric representation of a line which is fits to a set of given edge points. It takes as input the grey scale image, and produces as output an image showing the positions of tracked intensity discontinuities. The output of the edge detector defines where features are in the image, and the Hough transform determine what the features are and how many of them exist in the image. The main advantage of the Hough transform technique is that it is tolerant of gaps in feature boundary descriptions and is minimal unaffected by image noise. The result of the Hough transform is stored in a matrix that often is considered an accumulator (Fig. 6 (a)). One dimension of this matrix is the angles θ and the other dimension are the distances ρ, and each element has a value telling how many points/pixels are positioned on the line with parameters (ρ, θ). So the element with the highest value shows the line that is most represented in the input image.

After performing probabilistic Hough Transform, two lines are obtained determining and ensuring the right angle prism used to guide the incident light. Figure 5 shows in blue color the plot of the inclination angle of the prism through the digital processing. The main peaks are located in the Hough transform matrix. The left slope is calculated to be an apex angle of -45.38º and the right slope has an apex angle of 44.50º forming a total angle of 89,88º (Fig. 6 (b)), after that, an approximation was made to relate the prism rounding radius with the slope values and the contour of the vertex profile. The circle obtained used to achieve the radius of the prism vertex is showed in red color (Fig. 5 (b)). The prism base obtained has a width of 400 μm and the radius value obtained is 13.78 μm; this result could be affected by the pressure exerted to make the cut of the prism film.

3. PRISMATIC LIGHTGUIDE ANALYSIS

The lightguide structure analysed is a prismatic hollow tube with the thin polycarbonate film with right angle prism sections. The prism light guide transmits light by total internal reflection, which gives higher efficiency and homogeneous light distribution through the guide. Prismatic sheeting developments have provided further improvements in sunlight systems (Fig. 7), boosting efficiency for specific incidence angles with regard to the aluminum guides.

Light travels mainly in the hollow air space inside the guide and bounces off by total internal reflections (TIR) when the input light is highly collimated. This configuration has an angular acceptance cone, which is not an isotropic distribution in the space determined by the refractive index of the prismatic film, if the refractive index of the dielectric material is 1.5, as is the case of acrylic plastic, then the input light angle must be approximately less than 27.5º from the guide´s axial direction, even though the acceptance cone can be higher in the meridional plane.

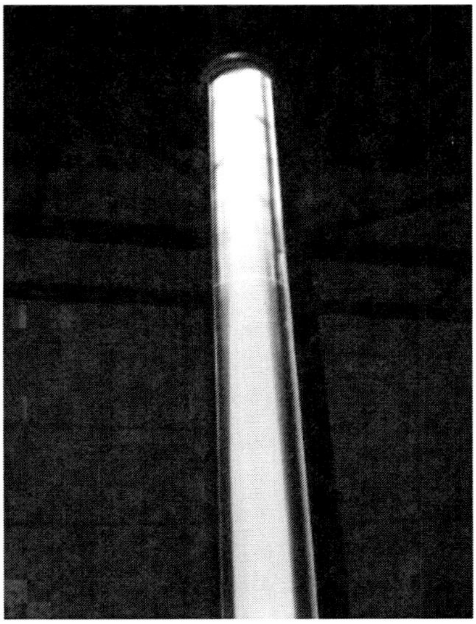

Figure 7. Prismatic light tubes are used for transporting and distributing natural light. Experimental setup in the School of Optics (Complutense University of Madrid)

Two different analysis have been done: Firstly, a theoretical simulation based on the specular reflection model in order to analyse the color characteristics to the output of two guides, rectangular and cylindrical of different lengths, one of them of aluminum material and the other one, internally coated of prismatic film structure this model have been developed by means of a mathematical software as Matlab. Secondly, the prismatic guides proposal evaluates with a ray tracing software through which we show the efficiency for a wavelength evaluated at the end of the guide, this model have been studied with Monte Carlo ray tracing.

3.1. Theoretical simulation

The analysis presented is based on the specular reflection model. There are two important objectives to evaluate with regard to the light quality as a function of the Spectral Power Distribution; these are Correlated Color Temperature (CCT) and the Color Rendering Index (CRI) of the light source, which we evaluate in this work using the guidelines of CIE 13.3.

Color rendering of an illuminant is the effect of the illuminant on the color appearance of objects by conscious or subconscious comparison with their color appearance under a reference illuminant. Colour Rendering serves to describe the effect of an illuminant on the color appearance of objects. The most fundamental natural source is daylight, thus the primary reference for comparing the rendering of a source should be the Standard Illuminant D65.

The Correlated Color Temperature is the temperature of the Planckian radiator whose perceived color most closely resembles that of a given stimulus at the same brightness and under specified viewing conditions (CIE 17.4, 1989). According to this definition, CCT can be calculated using one of the chromaticity diagrams. CIE still recommends to calculate CCT using the 1960 (u, v) chromaticity diagram (now deprecated). On the (u, v) diagram, find the point on the Planckian locus that is at the shortest distance from the given chromaticity point. CCT is the temperature of the Planck's radiation at that point. CCT can be calculated from CIE 1931 chromaticity coordinates x and y in many different ways. Complicated algorithms and simple equations alike have been proposed and used for several decades.

The simulated construction of the prismatic guides is based on a tube internally coated polycarbonate sheet whose outer face is composed of 90º micro-prisms; rays that enter on the guide with an angle suitable undergo total internal reflections. In this model, in order to include the round in corners and the absorption we estimate the spectral reflectance of the inner face of the prism sheet will be estimated in 0.99 throughout the spectrum analyzed. In this section two different types of light guides: circular and rectangular cross sections have been analyzed (Fig. 8).

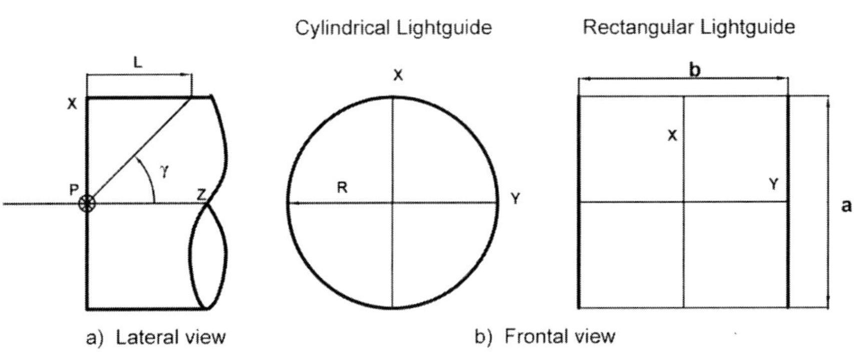

Figure 8. The Coordinate system defines the angles of incidence and observation in the guides. Longitudinal section (a) and transverse section (b).

The photometric profile of the source is an angular distribution adapted to the optimal transmission features of the prismatic film used, and the analysis was restricted to the spectral range between 380 and 780 nm. The CIE Illuminant D65 is the reference light source.

The spectral reflectance of aluminum is estimated as isotropic over the fence and has been experimentally determined using a Hitachi U-3400 spectrophotometer with special accessories for measuring the specular reflectance at 12° incidence. Experimental spectral reflectance measured of aluminum is shown in figure 9.

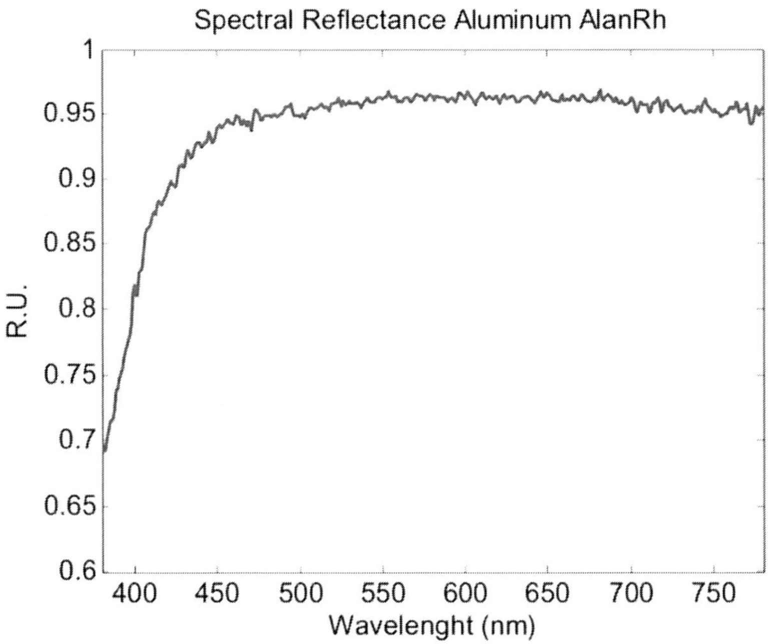

Figure 9. Aluminum spectral reflectance.

To determine the spectral distribution of radiation emerging from the rectangular guide, it is used the model developed by Whitehead (Whitehead, L.A., 1982). According to this model, the number of traversals (Δn) per unit length (Δz) of a ray of light undergoes inside the guide is given by the following expression:

$$\frac{\Delta n}{\Delta z} = \frac{\tan\theta}{Dr}, \tag{1}$$

where *Dr* is the average cross-sectional distance travelled by a ray in crossing the guide air space and θ is the angle by which any ray deviates from the guide's axial direction.

If the guide air space is rectangular, with dimensions *a* and *b* (fig.8), then

$$Dr = (a^{-1} + b^{-1})^{-1}, \tag{2}$$

If roughly circular with radius R

$$Dc = \frac{4R}{\pi}, \tag{3}$$

In the rectangular case, the dimension *a* is the same as *b* (0.8862 meters), the cylindrical guide aperture radius is 0.5 meters in order to maintain the same input area than rectangular guide, and the length of the guides (*L*) evaluated are 5, 10 and 15 meters.

The dependence of the radiant flux with the angle to the incident beam is given by equation 4

$$\varphi(\theta) = 2\pi I_0 \left[\cos\theta_2 - \cos\theta_1\right], \tag{4}$$

where I_0 is the total intensity power radiated by the source in a specific direction θ. Moreover, we obtain the spectral distribution of radiant flux incident S_λ, where S_λ is the primary illuminant spectral distribution. θ_1 and θ_2 are the angular limits and the intensity has I_0 value.

The spectral and angular distribution at the exit of the guide will be:

$$S' = \int_0^{\pi/2} P(\lambda,\theta)\rho(\lambda,\theta)^n d(\lambda,\theta), \tag{5}$$

ρ is the reflectance of the material for each wavelength studied and n is considering the number of internal reflections obtained for each incidence angle theta θ in the guide.

The prismatic guide has a higher transmission throughout the spectrum (Fig. 10) and the relative spectral power is constant. In the aluminum guide, the energy area of the shorter wavelengths obtained is reduced due to low spectral reflectance of the aluminum.

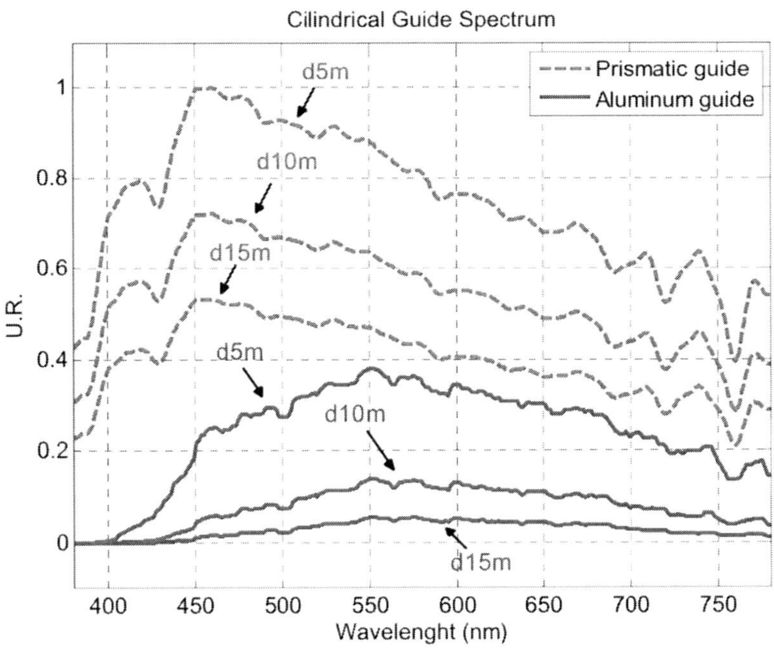

Figure 10. Spectral Power Distribution of the two types of guides for different lengths.

3.2. Color analysis for circular and rectangular lightguides

The spectral power distribution is used for the estimation of the spectral ratio resulting. The spectral ratio is used to compare the spectral distribution at the input and output of the guides. The spectral ratio (η_{λ}.) will be:

$$\eta(\lambda) = \frac{S'(\lambda)}{S(\lambda)} \tag{6}$$

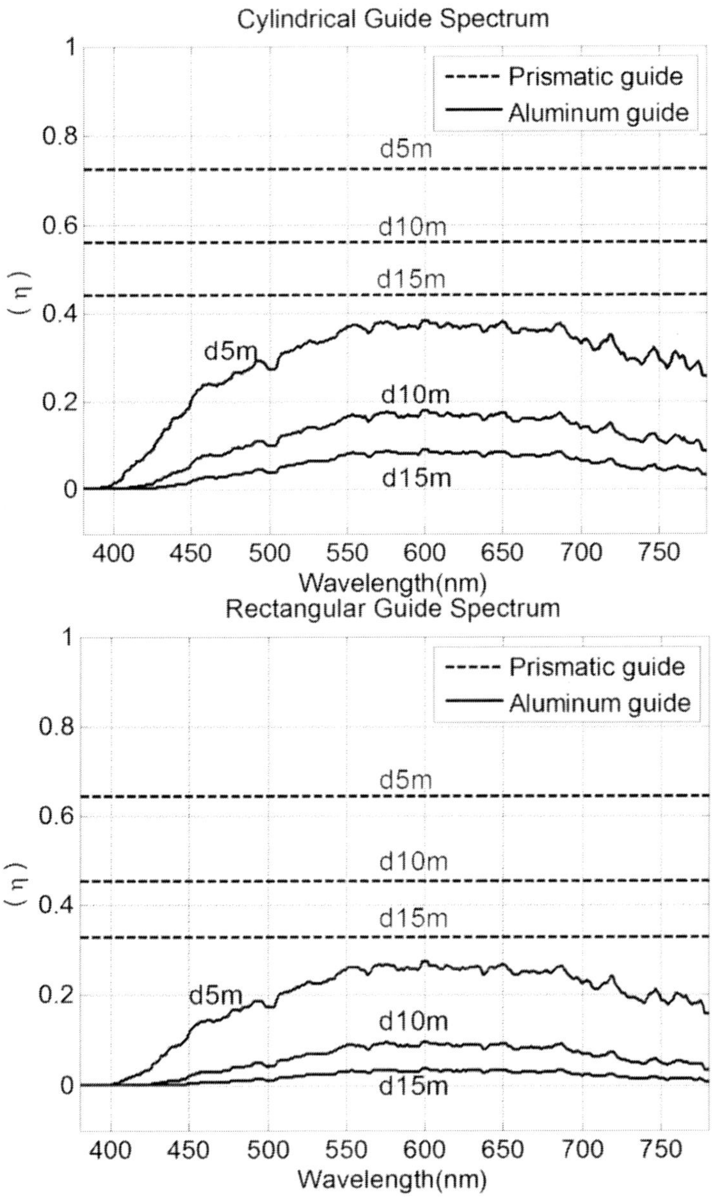

Figure 11. The ratio of spectral power distribution of the two types of guides for different lengths.

Considering $S(\lambda)$ the input spectral power distribution of the source and $S'(\lambda)$. Spectral power distribution ratio emergence from cylindrical guide is greater from rectangular guides with the same dimensions (figure 11). In addition, the ratio decreases significantly in 10 and 15 meters from aluminum guides. The prismatic guide has high transmission across the spectrum and the spectral power is constant. There is a downward trend in short wavelengths in aluminum guide energy due to the spectral reflectance characteristics of the material.

We define the spectral ratio, η_λ as the fraction of the incident spectral power distribution transmitted by the guide:

For comparing different behavior of light guide regarding color output flux it is used CIE 1931 diagram due to it is more spread on known color representation (CIE 13:3, 1995).

Color coordinates in CIE 1931 chromaticity diagram for studied lengths of aluminum and prismatic guides are show in figure 12. The chromatic coordinates with the light source (D65) chosen for different lengths of guide will be compared. The three consecutive black square points and blue circle points represent the aluminum guide in 5, 10 and 15 meters.

The three consecutive points (d5m, d10m and d15m) represent the aluminum guide. The results of the CIE 1931 chromaticity coordinates of the full spectrum transmitted in the prismatic guides differ from those transmitted in the aluminum guides approaching to the yellow zone. When the length of the aluminum guides increases, there are no changes in the results obtained for the prismatic guide (P), and the result are superposed with the illuminant D65 because the reflectance is maintained for the entire spectrum.

The CCT is obtained from the spectral distribution at the exit of the guide (Table 1). This calculation is carried out for different lengths of the two types of guides considered in this work. To determine the reference light source for a given test source, we must find the CCT of the test source. Once this data is known, the reference light source is a Plankian black body which has the same temperature.

Moreover, the Color Rendering Index (CRI) of the sources at the end of the guides it is calculated (fig.13). From the result obtained (Fig.10), it is possible to determine the significant color change in the aluminum fence, showing less color change difference for all lengths than the prismatic guides.

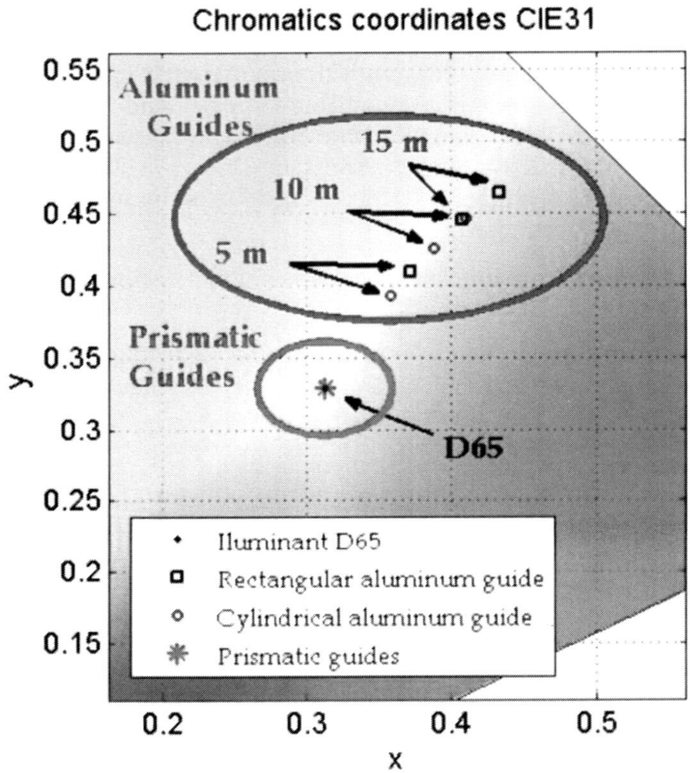

Figure 12. CIE 1931 chromatic coordinates of the studied lightguides.

Table 1. Correlated Color Temperature (CCT) and Color Rendering Index of each spectral power.

	Cylindrical Guide		Rectangular Guide	
	CCT (ºK)	CRI	CCT (ºK)	CRI
Iluminant D65	6503	99.97	6503	99.97
Prismatic guide (5m)	6503	99.97	6503	99.97
Prismatic guide (10m)	6503	99.97	6503	99.97
Prismatic guide (15m)	6503	99.97	6503	99.97
Aluminum guide (5m)	4707	89.81	4407	87.74
Aluminum guide (10m)	4126	85.45	3818	82.39
Aluminum guide (15m)	3781	81.98	3478	78.15

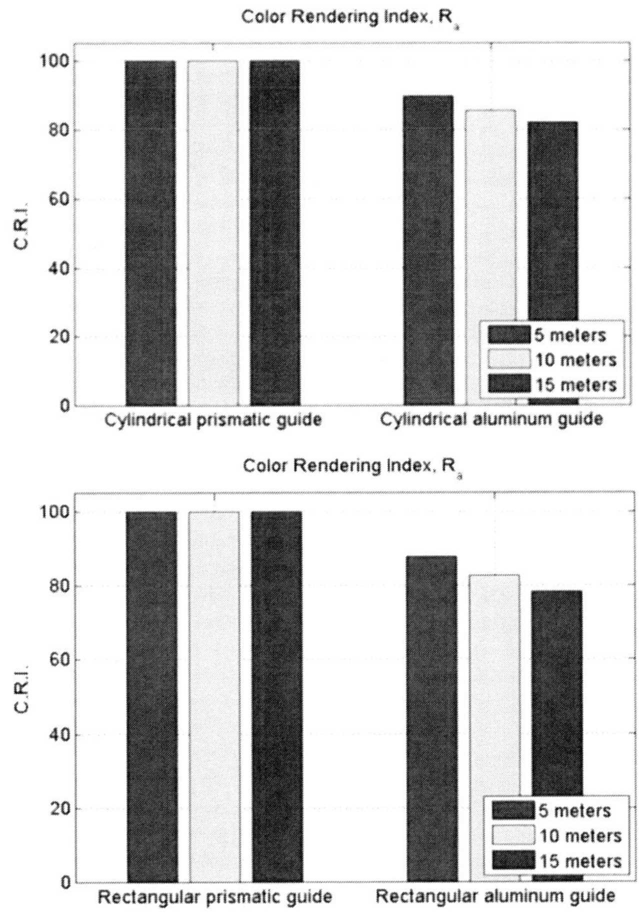

Figure 13. CRI of measured lightguides (cylindrical and rectangular).

3.3. Flux transfer analysis

The flux transmitted of the two types of prismatic guides for a wavelength is studied in order to analyze the efficiency of both systems. For this purpose, the distribution characteristics are calculated using a three-dimensional ray tracing. In this case, the lightguides have a 1.49 refractive index for the studied wavelength without absorption losses. Generally, the associated absorption loss in prismatic film is rather low, and taking into account that it is considered the model of light reflection, in this study this parameter will not be studied on the color quantity estimated.

The dimensions of the guides analyzed in the ray tracing study (Fig.14) are the same defined in the theoretical simulation (Fig.8) and the enlargement scale factor of the prism shape is on the order of 10. The light cone input is an extensive emitter with a random distribution of 30º semi angle (θ) which has the same size as the section of the light guide.

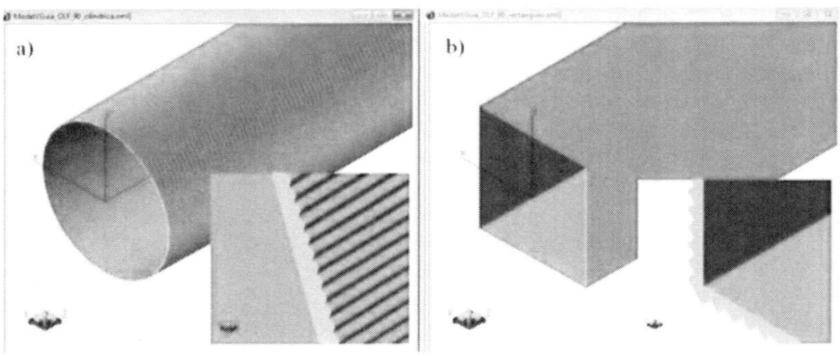

Figure 14. Perspective view of Ray trace simulation model of lightguides with a detail of the prismatic structure: cylindrical (a) and rectangular (b).

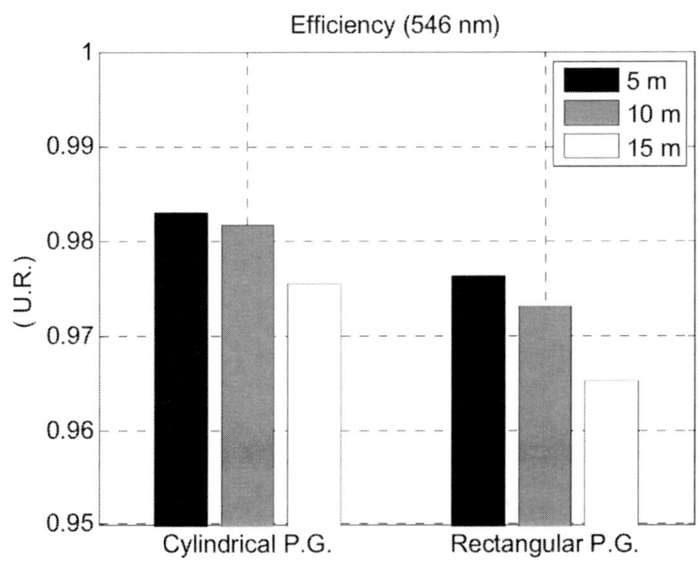

Figure 15. Output efficiency results for wavelength of 546 nm in Prismatic Guides.

The comparison of efficiency in rectangular and cylindrical prismatic guides is shown in Figure 15; the graphics indicates the influence of the shape in maintaining the transported flux and the efficiency for a wavelength (546 nm). The output flux efficiency of 15 meters cylindrical guide is 2% higher than rectangular guide; in addition, in the first meters the flux remains more constant.

4. HOLLOW PRISMATIC CPC

Compound parabolic concentrator (CPC) is an optical devices used in the solar energy related areas and also in other applications where radiant energy concentration is needed, being defined as one of the first devices that resulted from the practical application of nonimaging optics (Welford and Winston, 1978). Light from a defined range of angles of incidence is reflected by total internal reflection on the parabolic walls of the CPC and concentrated at the exit of the CPC.

The authors propose an innovative 3D hollow prismatic CPC (PCPC) in reverse mode made of a prismatic dielectric material, which has a high efficiency comparing it with aluminium CPC (ACPC). The basic idea is to use a hollow prismatic light guide with CPC shape. In figure 16 (up-left), we can observe the design in 2D geometry in the inverse mode proposal; all the rays entering at the focus of the parabola (F_1 and F_2) emerge through the exit aperture with the design angle θ. This paper reports 2D, 3D design (Fig. 16) and numerical analysis by ray-tracing software, furthermore experimental results are shown. A prototype has been developed and tested showed in figure 16 (down). The hollow PCPC in reverse mode has an entrance pupil that is small compared to the exit pupil depending on the design angle. This CPC design accepts light in 2π entering the entrance pupil and redirecting it in the CPC design angle. This new concept is made of a prismatic film; this dielectric layer accepts light not only in the entrance pupil (Entry 1) but also through the layer itself. This property allows an increase in efficiency compared with the ACPC.

In this section we analyzed the PCPC by raytracing software to determine the angular transmission, optical efficiency and irradiance distribution on the system exit aperture in a 3D system, later a comparison between ACPC and PCPC is presented.

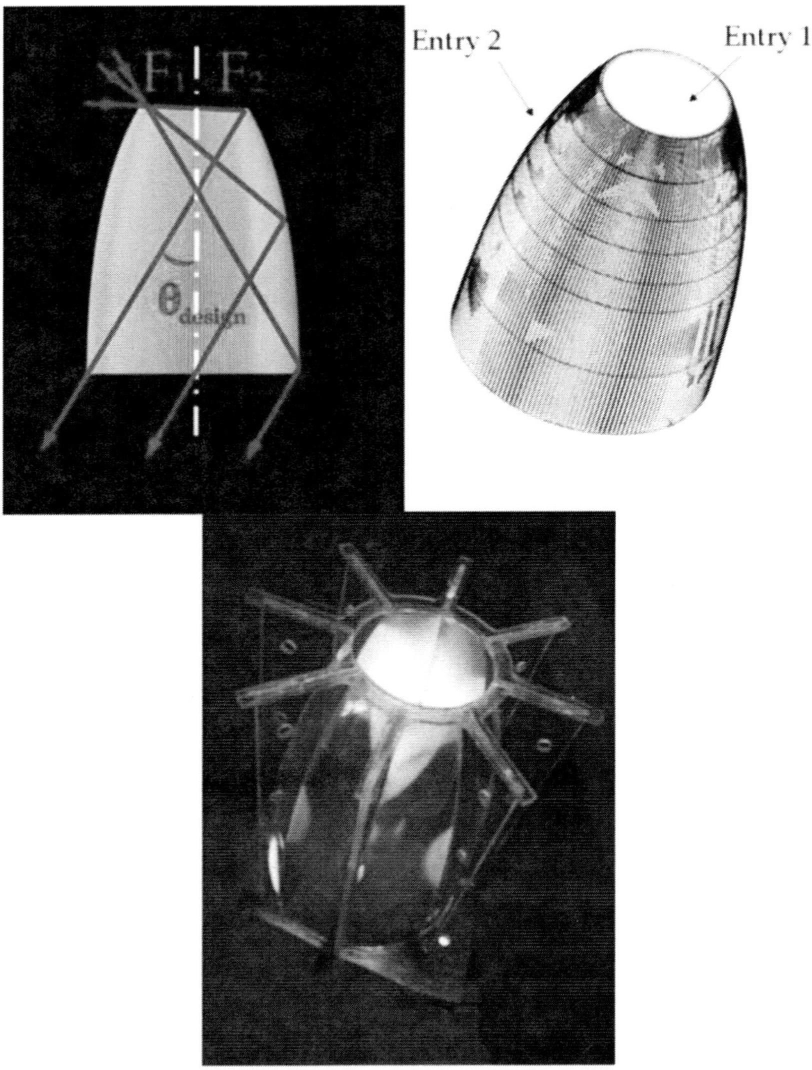

Figure 16. The CPC profile (up-left) with a ray tracing showing the design angle and the maximum input angle of design θ, 3D hollow HCPC 30º software design (up-right) and the experimental prototype (down) in which the reflector surface is a prismatic film supported by eight polycarbonate ribs.

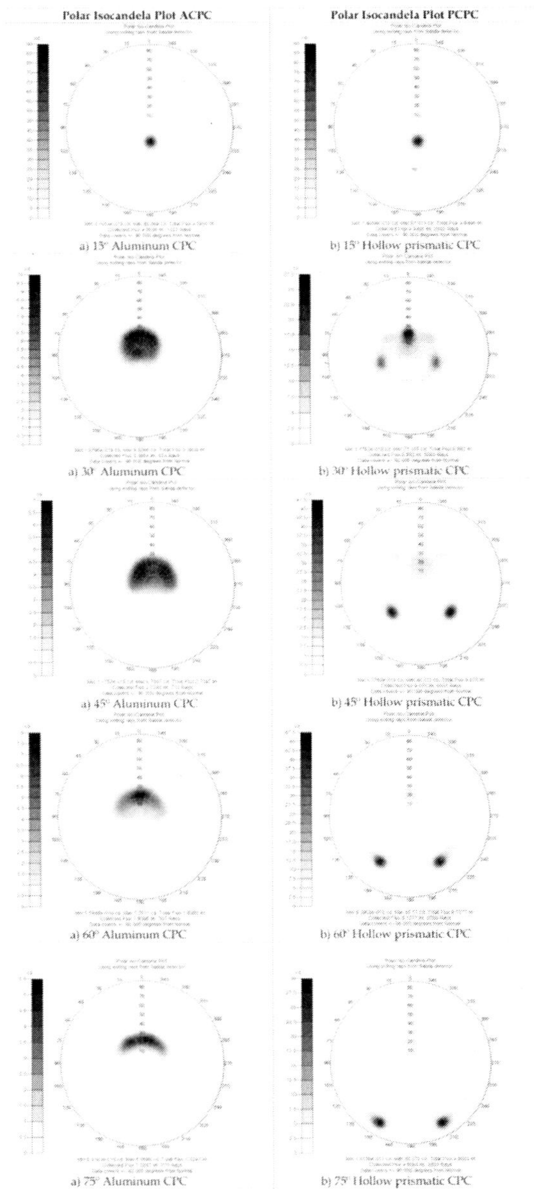

Figure 17. The polar intensity diagram provides the shape of the light distribution to the exit of both parabolic systems

4.1. Ray-tracing: Efficiency of the PCPC compared to ACPC

Ray-tracing is processed for two kinds of CPCs, one designed with standard aluminum with a reflectance of 1.0 and the other one, the prismatic CPC designed with PMMA.

4.1.1. Polar Isocandela Plot of ACPC and hollow PCPC

The polar intensity diagram provides the shape of the light distribution of both parabolic systems. Figure 17 (a) shows isocandela plot representing the ACPC intensity and angular distribution. The 17 (b) illustrations show the isocandela plot for PCPC when light enter through entry 1+2.

5. EFFICIENCY COMPARATIVE

Ray-tracing is processed for two kinds of CPCs, one designed with standard aluminum with a reflectance of 100 % and the other one, designed with dielectric prismatic film.

Computed efficiency is obtained by tracing collimated rays in 5º angle intervals for both systems and computing the obtained flux at the exit pupil of the system.

The comparison of PCPC and ACPC collectors is shown in figure 18; it indicates the effects of incidence angles and the efficiency of the PCPC. To compare both systems it is used collimated light at different angles. The flux obtained in the entry 1 (Φ) is used to calculate the final efficiency (η),

$$\eta = \frac{\phi_{entry1}}{\phi_{exitpupil}} \tag{7}$$

The PCPC accepts light out of this entry pupil Entry 1 so η can be higher that 100%.

When it is analysed the efficiency obtained in both systems adding the prismatic surface of the PCPC (entry1+entry2) to 85º, it is observed the improvement with regard to the ACPC, reaching a 600 % higher efficiency flux than an ACPC.

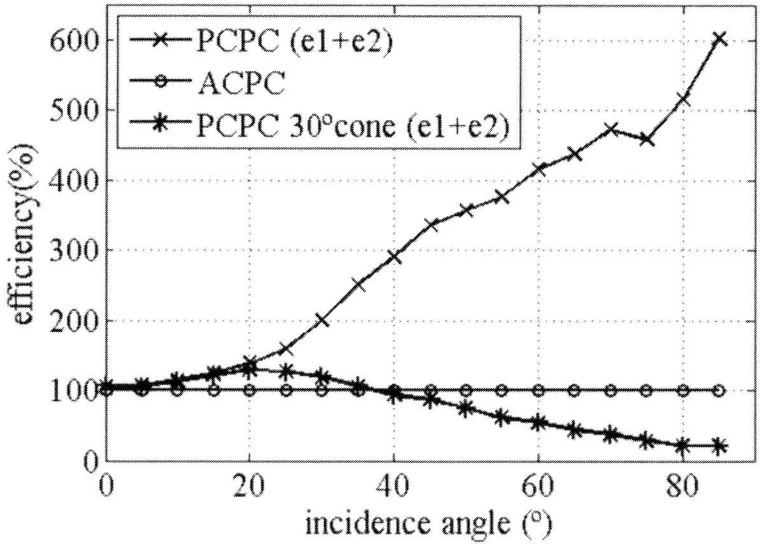

Figure 18. Output flux HPCPC VS output flux HPCPC 30º cone using entry1 and entry2

The efficiency of the PCPC to compare the outflow of 30º cone (*) (entry1+entry2) with regard to the efficiency of entry 1 it is evaluated. There is a clear profit for incidence angles ranging from 0º to 35º, though it is necessary to improve the efficiency for the higher incidence angles.

It is necessary to investigate how the PCPC is working when light enters in the entry pupil (entry 1) as it is done in ACPC.

The ACPC works better than PCPC as the incidence angle increases if we use only entry 1 in the prismatic PCPC. Figure 19 show this behaviour for different incidence angles. The TIR that suffers light beam in the outer surface of the PCPC is not happening when incidence angle increases, this behavior explains the decrease of efficiency according to the incidence angle increases The division of the beam showed in figure 20 (b) is due to the prismatic effect when light reach the film in the outer side as shown basic raytrace in figure 1a.

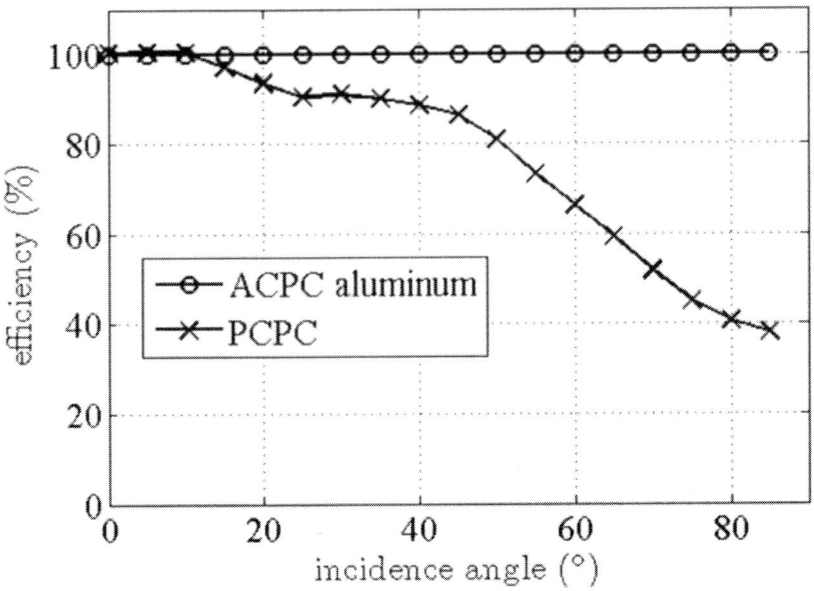

Figure 19. ACPC VS PCPC using entry 1.

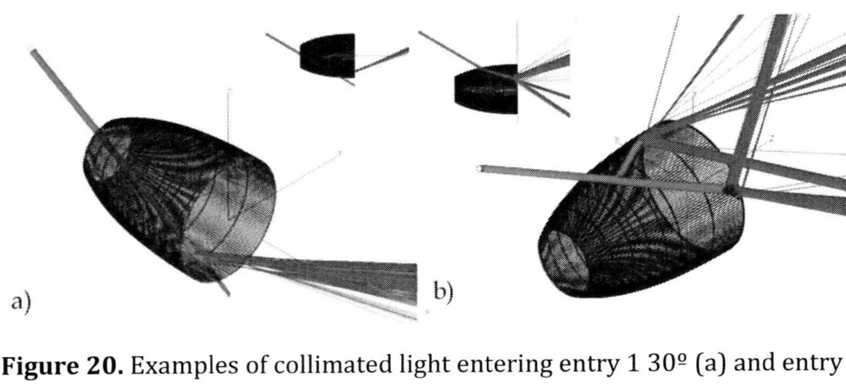

Figure 20. Examples of collimated light entering entry 1 30º (a) and entry 2 30º (b).

The light guiding after PCPC is a good alternative to light far away from the collecting system, the use of a hollow light guide is demonstrated as a good way to transport light, and new bending systems show a good way to reach all the lighting necessities.

6. EXPERIMENTAL MEASURES

A CCD video photometer (Radiant Imaging Prometric 1400) is used to measure output light distribution in exit pupil with a Lambertian screen (Fig.21). We measure output light in two experimental assemblies changing the incidence angle between 0º and 75º increasing the source angle in steps of 15º. Firstly, we evaluate entry 1, and secondly entry 1+2 is evaluated.

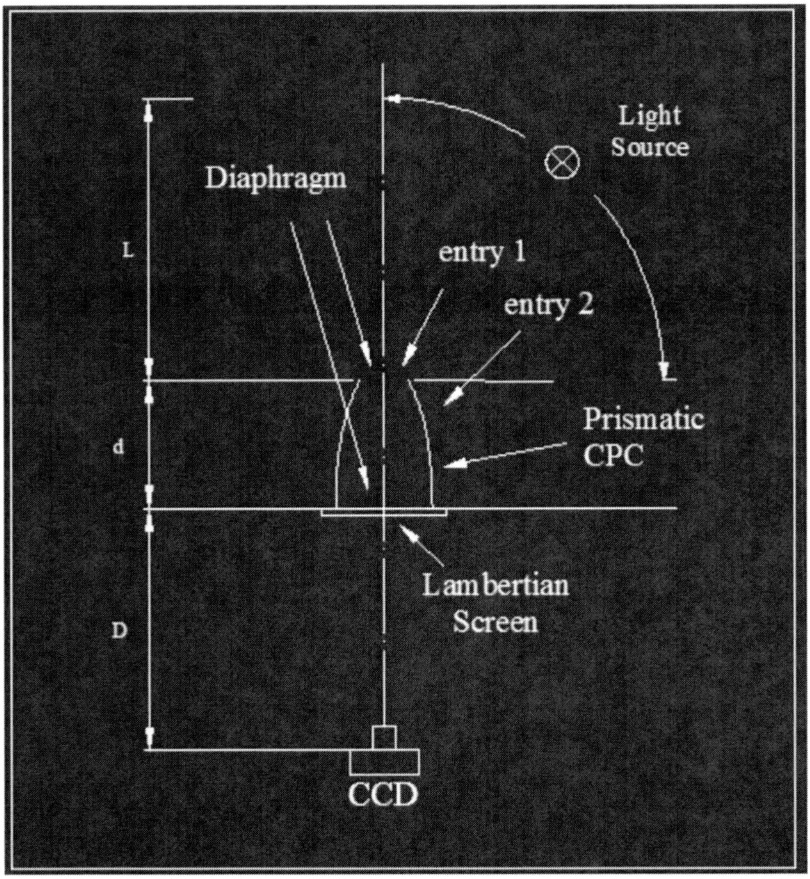

Figure 21. Schematic diagram of experimental setup The HPCPC has the following parameters: input aperture diameter (entry1): 88 mm; output aperture diameter (Lambertian screen): 187 mm, L: 4000 mm, d: 260 mm.

Normalized light distribution map onto exit pupil is shown in the figures 22 to 28. The (a) figure represents the illuminance map obtained with the ray tracing software, the (b) figure represents the map obtained with the experimental setup. Figures 22 to 24 show data for using entry 1 and 25 to 28 show data using entry 1+entry 2 for some light angles.

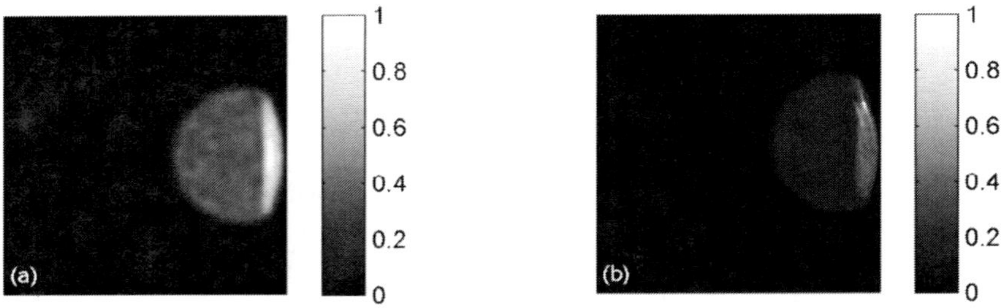

Figure 22. entry1. (a) Raytracing simulation, (b) Experimental measurement.

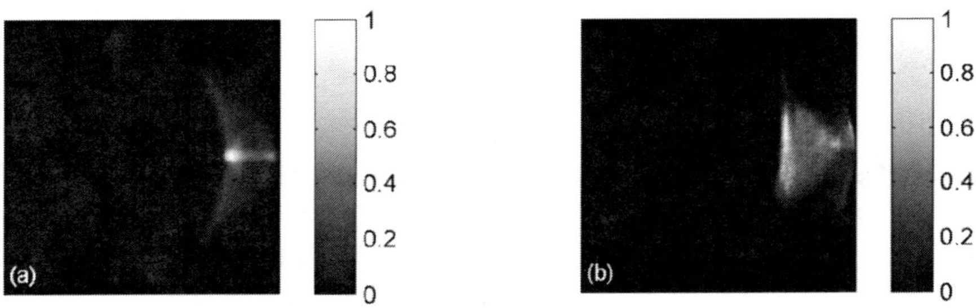

Figure 23. entry1. (a) Raytracing simulation, (b) Experimental measurement.

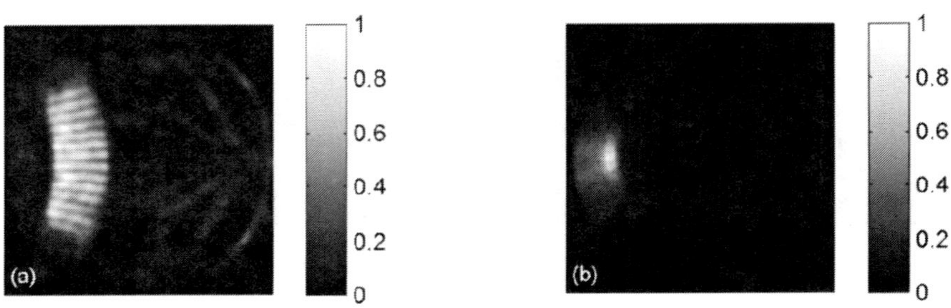

Figure 24. entry1. (a) Raytracing simulation, (b) Experimental measurement.

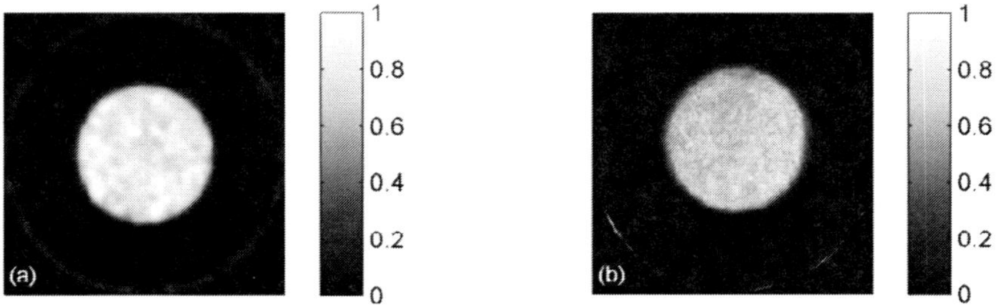

Figure 25. entry1+entry2. (a) Raytracing simulation, (b) Experimental measurement.

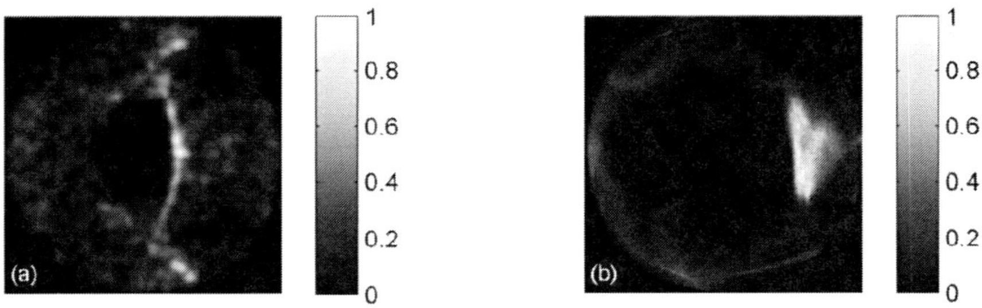

Figure 26. entry1+entry2. (a) Raytracing simulation, (b) Experimental measurement.

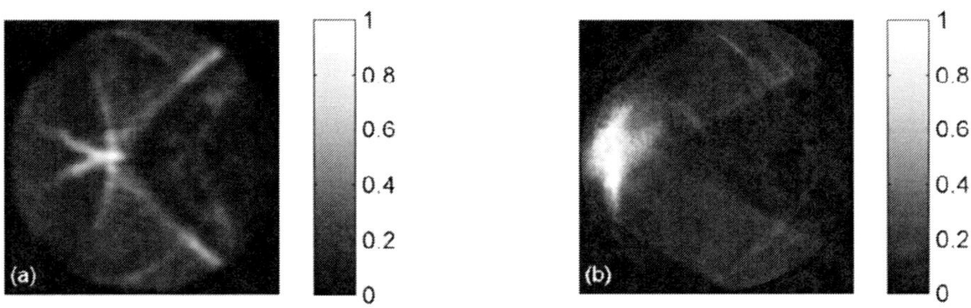

Figure 27. entry1+entry2. (a) Raytracing simulation, (b) Experimental measurement.

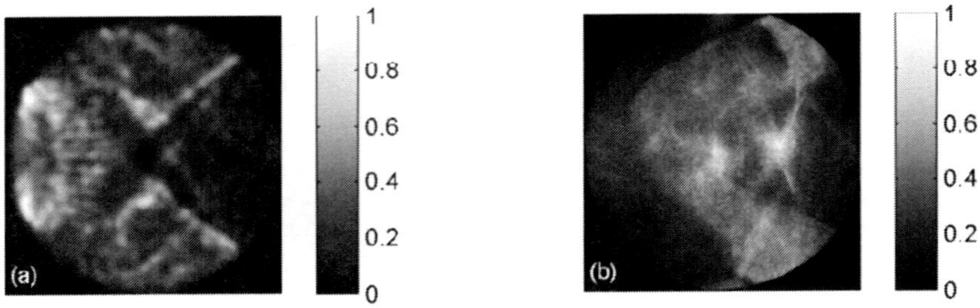

Figure 28. entry1+entry2. (a) Raytracing simulation, (b) Experimental measurement.

The circular peripheral wreath in figure 25 is due to the multiple internal reflections of the light inside the prismatic film. When we increase the angle of incidence, we can observe the division of the beam showed in figure 27 due to the prismatic film structure.

7. CONCLUSIONS

The present study shows the interesting combination of light transmission and reflection capabilities of dielectric prismatic film analysed in different guidance geometries and the approach of a collecting system for sunlight applications. Prismatic film can work as a perfect mirror or transparent material depending upon the angle that light strikes the material, it allow us to produce lighting products with unique properties.

It is important to analyse the micro-structure prism imperfections in the form of the surfaces: imperfect corners (apex and valley), surfaces which are not optically flat or which deviate from the expected angle, optical inhomogeneities in the material and the existence of the curved area on the peaks prism which modify the optical behaviour of the prism film; this imperfections modify the optical path and therefore the rays can be directed to other directions instead of undergoing total internal reflections. In order to check the importance of prism rounding, the apex is analysed through Hough Transform by digital image processing. The peak curved radio obtained is despised because of the radius size obtained by image processing is small with regard to the size of the prism structure. The prism are calculated to have an angle of 89.88º and the diameter of the curved region obtained in the prism´s peak is 27.56 μm, this result could be affected by the pressure exerted to make the cut of the film.

Dielectric prismatic guides have a high quality output flux regarding standard light guides based on specular material. Specific shapes such as a circular cross section show a soft increase in flux transmission, with the extent of the improvement dependent on the input position of the light ray.

When input light is in the admitted angle the prismatic light guides have a higher transmission in the entire spectrum than the reflective guides and their spectral efficiency is more constant. There is a downward trend in short wavelength in aluminum guide energy due to the spectral reflectance characteristics of the material.

Prismatic light guides turn out to be more robust in lighting quality maintenance than the aluminum guides, which are efficient but only capable of maintaining light quality distances lower than 5 meters. Cylindrical light guides have a 2% higher efficiency than rectangular although the rectangular shape could be more convenient in office buildings due to the occupied space and construction constraints.

In this study, the authors propose an innovative 3D hollow prismatic CPC (PCPC) working in reverse mode and made of a dielectric material, which has a high efficiency compared with aluminium CPC (ACPC). Transportation of daylight over longer distances requires an optimized collector, the PCPC is an appropriate design for natural light systems like skylights and collector guiding systems since it has properties as collimator to catch the light and to direct it and transport it long distances from a remote source with little attenuation. The hollow PCPC has an entrance pupil that is small compared to the exit pupil depending on the design angle. This CPC design accepts light in 2π entering the entrance pupil and redirecting it in the CPC design angle. This new concept is made of a prismatic structure film; this dielectric layer accepts light not only in the entrance pupil but also through the layer itself.

The results obtained shows that measured PCPC efficiency compared with standard aluminum is 600% higher at 85º incidence angle, a medium value of 300% increase is obtained in the range from 0 to 85º. There is a clear profit for incidence angles ranging from 0 º to 35 º, though it is necessary to improve the efficiency for the higher incidence angles.

The design of big structures in buildings is easier with this new system because of the minor weight of the plastic material which can be conformed in independent parts and development in moulding fabrication can improve the cost of the system.

REFERENCE

1. Fernandez. Alvarez-Balbuena, A. , D. Vázquez-Moliní, B. Garcia-Fernandez, A. Garcia-Botella, E. Bernabeu, 2009Skylight: a hollow prismatic CPC, Proceedings of the SPIE, 7423, 74230T.
2. Fernandez. Alvarez-Balbuena, A. , D. Vazquez-Moliní, B. García-Fernandez, L. García-Rodríguez, T. Galán-Cañestro, 2010Daylight illumination system by vertical Transparent Prismatic Lightguide for an office building. Colour and Light in Architecture. Knemesi, Verona, 360365
3. J. Canny, 1986A Computational Approach to Edge DetectionIEEE Trans. Pattern Analysis and Machine Intelligence, 86679698
4. CIE Publ. No.13.31995Technical Report- Method of measuring and specifying colour rendering properties of light source.
5. CIE Publ. No.17.41989IEC Pub. 50(845) International.Lighting Vocabulary.
6. M. D. Dubois, 2003Shading devices and daylight quality: an evaluation based on simple performance indicatorsLighting Research and Technology3516174
7. B. García-Fernández, D. Vázquez-Moliní, Fernández. Álvarez-Balbuena, A. , 2011Lighting quality for aluminum and prismatic light guidesProceedings of the SPIE 8170, 81700T.
8. A. A. García-Botella, A. Fernández-Balbuena, D. Vázquez-Moliní, E. Bernabéu, 2009Ideal 3D asymmetric concentratorSolar Energy113119Hough, P.V.C. (1962). Methods and means for recognizing complex patterns, U.S. Patent 3,069,654, Dec.
9. C. K. Hsieh, 1981Thermal analysis of CPC collectors Solar Energy, 2711929
10. J. T. Remillard, M. P. Everson, W. H. Weber, 1992Loss mechanisms in optical light pipes,Applied Optics313472327241
11. K. G. Kneipp, 1994Use of prismatic films to control light distributionInternational Lighting in Controlled Environments Workshop, NASA-CP-95-3309, 307318
12. M. S. Mayhoub, D. J. Carter, 2010Towards hybrid lighting systems: A reviewLighting Research and Technology425171

13. Wang M.W. & Tseng C.C.2009Analysis and fabrication of a prism film with roll-to-roll fabrication process.Optics Express17647184725

14. OLF2301Data sheet, www.3M.com/lightingproducts.

15. Swift P.D. & Smith G.B.1995Cilindrical mirror light pipes, Solar Energy Materials and Solar Cells, 362159168

16. TracePro® Opto-Mechanical DesignSoftware, http://www.lambdares.com/

17. D. Vázquez-Moliní, Fernández. Álvarez-Balbuena, A. , M. González-Montes, E. Bernabeu, A. García-Botella, L. García-Rodríguez, W. Pohl, 2009Guiding daylight into a building for energy-saving illuminationSPIE NewsroomProceedings of the SPIE DOI:

18. W. T. Welford, R. Winston, 1978The Optics of Nonimaging Concentrators: Light and Solar Energy.Academic Press, London.

19. R. Winston, 1975Development of the compound parabolic collector for photo-thermal and photo-voltaic applications Proceedings of the Society of Photo-Optical Instrumentation Engineers. Optics in Solar Energy Utilization, 6813644

20. R. Winston, J. C. Miñano, P. Benitez, 2005Nonimaging OpticsElsevier Academic Press, San Diego, CA.

21. L. A. Whitehead, 1981U. S. Patent 4,260,220. Prism Light Guide Having Surfaces which are in Octature.

22. L. A. Whitehead, 1982Simplified Ray Tracing in Cylindrical Systems,Applied Optics211935363538

INDEX